Stress Response Mechanisms in Fungi

Marek Skoneczny
Editor

Stress Response Mechanisms in Fungi

Theoretical and Practical Aspects

 Springer

Editor
Marek Skoneczny
Institute of Biochemistry and Biophysics
Polish Academy of Sciences
Warsaw, Poland

ISBN 978-3-030-13141-8   ISBN 978-3-030-00683-9   (eBook)
https://doi.org/10.1007/978-3-030-00683-9

This Springer imprint is published by the registered company Springer Nature Switzerland AG
The registered company address is: Gewerbestrasse 11, 6330 Cham, Switzerland

Preface

The concept of the stress response as a set of reactions of living organisms triggered by the adverse circumstances applied to them was formulated by Hungarian endocrinologist Hans Selye, the pioneer of studies on stress (Selye 1936). The first observations of the cell response to stress on a molecular level were perhaps made in the early 1960s when Ritossa observed characteristic transient activation (so-called puffing) of certain regions of the *Drosophila* salivary gland giant chromosomes in response to heat shock (Ritossa 1962). Stress is a common experience of any living creature, endangering its survival; however, on a longer evolutionary scale, it is a tool of selection pressure that selects those that endure it the best.

The title and subject of this book were selected not without a reason. Yeast and filamentous fungi are renowned for their capacity to survive under extremes and to exploit seemingly nonabsorbable nutrients. Thus, they (together with various branches of prokaryotes) are exposed to the widest variety of environmental stresses. They are also attractive model organisms in all life science studies at the molecular and cellular levels. They are relatively easy to cultivate and grow quickly, translating into short paths to results. They also have small genomes and are easily amenable to analysis and modification. *Saccharomyces cerevisiae*, the first eukaryote whose whole genomic sequence was revealed in 1996, occupies the dominant position as the preferred model organism. It is also a very convenient model in studies of the participation of individual gene products in the resistance to any stress condition owing to the availability of the collection of *S. cerevisiae* knock-out strains. More recently, numerous other yeast and fungal genomes have been sequenced, and their bearers have become convenient models in stress response mechanism studies. Studying stress response mechanisms with these organisms is also of great practical importance.

All these considerations have determined the structure of this book. Chapters 1, 2, 3, 4, and 5 describe the current knowledge on how yeast and filamentous fungi cope with various stresses of external or endogenous origin. Chapters 6 and 7 describe the practical—industrial, agricultural, and medical—aspects of yeast and filamentous fungi responses to stress. These organisms are widely used in all biotechnological processes, including the production of food, pharmaceuticals, and other compounds. The industrial environment is usually harsher and more unfriendly, and the substrates are quite often not optimal for growth. Thus, developing yeast and fungi strains highly resistant to all stresses they may encounter in the production process is economically desirable. Knowing fungal resistance mechanisms, especially fungal defenses to avoid death in the presence of fungicides, will help to develop better preparations that are fungi specific and will not affect humans, animals, and plants. This will contribute to the production of healthier foods and to lowering the economic loss in agriculture. Fungal pathogens are an obvious choice in the studies of human–host interactions. Fungal pathogens of humans, in addition to stresses experienced during their saprophytic life phase, are exposed to the stresses incurred by host defense mechanisms and antifungal medicines. As we know, they are very good at enduring these stresses, so studying their survival strategies will lead to better therapies.

The response to stress encompasses mounting defense mechanisms while minimizing already incurred damage. As we shall see, these approaches may be diverse and include programmed cell death (PCD), which clears the clonal cell population of individuals with irreparable damage to benefit the survivors. The stress response also prepares the cell for further stresses; thus, the cell adapts to stress, acquiring stress tolerance. Indeed, numerous observations have been reported that mild stress makes cells more resistant to severe stress of the same type, as well as to other stresses. In addition to the dynamic process of reacting to encountered stress, there are circumstances where cells must be constantly resistant to certain stresses, either because the stress acts too quickly and the cell has insufficient time to appropriately respond or because the stress conditions lead to the suspension of life processes. Therefore, some stress protecting agents must be present in the cells constitutively.

These mechanisms should be well orchestrated to assure the survival of an organism exposed to inhospitable conditions, that is, for most of its life. The objective of this book is to provide a concise yet up-to-date account of these mechanisms employed by the popular unicellular fungal models. This book will be a useful guide to this field, valuable for those interested in basic studies on stress and for those focused on its practical aspects.

Warsaw, Poland   Marek Skoneczny

References

Ritossa F (1962) A new puffing pattern induced by temperature shock and DNP in Drosophila. Experientia 18:571–573. doi: https://doi.org/10.1007/BF02172188

Selye H (1936) A syndrome produced by diverse nocuous agents. Nature 138:32

Contents

Editor and Contributors

About the Editor

Marek Skoneczny is a group leader at the Institute of Biochemistry and Biophysics, Polish Academy of Sciences in Warsaw, Poland. His research focuses on various topics of living cell response to environmental stresses using yeast as a model system and employing diverse methodologies including the genome-wide approaches. In his studies, he emphasizes the aspect of stress containment, i.e., compartmentalization of stress sources and remedies in organelles, such as peroxisomes.

Contributors

Razieh Karimi Aghcheh Department of Molecular Microbiology and Genetics, University of Goettingen, Goettingen, Germany

Sabina Bednarska Department of Biochemistry and Cell Biology, Faculty of Biology and Agriculture, University of Rzeszow, Rzeszow, Poland

Vasudha Bharatula Department of Biochemistry and Molecular Biology, Penn State College of Medicine, Hershey, PA, USA

Gerhard H. Braus Department of Molecular Microbiology and Genetics, University of Goettingen, Goettingen, Germany

James R. Broach Department of Biochemistry and Molecular Biology, Penn State College of Medicine, Hershey, PA, USA

Mafalda Cavalheiro Department of Bioengineering, Instituto Superior Técnico, University of Lisbon, Lisboa, Portugal

iBB - Institute for Bioengineering and Biosciences, Biological Sciences Re-search Group, Lisboa, Portugal

Yukio Kimata Graduate School of Biological Sciences, Nara Institute of Science and Technology, Nara, Japan

Kenji Kohno Institute for Research Initiatives, Nara Institute of Science and Technology, Nara, Japan

Kamil Krol Institute of Biochemistry and Biophysics, Polish Academy of Sciences, Warszawa, Poland

Magdalena Kwolek-Mirek Department of Biochemistry and Cell Biology, Faculty of Biology and Agriculture, University of Rzeszow, Rzeszow, Poland

Roman Maślanka Department of Biochemistry and Cell Biology, Faculty of Biology and Agriculture, University of Rzeszow, Rzeszow, Poland

Thi Mai Phuong Nguyen Graduate School of Biological Sciences, Nara Institute of Science and Technology, Nara, Japan

Adrianna Skoneczna Institute of Biochemistry and Biophysics, Polish Academy of Sciences, Warszawa, Poland

Marek Skoneczny Institute of Biochemistry and Biophysics, Polish Academy of Sciences, Warszawa, Poland

Miguel Cacho Teixeira Department of Bioengineering, Instituto Superior Técnico, University of Lisbon, Lisboa, Portugal

iBB - Institute for Bioengineering and Biosciences, Biological Sciences Re-search Group, Lisboa, Portugal

Renata Zadrąg-Tęcza Department of Biochemistry and Cell Biology, Faculty of Biology and Agriculture, University of Rzeszow, Rzeszow, Poland

Abbreviations

6-4PP	(6-4) pyrimidone photoproducts
8-oxoG	8-oxo-7,8-dihydroguanine
9-1-1	Rad9-Rad1-Hus1 complex
AAA	ATPases associated with various cellular activities
ABC	ATP-binding cassette superfamily
ABPA	Allergic bronchopulmonary aspergillosis
ACD	Accidental cell death
AMPK	AMP-activated protein kinase
AO	Alcohol oxidase
AP site	Abasic sites
AP-1	Activating protein-1
APX	Ascorbate peroxidase
ATP	Adenosine triphosphate
BER	Base excision repair
BiP	Immunoglobulin heavy chain binding protein
BIQ	1-[4-butylbenzyl]isoquinoline
BIR	Break-induced replication
BMDM	Bone marrow-derived macrophage
BZT	1,2-benzisothiazolinone
CAT	Catalase
CDC	Cell division cycle
CDK	Cyclin-dependent kinase
CFTR	Cystic fibrosis transmembrane conductance regulator
CMG	Cdc42/Mcm2-7/GINS helicase complex
CO	Crossover
CPA	Chronic pulmonary aspergillosis
CPD	Cyclobutane bi-pyrimidine dimers
CRD	Cysteine rich domains
CWI	Cell wall integrity
DDR	DNA damage response
dHJ	Double Holliday junction
dNTP	Deoxynucleotides
DRR	DNA replication stress response

DSB	DNA double-strand breaks
DSBR	Double-strand break repair
dsDNA	Double-stranded DNA
DTT	Dithiothreitol
eIF2α	Eukaryotic translation initiation factor 2α
ER	Endoplasmic reticulum
ERAC	ER-associated compartment
ERAD	ER-associated degradation
ERK	Extracellular signal-regulated kinase
ESCRT	Endosomal sorting complexes required for transport
ESR	Environmental stress response
FBS	Fetal bovine serum
FKBP	FK506 binding protein
GARP	Golgi-associated retrograde protein transport
GGR	Global genome repair
GOF	Gain-of-function
GPI	Glycosylphosphatidylinositol
GPX, CysGPx	Glutathione peroxidase
GR	Glutathione reductase
GRX	Glutaredoxin
GS	Glutathione synthetase
GSH	Glutathione
GSR	General stress response
GSSG	Glutathione disulfide
GST	Glutathione S-transferase
HDAC	Histone deacetylase
HHP	High hydrostatic pressure
HisRS	Histidyl-tRNA synthetase
HJ	Holliday junction
HOC	High oxygen consumption
HOG	High osmolarity glycerol
HR	Homologous recombination
HSP	Heat shock protein
HSR	Heat shock response
IA	Invasive aspergillosis
ICL	Inter-strand cross-links
IDP	Intrinsically disordered proteins
INQ	Intranuclear quality-control compartment
IPOD	Insoluble protein deposit
JUNQ	Juxtanuclear quality-control compartment
KHMTase	Lysine histone methyltransferase
LBER	Long-patch BER
LOC	Low oxygen consumption
MAPK	Mitogen-activated protein kinases
MDR	Multidrug resistance

MEC	Mitotic exit checkpoint
MEK	MAPK/ERK kinase
MFS	Major facilitator superfamily
MMR	Mismatch repair
MRX	Mre11-Rad50-Xrs2 complex
MVB	Multivesicular body
NADPH	Nicotinamide adenine dinucleotide phosphate
NCO	Non-crossover
NCR	Nitrogen catabolite repression
NEF3	Nucleotide excision factor 3
NER	Nucleotide excision repair
NES	Nuclear export sequence
NHEJ	Non-homologous end joining
NLS	Nuclear localization sequence
NO	Nitric oxide
NOX	NADPH oxidase
OSR	Oxidative stress response
PAF-AH	Platelet-activating factor acetylhydrolase
PAPS reductase	Reductase 3$'$-phosphoadenosine-5$'$-phosphosulfate
PCD	Programmed cell death
PDS	Post-diauxic shift
PGPF	Plant growth promoting fungi
PGPR	Plant growth promoting rhizobacteria
PIKK kinase	Phosphoinositide 3-kinase-related kinase
PKA	Protein kinase A
PM	Plasma membrane
Pol	DNA polymerase
polyQ	Polyglutamine
PP pathway	Pentose phosphate pathway
PRX, TPx	Peroxiredoxin
PTM	Post translational modification
RER	Ribonucleotide excision repair
RESS	Repression under secretion stress
RiBi	Ribosomal biogenesis
RIDD	Regulated IRE1-dependent decay of mRNA
RNR	Ribonucleotide reductase
RNS	Reactive nitrogen species
ROS	Reactive oxygen species
RP	Ribosomal protein
RPA	Replication protein A complex
SAC	Spindle assembly checkpoint
SAM	S-adenosyl methionine
SAP	Secreted aspartyl proteinase
SAPK	Stress-activated protein kinase
SAPK cascade	Stress-responsive MAPK cascade

SBER	Short-patch BER
SDSA	Synthesis-dependent strand annealing
SM	Secondary metabolite
SmF	Submerged fermentation
SOD	Superoxide dismutase
SREBP	Sterol regulatory element-binding proteins
SSA	Single-strand annealing
SSB	Single-strand break
ssDNA	Single-stranded DNA
SSF	Solid-state fermentation
STRE	Stress response element
t-BOOH	*tert*-Butyl hydroperoxide
TCA	Tricarboxylic acid cycle
TCR	Transcription coupled repair
TF	Transcription factor
TLS	Trans-lesion synthesis
TRR	Thioredoxin reductase
TRX	Thioredoxin
Ub	Ubiquitin
UGGT	UDP-Glucose:glycoprotein glucosyltransferase
UPR	Unfolded protein response
UPRE	Unfolded protein response element
UPS	Ubiquitin-dependent proteasome system
UV	Ultraviolet
V-ATPase	Vacuolar H+-ATPase
VOC	Volatile compound
WOA	Weak organic acids
YMC	Yeast metabolic cycle
γ-GCS	γ-Glutamylcysteine synthetase

Response Mechanisms to Oxidative Stress in Yeast and Filamentous Fungi

Renata Zadrąg-Tęcza, Roman Maślanka, Sabina Bednarska, and Magdalena Kwolek-Mirek

Abstract

Oxidative stress is defined as a disturbance in the balance between the production of reactive oxygen species (ROS) and antioxidant defences, and thereby the ROS level exceeds the antioxidant capacity needed to maintain the intracellular redox environment in a reduced state. ROS are produced as a result of normal cellular metabolism as well as environmental factors. They display stronger reactivity than molecular oxygen and have the capacity to damage cellular components, which leads to disturbance in cellular homeostasis. The ability to detect environmental signals, perform signal transduction and trigger the response is a fundamental feature of all cells and organisms. This chapter reviews factors causing oxidative stress and the complex system of stress response including not only rapid transcriptional changes but also activation of the response to damage, which enables the cell to survive or trigger the cell death pathway.

1.1 Introduction

Oxygen is one of the most important elements for living organisms, currently making up almost 21% of the Earth's atmosphere. The emergence of oxygen and gradual increase of its content in the atmosphere became an important factor for evolutionary changes enabling the development of complex life forms (Reinhard et al. 2016; Taverne et al. 2018). This is mainly related to specific chemical and thermodynamic properties of oxygen enabling the development of aerobic respiration. Yet aerobic respiration is not only connected with higher efficiency of energy

R. Zadrąg-Tęcza (✉) · R. Maślanka · S. Bednarska · M. Kwolek-Mirek
Department of Biochemistry and Cell Biology, Faculty of Biology and Agriculture, University of Rzeszow, Rzeszow, Poland
e-mail: retecza@ur.edu.pl

© Springer Nature Switzerland AG 2018
M. Skoneczny (ed.), *Stress Response Mechanisms in Fungi*,
https://doi.org/10.1007/978-3-030-00683-9_1

production but also provides intermediate metabolites that are vital for maintaining internal homeostasis of the cell (Jiang et al. 2010). The role of oxygen, however, has been raising a lot of controversy and is constantly a subject of scientific discussion. The controversy is primarily related to the fact that aerobic metabolism leads to production of toxic by-products such as reactive oxygen species (ROS). ROS is a collective term describing free radicals, including the superoxide anion and hydroxyl radical as well as nonradical but highly reactive molecules formed upon incomplete reduction of oxygen, such as hydrogen peroxide (Perrone et al. 2008). ROS have been traditionally viewed as harmful because they manifest a stronger reactivity than molecular oxygen and have the capacity to damage cellular components. As a consequence, ROS cause disturbance of cellular homeostasis, which may lead to cell ageing or death. However, recent data have emphasised that a low level of ROS is necessary for important biological processes such as cell proliferation or differentiation (Theopold 2009; Bigarella et al. 2014). ROS may act as signalling molecules in many physiological processes via oxidation thiol groups (–SH) of cysteine residues in protein kinases (Schieber and Chandel 2014; Zhang et al. 2016). ROS can also be an important element responsible for ensuring of the cellular redox homeostasis which is crucial for maintaining efficiency of many intracellular processes, and their disorders influence regulation of cell growth or ageing process (Ayer et al. 2013). Considering both the beneficial and the toxic nature of ROS, the key issue is the type and content of ROS. Low levels of ROS induce signalling pathways, thus supporting physiological processes, whereas high levels of ROS can be toxic. Therefore, in a physiological state, cellular ROS level is maintained in a dynamic equilibrium by cellular processes responsible for ROS generation and elimination. ROS are produced under cellular metabolic processes and by environmental factors. Systems responsible for removal or detoxification of ROS include superoxide dismutase, catalase, glutathione peroxidase, components of glutathione/glutaredoxin and thioredoxin. Disturbance in the balance between the production of reactive oxygen species and antioxidant defences is referred to as oxidative stress. This is the state when the ROS level exceeds the antioxidant capacity needed to maintain the intracellular redox environment in a reduced state (Brown et al. 2017).

Activation of stress response mechanism, which has a dose-dependent nature, is essential for survival in stress conditions. The biological purpose of oxidative stress response in the first place is to restore the cellular redox homeostasis. This requires transcriptional changes regulated via Yap1p and Skn7p factors responsible for the expression of key genes coding enzymes required for ROS detoxification (Lee et al. 1999; Ayer et al. 2013). It is considered that transcriptional activation is preceded by the metabolic flux from glycolysis to the pentose phosphate pathway, which promotes generation of reducing power in the form of NADPH (Ralser et al. 2007). Effectiveness of these activities defines the next steps of cellular stress responses. Increase in the ROS level leads to transiently delayed cell cycle progression to enable DNA-damage repair and activation of the cellular degradation system. These elements of the stress response mechanism are intended to maintain cell vitality and ensure its survival. In the case of higher doses of ROS, mechanisms responsible for triggering the apoptotic cell death programme are activated.

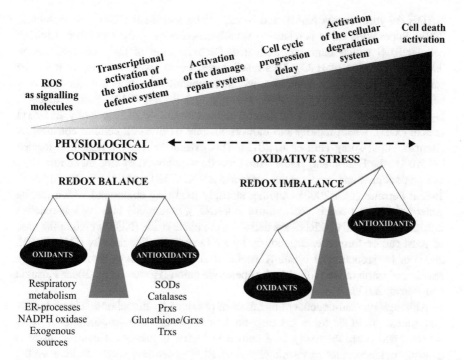

Fig. 1.1 The subsequent steps of the complex response mechanism to oxidative stress in yeast

The mechanism of response to oxidative stress is complex and adjusted to the ROS level (Fig. 1.1). Basic principles of this mechanism are generally conserved among eukaryotic organisms as well as different species of fungi. Nevertheless, there are also some differences arising, for instance, from fungal life strategies or ecological conditions. The primary goal of this chapter is to describe the principal mechanisms of response to oxidative stress on the basis of the data related to the *Saccharomyces cerevisiae* yeast and to identify differences between *S. cerevisiae* and other selected fungi presented in the last paragraph of this chapter.

1.2   Sources of Reactive Oxygen Species

Cellular metabolism is strictly connected with electron transfer operations; in most of such operations, oxygen acts as an electron acceptor. Hence, metabolic reactions may result in formation of ROS. Mitochondria were generally considered as the main source of ROS, also in the *S. cerevisiae* yeast (Longo et al. 1996). A primary ROS is superoxide anion ($O_2^{\bullet-}$), which is generated by electron leakage in the mitochondrial transport chain and one-electron reduction of molecular oxygen (Bartosz 2009). In the case of yeast, the main sources of one-electron leakage in the respiratory chain are complex III (the cytochrome bc1 complex) and the external

NADH dehydrogenases Nde1p and Nde2p (Fang and Beattie 2003; Herrero et al. 2008). Superoxide anion is relatively weakly reactive but is the precursor of further cellular ROS which must be eliminated. Detoxification of $O_2^{\bullet-}$ via dismutation leads to generation of H_2O_2. This reaction can occur spontaneously, or it can be catalysed by superoxide dismutase, whose enzymatic activity is a major source of cellular H_2O_2 (Herrero et al. 2008; Farrugia and Balzan 2012). Similar to $O_2^{\bullet-}$, H_2O_2 stimulates generation of several more harmful ROS, especially hydroxyl radical ($^{\bullet}OH$), which rapidly and indiscriminately oxidises all cellular constituents (Bartosz 2009; Aung-Htut et al. 2012). This extremely reactive radical is formed in vivo via the Fenton and Haber-Weiss reactions catalysed by free metal ions such as iron or copper and involves the combined action of H_2O_2 and $O_2^{\bullet-}$ (Biliński et al. 1985a; Perrone et al. 2008). Another strongly oxidising compound which can be generated in yeast cell is peroxynitrite ($ONOO^-$), commonly classified as reactive nitrogen species (RNS). RNS are derived from nitric oxide (NO), which in the case of yeast can be formed endogenously by a NO synthase-like activity (Osório et al. 2007) or by reduction of nitrite (Castello et al. 2006). Peroxynitrite is formed by reaction of nitric oxide radical with superoxide anion (Herrero et al. 2008; Farrugia and Balzan 2012).

Although oxygen-dependent metabolism provided by mitochondria is an important source of ROS, there are also no less important extra-mitochondrial ROS sources, which are relatively well known in higher organisms. However, there is growing evidence that extra-mitochondrial ROS generation seems to be essential also in the case of yeast (Rinnerthaler et al. 2012; Leadsham et al. 2013; Busti et al. 2016; Maslanka et al. 2017). It was confirmed by results showing that yeast growing in the medium with high glucose concentration where mitochondrial respiration is repressed has a high $O_2^{\bullet-}$ level (Reddi and Culotta 2013; Maslanka et al. 2017). The main compartment involved in extra-mitochondrial ROS generation is endoplasmic reticulum (ER). In the ER-lumen, oxidative protein folding takes place, of which the side effect is the formation of H_2O_2 and other ROS. Their generation is connected especially with disulfide bond formation, during which two electrons are transferred to the protein disulfide isomerase (Pdi1p), flavin-containing oxidoreductase (Ero1p) and oxygen, the latter acting as the terminal electron acceptor. Furthermore, ER stress caused by accumulation of misfolded proteins in the ER-lumen induces production of ROS including $O_2^{\bullet-}$ (Perrone et al. 2008; Aung-Htut et al. 2012). The connection between ER and extra-mitochondrial ROS production is also strengthened by studies demonstrating the existence and activity of Yno1p, which is the only NADPH oxidase discovered so far in the *S. cerevisiae* yeast (Rinnerthaler et al. 2012; Leadsham et al. 2013). NADPH oxidases are the ER-located enzymes producing $O_2^{\bullet-}$ by a one-electron reduction of oxygen in a NADPH-dependent fashion. It is worth underlining that in contrast to the above sources, $O_2^{\bullet-}$ generated by NADPH oxidases is formed as the main product of the reaction, not as a side effect of their action (Rinnerthaler et al. 2012). The activity of NADPH oxidases is directly dependent on the availability of cytosolic NADPH formed mainly in the pentose phosphate (PP) pathway, which generates two important issues: (1) formation of extra-mitochondrial ROS may be directly connected with the cellular biosynthesis processes (Weinberger et al. 2010; Maslanka et al. 2017) and (2) - pro-oxidant properties of NADPH (Maslanka et al. 2017; Xiao et al. 2018) may stand in opposition to its generally accepted antioxidant role (Izawa et al. 1998; Xiao et al. 2018). ER is also involved in detoxification of xenobiotics. This task is fulfilled by ER detoxifying enzymes, including cytochrome P-450 isozymes, which are able to reduce molecular oxygen and produce $O_2^{\bullet-}$ and H_2O_2 (Crešnar and Petrič 2011). However, the precise location of the actual cellular compartments responsible for ROS production may be difficult to determine because there is a connection between mitochondrial and ER-depended ROS generation (Perrone et al. 2008; Leadsham et al. 2013; Busti et al. 2016). Interorganellar signalling may be assured by using Ca^{2+} as a secondary signalling molecule (Perrone et al. 2008; Busti et al. 2016). Mitochondrial dysfunction may lead to increased production of ROS from the ER-localised NADPH oxidase Yno1p (Leadsham et al. 2013). In turn, severe ER stress leads to an increased release of Ca^{2+} from ER to the cytosol, which may elevate mitochondrial matrix Ca^{2+} levels and result in an increased production of ROS by the respiratory chain (Perrone et al. 2008). On the other hand, depletion of Ca^{2+} induces ER stress and causes an increased generation of ROS, which is, however, independent of mitochondrial activity (Busti et al. 2016). Other extra-mitochondrial cellular compartments of ROS production are peroxisomes. In yeast, these organelles contain several enzymes crucial for the β-oxidation of fatty acids or oxidative deamination of D-amino acids, which lead to generation of H_2O_2 and other ROS as a by-product (Herrero et al. 2008).

The intracellular generation of ROS in the case of yeast is strictly connected with type and availability of nutrients or culture conditions. The level of ROS may differ depending both on carbon source, e.g. glucose, fructose, glycerol, ethanol and growth phase of the culture (Drakulic et al. 2005; Perez-Gallardo et al. 2013; Semchyshyn et al. 2014). Additionally, several environmental stressors such as UV radiation (Kozmin et al. 2005), heavy metals such as cadmium (Brennan and Schiestl 1996) and arsenic (Du et al. 2007), acetic acid (Ludovico et al. 2001), salt (Koziol et al. 2005), ethanol (Charoenbhakdi et al. 2016) and heat stress (Moraitis and Curran 2004) are connected with ROS production and oxidative stress induction. Oxidative stress may also be generated by addition of redox-cycling xenobiotics, of which paraquat and menadione are well-known ROS inducers used in yeast studies. Their activity is based on stimulation of $O_2^{\bullet-}$ by accepting electrons from cellular reducers and transferring them to molecular oxygen (Herrero et al. 2008).

1.3 Antioxidant Defence and Thiol Redox Homeostasis

High reactivity and oxidising properties of ROS force cells to control their ROS levels and develop protective mechanisms. The purpose of these mechanisms is to (1) prevent reaction of ROS with cellular components, (2) interrupt the free-radical chain reactions and (3) remove the effects of ROS reactions with biomolecules. Effective defence mechanisms allow the cell to maintain redox homeostasis, i.e. the

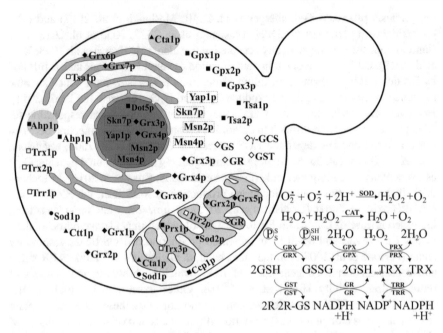

Fig. 1.2 The intracellular localisation of the transcription factors and the main elements of the antioxidant defence and thiol redox homeostasis. The proteins within the same group were indicated by appropriate markers: (filled dots) superoxide dismutases, (filled triangles) catalases, (filled squares) peroxidases, (open diamonds) glutathione/(filled diamonds) glutaredoxin pathway, (open square) thioredoxin pathway. The transcription factors were outlined

balance between oxidation and reduction reactions. Maintaining redox homeostasis requires the presence of a well-coordinated system of proteins and low molecular weight compounds. An important role in the defence against ROS is played by antioxidant enzymes, which catalytically reduce ROS through electron transfer, often via hydrogen (Fig. 1.2). An antioxidant enzyme that catalyses the two-step disproportionation of superoxide anion to hydrogen peroxide and oxygen molecule (dioxygen) is superoxide dismutase (SOD). In the *S. cerevisiae* yeast, there are two SOD isoenzymes—the copper- and zinc-containing enzyme (CuZnSOD, SOD1) occurring in the cytosol and in the intermembrane space of mitochondria and the manganese-containing enzyme (MnSOD, SOD2) located in the mitochondrial matrix (Sturtz et al. 2001; Culotta et al. 2006). The Sod2p and the mitochondrial fraction of Sod1p are involved in the disproportionation of $O_2^{\bullet-}$ generated by the mitochondrial respiratory chain, while the cytosolic fraction of Sod1p seems to protect yeast cells against $O_2^{\bullet-}$ formed upon other metabolic reactions and xenobiotic biotransformation (Sturtz et al. 2001; Herrero et al. 2008; Klöppel et al. 2010). Sod1p also plays a role in signalling through H_2O_2, a product of reaction of this enzyme. In response to elevated endogenous and exogenous ROS, Sod1p may rapidly relocate into the nucleus, then bind to the promoters and regulate the expression of genes involved in antioxidant defence and DNA-damage repair (Tsang et al. 2014). Deletion of *SOD1* gene causes a number of oxidative stress-related alterations in cells, among others a lower growth rate in air, inability to grow in an atmosphere of 100% oxygen in a rich cultivation medium, auxotrophy with respect to lysine and methionine (Biliński et al. 1985b), elevated free iron concentration (Srinivasan et al. 2000) and inactivation of proteins containing the 4Fe–4S groups (Wallace et al. 2004). Moreover, the Δ*sod1* mutant exhibits an increased sensitivity to oxidative stress-inducing agents that either generate ROS, such as paraquat and menadione (Biliński et al. 1985b; Gralla and Valentine 1991), or decrease the level of reduced glutathione (GSH), e.g. dithiopyridine (López-Mirabal et al. 2007) and acrolein (Kwolek-Mirek et al. 2009). Depletion of *SOD1* was also shown to increase intracellular ROS content (Drakulic et al. 2005; Kwolek-Mirek et al. 2011). Furthermore, the loss of solely *SOD1* or both *SOD1* and *SOD2* dramatically reduced the chronological and replicative lifespans of the yeast (Longo et al. 1996; Barker et al. 1999; Wawryn et al. 1999).

The level of H_2O_2 increasing inter alia as a result of $O_2^{\bullet-}$ disproportionation is reduced by antioxidant enzymes such as catalase and peroxidase. Catalase (CAT) is an iron-containing enzyme with a heme reactive group located inside the enzyme structure, which catalyses the reduction of H_2O_2 into O_2 (dioxygen) and H_2O molecules. In yeast, there are two catalase isoenzymes—catalase T (CTT1) in the cytosol and catalase A (CTA1) in peroxisomes and mitochondria (Petrova et al. 2004). Ctt1p activity is important during the oxidative stress response and in protecting proteins against oxidative inactivation (Wieser et al. 1991; Nishimoto et al. 2016). *CTT1* gene expression is induced by heat, osmotic, starvation and H_2O_2 stress (Grant et al. 1998). On the other hand, Cta1p is involved in the detoxification of H_2O_2 formed during fatty acids β-oxidation in peroxisomes and as a result of SOD reaction in mitochondria (especially in respiratory conditions) (Petrova et al. 2004). Surprisingly, a loss of either individual catalase or both *CTT1* and *CTA1* had no effect on H_2O_2 sensitivity (Izawa et al. 1996; Grant et al. 1998). These results demonstrate that there are other enzymes—peroxidases—that can effectively remove H_2O_2 in cell. The yeast peroxidases, unlike superoxide dismutases and catalases, are not metalloenzymes (except cytochrome c peroxidase). They reduce hydrogen peroxide or other organic peroxides using thiols as an electron donor. There are three families of peroxidases: glutathione peroxidase (GPX, CysGPx), thioredoxin-dependent peroxidase (peroxiredoxins; PRX, TPx) and cytochrome c peroxidase (Herrero et al. 2008). Glutathione peroxidases Gpx1p, Gpx2p and Gpx3p catalyse reduction of H_2O_2 or phospholipid hydroperoxides to H_2O or corresponding alcohol using reduced glutathione (Inoue et al. 1999; Avery and Avery 2001), although for Gpx2p the thioredoxin system is the preferable electron donor in vivo (Tanaka et al. 2005). Gpx3p has also been identified as a redox transducer of Yap1p, and this activity is strictly dependent on the thioredoxin system (Delaunay et al. 2002). Furthermore, in yeast there are five peroxiredoxins, namely, Ahp1p, Tsa1p, Tsa2p, Dot5p and Prx1p, which detoxify H_2O_2 or alkyl hydroperoxides to H_2O or respective alcohol. The oxidised peroxiredoxins are then reduced by thioredoxins. Peroxiredoxins are not redundant enzymes because they differ in subcellular localisation and peroxide specificities. Tsa1p, Tsa2p and Ahp1p are cytosolic enzymes, although Ahp1p contains a peroxisomal targeting signal. In contrast, Dot5p is nuclear, whereas Prx1p is mitochondrial (Park et al. 2000; Fourquet et al. 2008). Tsa1p may be also presented in the form associated with ribosomes while providing a protective function in protein translation processes (Trotter et al. 2008). Deletion of three genes coding glutathione peroxidases and five genes coding peroxiredoxins (Δ8 mutant) leads to increased mutational load and decreased vitality (Kaya et al. 2014). Furthermore, in the mitochondrial transmembrane space of yeast, there is cytochrome c peroxidase (Ccp1p), which catalyses the reduction of H_2O_2 to water by ferrocytochrome c (Jiang and English 2006). Besides its antioxidative role, Ccp1p performs a signalling function by transmitting the stress signal to the transcription factor Skn7p (Charizanis et al. 1999; Jiang and English 2006) and participates in heme turnover in cells (Kathiresan and English 2016).

The thiol redox system, which consists of the glutathione/glutaredoxin and thioredoxin pathways, uses the cysteine residues to catalyse thiol-disulfide exchange reactions, thereby controlling the redox state of cytoplasmic cysteine residues and regulating their biological functions. The yeast glutathione/glutaredoxin pathway consists of glutathione, enzymes associated with its metabolism and two classes of glutaredoxins (Toledano et al. 2007). Glutathione (GSH) γ-L-glutamyl-L-cysteinylglycine is a tripeptide whose thiol group participates in redox reactions. GSH is synthesised in a two-step ATP-mediated reaction in cytosol. In the first step, γ-glutamylcysteine is formed from glycine and cysteine; this reaction is catalysed by γ-glutamylcysteine ligase (γ-GCS, γ-glutamylcysteine synthetase, product of *GSH1* gene). In the second step, the reaction of γ-glutamylcysteine and glycine is catalysed by glutathione synthetase (GS, product of *GSH2* gene) (Penninckx 2002). Deletion of the *GSH1* gene results in the inability to grow in a minimal cultivation medium without GSH and hypersensitivity to oxidative stress (Lee et al. 2001). Yeast can also take up glutathione from the cultivation medium via specific Hgt1p transporter (Bourbouloux et al. 2000). The formation of double bond between the thiol groups of two glutathione molecules leads to formation of glutathione disulfide (GSSG). GSH oxidation may occur in a non-enzymatic way or via glutathione peroxidase, if the oxidising agent is H_2O_2 or organic hydroperoxides. In turn, GSSG is reduced to GSH in a reaction catalysed by glutathione reductase (GR, product of *GLR1* gene) using NADPH as the reducing agent (Grant 2001; López-Mirabal and Winther 2008). Deletion of the *GLR1* gene results in more oxidative intracellular redox potential and hypersensitivity to oxidative stress (Grant et al. 1996; López-Mirabal et al. 2007). In yeast, the reduced form of glutathione occurs in high concentrations (2–10 mM) (Penninckx 2002), whereas its oxidised form is presented in much lower concentrations (4 µM) (Østergaard et al. 2004). The GSH/GSSG ratio in the total yeast cell extracts ranges from 30 to 100, which corresponds to the redox potential of GSH/GSSG couple from −220 to −232 mV (Schafer and Buettner 2001). Both the glutathione concentration and the redox potential of GSH/GSSG couple differ between cellular compartments. The GSH/GSSG couple is considered to be the main cellular redox buffer that may affect the redox status of proteins (Penninckx 2002; López-Mirabal and Winther 2008). Glutathione plays an important role in the response to oxidative stress. Cytosolic glutathione represents the first major pool of thiols which may be the target of oxidation in response to exposure to an exogenous oxidant. In turn, mitochondrial glutathione pool is crucial for oxidant tolerance (Gostimskaya and Grant 2016). Furthermore, glutathione also plays a role in cellular defence against reactive electrophiles, in particular xenobiotics. Many xenobiotics can react either spontaneously with GSH to form glutathione S-conjugates or via glutathione S-transferases (GST). Glutathione S-conjugates are transported to vacuole by Ycf1p or outside the cell (Zadziński et al. 1996). Glutaredoxins (GRXs) are another element of glutathione/glutaredoxin pathway in the thiol redox system. These enzymes are thiol-disulfide oxidoreductases that reduce protein disulfides or glutathione-protein mixed disulfides employing GSH as electron donor. The oxidised form of GRXs is transferred to the reduced form via GR by receiving electrons from NADPH to regenerate GSH from GSSG formed during the GRXs reaction (Herrero et al. 2010). Due to a number of catalytic thiol groups (number of cysteine residues in their active site), GRXs are divided into two classes: dithiol and monothiol. There are three dithiol GRXs in yeast, namely, Grx1p, Grx2p and Grx8p. These are cytosolic proteins, but the Grx2p fraction is also found in mitochondria (Pedrajas et al. 2002; Deponte 2013). Dithiol GRXs contribute to the regulation of redox signalling by modulation of the redox status of Trx's target proteins (Herrero et al. 2008; Matsuzawa 2017). Moreover, Grx1p and Grx2p have partially overlapping functions with TRX, GPX and GST (Herrero et al. 2008; Deponte 2013). Mutant $\Delta grx1$ is sensitive to oxidative stress induced by $O_2{}^{\bullet-}$, whereas mutant $\Delta grx2$ is sensitive to H_2O_2 (Grant 2001). Other yeast glutaredoxins are monothiol GRXs (Grx3-7p). Grx3p and Grx4p are located both in cytosol and nucleus. They participate in the regulation of the transcription factor Aft1p, which induces the expression of the genes related to iron assimilation. Deletion of *GRX3* and *GRX4* genes leads to disturbance of iron homeostasis and its accumulation in cell and increases sensitivity to oxidants (Pujol-Carrion et al. 2006; Herrero et al. 2008). Grx5p is a mitochondrial protein and is involved in biosynthesis of Fe–S groups. Deletion of the *GRX5* gene causes slow growth, inability to respire, iron accumulation, inactivation of mitochondrial enzymes containing Fe–S groups, high content of carbonyl group in proteins and hypersensitivity to oxidants (Rodríguez-Manzaneque et al. 2002; Herrero et al. 2008). Yeast cells also contain membrane-associated glutaredoxins Grx6p and Grx7p structurally and enzymatically similar to dithiol GRXs, which may act as redox regulators at the early stages of the protein secretory pathway (Izquierdo et al. 2008).

Thioredoxin pathway in the thiol redox system consists of thioredoxin (TRX), thioredoxin reductase (TRR), NADPH coenzyme and regulatory proteins of TRX (Pedrajas et al. 1999; Matsuzawa 2017). TRX has two thiol moieties in redox-active cysteine residues, Cys32 and Cys35 as Trx1p active sites, which are important for the activity and conformation of TRX. In the inactive and oxidised form of TRX, disulfide bonds are formed between two conserved thiol moieties. The oxidised TRX is transformed to the active and reduced form via TRR by receiving electrons from NADPH. Reduced TRX may transfer reducing equivalents to many substrates. The active TRX reduces proteins with disulfide bonds by transferring electrons from its reactive thiol moieties. In yeast, there are two cytosolic isoenzymes of thioredoxin, Trx1p and Trx2p, and one mitochondrial, Trx3p. In addition, yeast has two isoenzymes of thioredoxin reductase—Trr1p in the cytosol and Trr2p in the mitochondrion (Pedrajas et al. 1999). Trx1p and Trx2p reduce inter alia ribonucleotide reductase (RNR; Rnr1p-Rnr4p) involved in the maintenance of an appropriate pool of nucleotides for DNA synthesis, reductase $3'$-phosphoadenosine-$5'$-phosphosulfate (PAPS reductase; Met16p) necessary for sulfur assimilation and methionine biosynthesis and cytosolic peroxiredoxin Tsa1p involved in the detoxication of hydroperoxides (Garrido and Grant 2002; Vignols et al. 2005; Herrero et al. 2008). On the other hand, Trx3p has been implicated in the defence against ROS generated during respiration metabolism (Pedrajas et al. 1999). Deletion of either *TRX1* or *TRX2* has no effect on cell growth or morphology, but deletion of both genes affects the cell cycle and leads to methionine auxotrophy. The Δ*trx1*Δ*trx2* mutant shows a 33% increase in generation time and a significant increase in cell size (Muller 1991). Moreover, this double mutant is hypersensitive to H_2O_2, *t*-BOOH, DTT and acrolein (Trotter and Grant 2002; Vignols et al. 2005; Kwolek-Mirek et al. 2009).

1.4   Transcriptional Regulation of Antioxidant Defence

Stress response on the transcriptional level is a programme of modifications in gene expression to induce the production of a number of cytoprotective proteins to counteract oxidative stress. This programme includes direct or indirect activation of transcription factors as well as binding of such factors to specific DNA regions and induction of transcription of target genes mainly associated with antioxidant defences and thiol redox homeostasis maintenance. Several transcription factors, such as Yap1p, Skn7p, Msn2p and Msn4p, attend the response to oxidative stress (Fig. 1.2); however, Yap1p is the major regulator of oxidative stress response in the *S. cerevisiae* yeast.

The crucial role of Yap1p in oxidative stress sensing and response depends on the redox reactivity of its cysteine thiol residues. Yap1p is a transcription factor distributed mainly in the cell cytosol and in the nucleus under non-stress conditions and is rapidly transported to the nucleus upon oxidation of respective thiol groups (Kuge et al. 1997). Yap1p consists of basic leucine zipper (b-Zip) domain and two cysteine-rich domains (CRD) containing six redox-active cysteines. The N-CRD region constitutes the nuclear localisation sequence (NLS), while C-CRD constitutes the nuclear export sequence (NES). NLS binds Pse1p, which mediates the transfer of Yap1p into the cell nucleus (Isoyama et al. 2001), and NES binds exportin Crm1p (Yan et al. 1998). The interaction of Yap1p with Crm1p and the resulting Yap1p export from the nucleus is inhibited when the thiol groups of cysteine in CRDs become oxidised. Accumulation of Yap1p in the nucleus causes binding to specific DNA regions (Yap1 response element, YRE) (Kuge et al. 2001; Nguyên et al. 2001) and target gene expression. YRE sequence was found in the promoters of many genes (~70), in particular the genes responsible for glutathione synthesis (*GSH1*), thioredoxin system (*TRX2*, *TRR1*) and antioxidant enzymes (*SOD1*, *CTT1*) (Lee et al. 1999; Temple et al. 2005). Yap1p may be activated in a Gpx3p-dependant or Gpx3-independent manner (Azevedo et al. 2003). The former is characteristic of H_2O_2 or other peroxides (Delaunay et al. 2002). Gpx3p (also called Orp1p, oxidant receptor protein) is a sensor for peroxides which directly oxidise catalytic thiol group of Gpx3p to sulfenic acid, which subsequently condenses transiently with a thiol group in Yap1p, thus forming intramolecular trans-domain disulfide bonds between Cys303–Cys598 and Cys310–Cys629 (Wood et al. 2003). An additional Ybp1p protein (Yap1-binding protein) is required for the signal transduction from Gpx3p to Yap1p (Veal et al. 2003). The Gpx3p-independent way of Yap1p activation involves direct covalent modifications of thiol cysteine residues only in the C-CRD domain (C598, C620 and C629) and does not require Ybp1p (Azevedo et al. 2003). This way of activation is typical for thiol-reactive electrophilic compounds like N-ethylmaleimide, 4-hydroxynonenal, cadmium, menadione, diamide (Azevedo et al. 2003), acrolein (Kwolek-Mirek et al. 2009; Ouyang et al. 2011) and allicin (a natural garlic bioactive substance) (Gruhlke et al. 2017). The disulfides in Yap1p are back-reduced by thioredoxin Trx2p, and the interaction of Yap1p with exportin Crm1p is allowed to move Yap1p again to cell cytosol (Delaunay et al. 2000; Mason et al. 2006). Deletion of both thioredoxin genes results in constitutively activated Yap1p (Izawa et al. 1999).

Skn7p is a constitutive nuclear protein, and its expression does not change in the presence of oxidants (Lee et al. 1999; Mulford and Fassler 2011). Skn7p cooperates with Yap1p in the response to peroxides. These transcription factors bind to different sites in DNA. Skn7p has the N-terminal DNA-binding domain similar to heat shock factor (Hsf1p) and the C-terminal receiver domain containing aspartic acid residue required for phosphorylation by Sln1p and is implicated in osmotolerance (Ketela et al. 1998). Approximately half of the genes controlled by Yap1p are regulated by both Yap1p and Skn7p (including superoxide dismutases *SOD1* and *SOD2*, peroxiredoxins *TSA1*, *AHP1*, thioredoxin *TRX2* genes), while the other half is regulated in a Skn7p-independent manner (glutathione synthesis *GSH1*, PP pathway) (Lee et al. 1999). Skn7p binds Hsf1p in the H_2O_2-dependent induction of heat shock genes (Raitt et al. 2000).

Msn2p and Msn4p are homological proteins taking part in the response to many environmental factors, including heat, osmotic, ethanol, low glucose or pH stress (Causton et al. 2001). These transcription factors bind to the STRE sequence, and their function is to control the general stress response; however, they also regulate some oxidative stress-responsive genes (e.g. *CTT1*), and the yeast mutant lacking both *MSN2* and *MSN4* genes is hypersensitive to oxidative stress (Martínez-Pastor et al. 1996). The purpose of general stress response mediated by Msn2/4p is activation of the cellular defence mechanism, which in the case of mild stress condition may lead to the two phenomena: adaptive stress response and cross-stress resistance. These phenomena concern resistance to a second, strong stress after exposure to mild-primary stressors, which in the adaptive stress response related to the same type of stressors while in the cross-stress resistance related to the different type of stressors (Swiecilo 2016; Auesukaree 2017).

1.5   Oxidative Modifications of Cellular Macromolecules

Activation of stress response aims to the restoration of the redox homeostasis and thereby prevention of the oxidative damages. However, when the level of ROS exceeds the antioxidant capacity of the cell, ROS can react with cellular macromolecules, including proteins, lipids and nucleic acids. Almost 70% of all the oxidised macromolecules are of proteinaceous nature, indicating that proteins are the most prominent in vivo targets of oxidants (Dahl et al. 2015). Protein oxidative modifications, also known as protein oxidation, can be generally classified into two categories, reversible and irreversible, both of which can be selectively induced by ROS and RNS. Most of the modifications are related to cytosolic proteins; others are connected with nuclear compartment (Cai and Yan 2013; Jung et al. 2014). Reversible oxidative modifications are related to proteins with an amino acid containing the thiol group. Cysteine and methionine are the only known amino acids that can be repaired in an enzymatic way. Modifications of cysteine involve S-sulfenation, S-nitrosylation, S-glutathionylation and disulfide formation. In turn, methionine is readily oxidised to methionine sulfoxide (Cai and Yan 2013; Dahl et al. 2015). All of these cysteine and methionine oxidation modifications play beneficial roles because they may protect the target proteins from further oxidation that would otherwise permanently damage the target proteins. These modifications also play a role in redox signalling cascades that boost cellular defence systems to better counteract stress conditions (Cai and Yan 2013). Moreover, interconversion of methionine and methionine sulfoxide may regulate biological activity of proteins through alteration of their catalytic properties and efficiency and through modulation of surface hydrophobicity of proteins (Levine et al. 2000). The systems responsible for reducing oxidative protein modification include both glutathione/glutaredoxin and thioredoxin pathways and methionine sulfoxide reductase (Levine et al. 2000; Jung et al. 2014). In yeast, S-glutathionylation of several proteins associated with glycolysis and biosynthesis under H_2O_2 was identified (Shenton and Grant 2003). On the other hand, irreversible oxidative modifications of proteins include mainly protein carbonylation but also oxidation of histidines and tryptophans, nitration of tyrosine, formation of dityrosines and sulfinic/sulfonic acids intermediates (Dahl et al. 2015). In the *S. cerevisiae* yeast, the carbonyl groups formed on several amino acids residues, including arginine, histidine, lysine, proline, tryptophan and cysteine, are the most widely used biomarkers of protein oxidation and oxidative stress level. They can be generated both during physiological conditions such as carbon and nitrogen starvation (Aguilaniu et al. 2001), during replicative and chronological lifespans (Reverter-Branchat et al. 2004) under exposure to, e.g. H_2O_2, menadione (Cabiscol et al. 2000) or acrolein (Kwolek-Mirek et al. 2015). While some ROS-/RNS-mediated posttranslational modifications are intentional and reversible and belong to redox-regulated processes, irreversible protein modifications are typically destabilising and capable of triggering of a major secondary stress on the proteostasis of the cell. These oxidative modifications may lead to oligomerisation, fragmentation, destabilisation, aggregation and/or enhanced degradation of proteins (Dahl et al. 2015).

Other cellular macromolecules that may be oxidised by ROS are lipids. Lipid peroxidation is a process under which oxidants attack lipids containing C–C bond(s), especially polyunsaturated fatty acids. Due to the fact that yeast does not contain such fatty acids, there is no lipid peroxidation process in its cells. However, there are many data where this parameter is determined and its level changes under the oxidative stress (Castro et al. 2008; Kwolek-Mirek et al. 2009; Braconi et al. 2010). These results are most likely the effect of non-specific reaction of the thiobarbituric acid (the most commonly used compound for determination of lipid oxidation) with aldehydes originating from sources other than lipid peroxidation (Dennis and Shibamoto 1989; Miyake and Shibamoto 1999).

Nucleic acids are more stable than proteins and lipids and therefore not easily transformed into a free-radical state. $O_2^{\bullet-}$ and H_2O_2 do not damage nucleic acids in contrast to $^{\bullet}OH$ where reaction with nucleic acids results in damage to pyrimidines and purines as well as disruption of phosphodiester bonds (single- and double-strand breaks). More than 100 different types of oxidatively damaged bases have been identified (Cadet and Wagner 2013). The guanine occurring in the DNA or RNA is the most prone to oxidation. The oxidatively derived products of guanine were consistently observed in high yields from $^{\bullet}OH$, carbonate radical or singlet oxygen oxidations. The major guanine base oxidation product is 8-oxo-7,8-dihydroguanine (8-oxoG), which serves as a marker of oxidative damage of nucleic acids. The 8-oxoG can pair with cytosine and adenine nucleotides with an almost equal efficiency causing transversion mutation (Kaya et al. 2014). In the *S. cerevisiae* yeast, these damages are induced among others by UVB-photosensitised titanium dioxide (Pinto et al. 2010) or lack of thiol-based redox control ($\Delta 8$ mutant) (Kaya et al. 2014) (For more details, see Chap. 3 of this book dedicated to yeast and other fungi response mechanisms to genotoxic stress).

1.6   Cell Cycle Progression Delay

Oxidative stress response mechanism of the yeast cells has a complex nature and includes a number of mutually complementing pathways activated in the right order depending on the type, time of exposure or the stress level. Activation of these pathways involves a high costs and therefore has repercussions on the proliferative activity of the cell. The molecular mechanism of cell cycle control is sensitive to many environmental conditions and stressors including oxidative stress, which may disturb intracellular homeostasis, e.g. disorder of redox control can result in cell entering the quiescent state (Chiu and Dawes 2012). The key point in cell cycle control is START occurring in the G1 phase, whereas its transition initiates DNA synthesis in S phase. START coordinates the cell cycle with cell growth. Additional checkpoints of the cell cycle located at the G1/S and G2/M boundaries are regarded as internal regulatory system that may arrest the cell cycle (Kraikivski et al. 2015). Yeast may respond to ROS exposure by delaying or even arresting cell cycle progression via different molecular mechanisms. The aim of cell cycle progression delay is to activate an adequate repair system (Morano et al. 2012). Depending on the type of ROS, cell cycle delay may take place at different checkpoints, e.g. treatment with H_2O_2 leads to cell cycle arrest in G2 phase via Rad9p-dependent pathway (Burhans and Heintz 2009), whereas treatment with $O_2^{\cdot-}$ generator such as menadione or paraquat leads to a G1 phase delay in cell cycle progression before the START checkpoint, independent on the Rad9p activation (Flattery-O'brien and Dawes 1998). The pathway induced by H_2O_2 involving Rad9p protein is activated mainly in response to DNA damage. Rad9p serves as the checkpoint protein for DNA damage and acts by recognition of damage and initiating a signal transduction cascade that activates Mec1p and Rad53p. It is considered that the *MEC1* and *RAD53* genes are the central transducers of the DNA checkpoint pathway; in turn, yeast Rad9p protein is subjected to Mec1-/Tel1-dependent hyperphosphorylation and interacts with Rad53p after DNA damage (Sun et al. 1996; Vialard et al. 1998). In turn, cell cycle arrest in the G1 phase induced by $O_2^{\cdot-}$, independent of Rad9p, is mediated by inhibition of transcription of *CLN1* and *CLN2* genes encoding cyclins required for passage through the G1 checkpoint. This type of cell cycle arrest is mediated by the Swi6p transcription factor that activates the expression of G1 phase cyclins (Fong et al. 2008). Swi6p has reactive Cys404. Its oxidation to the sulfenic acid leads to destabilisation of the Swi6p complexes with Swi4p or Mbp1p, which inhibits their recruitment to the promoters of G1 cyclin genes, thereby altering transcription of the cyclin genes required for triggering of the S phase. Swi6p regulates cell cycle delay by acting as a sensor and transducer for oxidative stress and is therefore considered to be a central modulator of cell cycle regulation in response to oxidative stress (Chiu et al. 2011). Deletion of *SWI6* leads to inability of cells to arrest the cell cycle in the G1 phase (Fong et al. 2008).

1.7 Induction of Cellular Degradation Systems in Response to Oxidative Stress

The mechanisms that aim at detecting, repairing and/or removing oxidatively damaged macromolecules are necessary for cell survival in stress conditions. Eukaryotic cells possess two degradation mechanisms that remove irreparably oxidised biomolecules: the ubiquitin-dependent proteasome system (UPS) and autophagy. They both play a pivotal role in the removal of non-functional or unnecessary proteins, thereby maintaining cellular proteostasis. Nevertheless, autophagy accounts for elimination of damaged organelles, including selective degradation of the mitochondria.

The UPS is a highly conserved mechanism of selective degradation of misfolded, damaged or currently unnecessary intracellular proteins. Under physiological conditions, functional UPS is responsible for cell proliferation, as many short-lived proteins involved in cell cycle progression are substrates for proteolysis (Wolf and Hilt 2004). In the growing *S. cerevisiae* yeast, more than 90% of cellular protein degradation is ensured by UPS, and about 80% of the 26S proteasome is located in the nucleus (Russell et al. 1999). Selection of degraded proteins is precisely controlled. Substrate proteins are tagged by attachment of ubiquitin

(Ub) via isopeptide bond between the ε-amino group of a substrate lysine residue and the C-terminal glycine of Ub (Hershko and Ciechanover 1998). The topology and length of the attached ubiquitin chains are particularly important. Substrates modified by a polyubiquitination through Lys48 of Ub are targeted to proteolysis in the proteasome. The process is a cascade mechanism involving three types of enzymes: ubiquitin-activating enzyme (E1), ubiquitin-conjugating enzymes (E2) and the specific ubiquitin-protein ligase (E3) (Finley et al. 2012). The yeast cell has a single E1 enzyme (encoded by the *UBA1* gene), 11 E2 enzymes and a large family of E3s (about 60–100) (Finley et al. 2012). When the substrate proteins are tagged by Ub, they undergo degradation by the proteasome, a large macromolecular complex of proteins. It generally occurs in the form of 26S proteasome composed of two subcomplexes: a catalytic core complex (20S proteasome) and regulatory complex (19S complex) (Groll et al. 1997; Finley et al. 2012). The 20S proteasome is the enzymatic core of the proteasome system. It is formed by 28 subunits arranged in four heptameric rings: two outer α-rings and two inner β-rings. The α-subunits are responsible for the gating of the 20S proteasome proteolytic chamber, while the β-subunits are responsible for proteolysis. However, only three of the β-subunits (β1, β2, β5) are catalytically active and responsible for caspase-like, trypsin-like and chymotrypsin-like activities of the proteasome, respectively (Groll et al. 1997; Kisselev et al. 2006; Finley et al. 2012). The activity of the specific proteolytic site of proteasome depends on the amino acid composition of the degraded protein. In yeast, inactivation of chymotrypsin-like sites (which preferentially cleave the proteins after hydrophobic residues) impairs cell growth and resistance to environmental stress. Hence, the chymotrypsin-like activity is suggested as the main determinant of the rate of protein breakdown by the proteasome (Kisselev et al. 2006). The protein degradation by the 26S proteasome in ATP-/ubiquitin-dependent manner is characteristic of physiological conditions. Nevertheless, under oxidative stress damage of proteins can more or less affect cellular proteolysis processes. Although degradation of oxidatively damaged proteins may occur through the 26S-dependent mechanism, several studies demonstrated that the free form of the 20S proteasome is crucial for removing these proteins (Aiken et al. 2011; Demasi et al. 2013; Jung et al. 2014). This phenomenon seems to be connected with several factors: (1) oxidative stress causes disassembly of the 26S proteasome, (2) the 20S proteasome is more resistant to oxidative stress than both the 26S proteasome and enzymes involved in ubiquitylation and (3) protein degradation by the 20S proteasome is ubiquitin-/ATP-independent (Wang et al. 2010; Demasi et al. 2013; Jung et al. 2014). In a free 20S proteasome, the entrance into catalytic chamber is blocked by the interaction of the N-terminal domains of the α-subunits (Chondrogianni et al. 2014). However, it has been noted that hydrophobic patches exposed in oxidatively damaged proteins induce conformational changes and gate opening in the 20S proteasome (Aiken et al. 2011). In addition, the oxidative imbalance results in S-glutathionylation of cysteine residues in the 20S proteasome, which may prevent irreversible oxidation of the cysteine thiol group but also enables gate opening of the 20S proteasome and increases degradation of oxidised proteins (Demasi et al. 2013). Nevertheless, oxidative damage of protein is not a defined state but a continuous process transforming native and functional proteins to covalently cross-linked aggregates. Changes in the amount of oxidative modifications affect proteasomal degradation. Undamaged, properly folded proteins are usually not recognised as substrates for proteasomes. Slight oxidative damages induce conformational changes which can expose degradation signals, such as hydrophobic sequences or the N-terminal amino acid residue of the protein. At this point, proteolytic activity of both the 26S proteasome (connected with higher level of ubiquitination) and the 20S proteasome increases. Further oxidative modifications disrupt activity of proteins leading to their complete unfolding. At the same time, the 26S proteasome is dissembled, which may lead to accumulation of ubiquitinated substrates, and oxidatively damaged proteins are removed only by the 20S proteasome activity (Wang et al. 2010; Jung et al. 2014). Further conditions of oxidative stress lead to creation of protein aggregates. At first, they are hydrophobic protein aggregates that can still be degraded, but when proteins in aggregates are covalently cross-linked, they become resistant to proteasomal degradation (Jung et al. 2014). Moreover, proteasome is also a target of oxidative modifications, including carbonylation and crosslinking, which may inhibit its function (Aiken et al. 2011; Andersson et al. 2013).

The other mechanism of removal of damaged biomolecules in the cell is autophagy. In contrast to UPS, autophagy is the only way of degradation for cellular organelles. Cellular compartments of the autophagy-dependent degradation are lysosomes; in the case of yeast cells, their role is played by vacuoles. Depending on the way of delivery of degradable substrates into the vacuoles, three types of autophagy are distinguished: macroautophagy, microautophagy and chaperone-mediated autophagy (not observed in yeast cells) (Klionsky et al. 2010). Macroautophagy (commonly referred to as autophagy) occurs through non-selective sequestration of large regions of cytoplasm into double-layered vesicles, called autophagosomes, which then fuse with vacuoles, resulting in degradation and recycling of internal contents. Biogenesis of autophagosome is a multi-stage process in which autophagy-related proteins (Atg proteins) play a role in autophagy induction, cargo recognition, formation of phagophore, its elongation and creation of autophagosome (Farré et al. 2009; Hecht et al. 2014). ATG genes were discovered in the S. cerevisiae yeast; so far genetic screens have identified 36 ATG genes required for autophagosome formation and development (Tsukada and Ohsumi 1993; Pollard et al. 2017). Although details of the macroautophagy pathway are still being explained, Atg proteins can be divided into four main complexes with specific functions. The complex of Atg1, Atg13, Atg17, Atg29 and Atg31 proteins is responsible for initiation of autophagosome creation. The next complex, composed of phosphatidylinositol 3-kinase (Vps34), Atg6, Vps15 and Atg14 or Vps38 proteins, is mainly responsible for delivery of a specific phospholipid to the forming membranes, namely, phosphatidylinositol triphosphate. The third and fourth complexes, i.e. Atg12-Atg5 and Atg8-phosphatidylethanolamine (PE), are ubiquitin-like conjugation systems which participate in phagophore elongation (Farré et al. 2009; Pollard et al. 2017). Degradation of sequestered biomolecules occurs in vacuoles, which contain several proteases, e.g. proteinase A (Pep4p), proteinase B (Prb1p) or carboxypeptidase Y (Li and Kane 2009; Hecht et al. 2014). There is a strict connection between autophagic process and vacuolar acidification. Vacuolar pH in the yeast cell is generally ca. 5–6.5 and determines the cell's proteolytic activity (Li and Kane 2009). In the case of yeast cells, autophagy is involved in maintaining the intracellular homeostasis, and it plays an important role as a pro-survival pathway in stress conditions, mainly during nutrient starvation (Cebollero and Reggiori 2009; Hecht et al. 2014). One may also observe the protective role of autophagy against ROS generation and oxidative stress in yeast cells (Kissová et al. 2006; Bhatia-Kiššová and Camougrand 2010; Suzuki et al. 2011). It has been shown that intracellular ROS regulate autophagy (Kissová et al. 2006; Marques et al. 2006; Scherz-Shouval et al. 2007). ROS (especially H_2O_2) can modify thiol groups of Atg proteins or up-regulate vacuolar proteases, such as Pep4p (Marques et al. 2006; Scherz-Shouval et al. 2007). In comparison to UPS, autophagy occurs only in the cytoplasm, but it can degrade larger pool of substrates, including protein complexes, oxidatively damaged protein aggregates and ubiquitylated proteins accumulated in the situation of proteasome impairment (Korolchuk et al. 2010). The connection between UPS and autophagy is important: firstly, because proteasomal components are degraded by autophagy and, secondly, because impairment of one pathway affects the other one (Korolchuk et al. 2010; Marshall et al. 2016). It has been shown that impairment of UPS leads to upregulation of autophagy; however, inactivation of autophagy results in impairments of UPS and accumulation of ubiquitylated proteins (Korolchuk et al. 2010; Lu et al. 2014).

An unprecedented advantage of autophagy is the possibility of degradation of whole organelles. This may take place by a non-selective as well as selective type of autophagy. An important selective type of autophagic degradation is the removal of dysfunctional or redundant mitochondria, known as a mitophagy (Lemasters 2005; Kraft et al. 2009). In yeast cells, mitophagy can be observed both under physiological and stress conditions (Zhang et al. 2007; Kanki and Klionsky 2008; Bhatia-Kiššová and Camougrand 2010). When availability of nutrients changes, mitophagy regulates the number of mitochondria (Tal et al. 2007; Kanki and Klionsky 2008; Bhatia-Kiššová and Camougrand 2010). Mitophagy also plays a role of the mitochondria quality control mechanism, which is particularly important due to (1) generation of ROS in the mitochondria, (2) lack of effective repair mechanism in the mitochondria and (3) induction of cell death by accumulation of damaged mitochondria (Kraft et al. 2009; Breitenbach et al. 2014). Hence, the disorder of mitophagy can lead to excessive generation of ROS, oxidative stress, disruption of cell homeostasis and release of pro-apoptotic factors causing cell death (Farré et al. 2009). Mitophagy occurs and is regulated independently of non-selective macroautophagy. Studies performed on yeast cells identified several proteins directly connected with mitophagy, including Atg32p, Aup1p and Uth1p (Kissová et al. 2004; Tal et al. 2007; Kanki et al. 2009).

1.8   Activation of the Programmed Cell Death Pathway

Cell cycle progression delay and activation of the damage repair system aim to restore an internal homeostasis disturbed by stress. The effectiveness of these activities will have a crucial impact on unlocking of the cell cycle and return to reproduction. For yeast as a unicellular organism, maintenance of the ability to reproduce is very important. A unicellular organism should be considered in two categories: as a cell and as an organism. Maintenance solely a proper physiological fitness, which is defined as cell vitality (Kwolek-Mirek and Zadrag-Tecza 2014), is beneficial for cell, but its inability to reproduce (cell viability) may lead to population regression. Hence, maintaining the population in its environmental niche is undoubtedly related to the reproductive capacity (the number of buds produced) of its cells. Reduction of the yeast cell's reproductive capacity is observed inter alia under oxidative stress conditions. Deletion of genes encoding the important elements of antioxidant defence system causes reduction in the reproductive capacity and shortening of the chronological lifespan (Wawryn et al. 1999; Schroeder and Shadel 2014). Reduction of the yeast cell's reproductive capacity under oxidative stress conditions is not directly connected with the ROS level but can be a side effect of mechanism of cell cycle progression delay (Zadrag-Tecza et al. 2009). On the one hand, cell cycle progression delay allows for repair of damages caused by ROS; on the other hand, it can lead to a faster achievement by the cell of a size that prevents further reproduction, which may lead to terminal cell cycle arrest and thus shorten the replicative lifespan (Zadrag-Tecza et al. 2009).

Cell cycle regulation, either in physiological or stress conditions, is one of the most important issues for cell, especially in the case of unicellular organisms such as the budding yeast. At cell cycle checkpoints, cells have the opportunity of arresting and making important life and death decisions via activation of various damage response mechanisms. This crosstalk between cell cycle regulation and the programmed cell death is vital, because the decision on choosing the cell death instead of its survival is irreversible. Cell death means plasma membrane breakdown or cellular fragmentation. Usually this is the final element of the cellular stress response activated when the pro-survival response mechanisms and antioxidant defences have been insufficient. In the case of yeast, cell death may occur via at least two pathways: as an accidental cell death (ACD) which is a rapid, uncontrollable and unavoidable form of death and as programmed cell death (PCD) which depends on the activation of the molecular pathways (Carmona-Gutierrez et al. 2018). PCD such as apoptosis in the case of unicellular organisms may be less obvious than in multicellular organisms. However, it is known that in physiological conditions, *S. cerevisiae* yeast cells form colonies and have also the ability to form biofilms when nutrients are depleted (Vachova and Palkova 2005). Therefore, it is considered that apoptosis may be a mechanism to clear old or damaged yeast cells. In addition, redistribution of nutrients from aged or stressed cells enhances the chances for survival of young cells (Fabrizio et al. 2004; Buttner et al. 2006). PCD such as apoptosis is mediated by two signalling pathways, the extrinsic, induced by death receptor, and the intrinsic, including the mitochondrial signalling pathway

(Eisenberg et al. 2007). Yeast cells showed key morphological features of apoptosis (similar to mammalian cells) such as phosphatidylserine externalisation, chromatin condensation, DNA fragmentation, reduction of cellular volume and formation of apoptotic bodies (Madeo et al. 1997). Apoptotic cells often indicate the outer membrane permeabilisation and then the release of mitochondrial cytochrome c and loss of the mitochondrial transmembrane potential (Ludovico et al. 2002; Sapienza et al. 2008). The intrinsic pathway of apoptosis is induced by a lot of signals such as DNA damage or starvation but predominantly by oxidative stress. In yeast, similar to mammalian cells, ROS act as crucial modulators of apoptosis (Madeo et al. 1999). High levels of ROS act as a pro-apoptotic signal. It was shown that other stress conditions which are accompanied with ROS accumulation may also induce the cells apoptosis, e.g. salt stress (Wadskog et al. 2004), osmotic stress (Silva et al. 2005), heat stress (Lee et al. 2007), acetic acid (Ludovico et al. 2001), actin cytoskeletal defects (Gourlay and Ayscough 2006), disruption of iron homeostasis (Almeida et al. 2008) and acrolein (Kwolek-Mirek et al. 2015). There are a number of important yeast cell death regulators such as cytochrome c, apoptosis-inducing factor (Aif1p), endonuclease G (Nuc1p) or metacaspase (Yca1p) (Madeo et al. 2002; Buttner et al. 2007; Carmona-Gutierrez et al. 2010). Some of these factors are connected with the mitochondrial pathway of apoptosis, which has been linked to complex mitochondrial processes, such as cytochrome c and Aif1p release, depolarisation of mitochondrial membrane potential and mito-chondrial fragmentation. These mitochondrial yeast apoptosis events can be observed in the case of apoptosis induced during chronological and replicative ageing as well as apoptosis induced by acetic acid or H_2O_2 (Ludovico et al. 2001; Severin and Hyman 2002; Madeo et al. 2004). Generally, mitochondria are very dynamic structures undergoing constant fusion and fission. A balance between these activities is required for normal functioning of the cell and minimisation of ROS generation (Westermann 2008). Extensive mitochondrial fragmentation is a com-mon feature observed under oxidative stress that has been associated with the release of apoptotic factors. The release of pro-apoptotic factors such as cytochrome c and Aif1p is connected with the loss of inner and outer mitochondrial membrane integrity (Strich 2015). The key element of the cell death via apoptotic pathway is caspases. Yeast cells possess the single *YCA1* gene encoding the type I metacaspase (analogous to mammalian caspases), which participates in the execution of oxidative stress-induced PCD (Madeo et al. 2002). Yeast cell death induced by ROS-associated stress stimuli can occur via two alternative pathways: Yca1-dependent apoptosis or Yca1-independent apoptosis (Madeo et al. 2009). These two pathways differ from each other because Yca1-independent apoptosis occurs without cytochrome c release, which requires Yca1p (Perrone et al. 2008). Cell death occurring in an Yca1-independent manner requires the activity of proteasomes and of caspase-like proteases (Wilkinson and Ramsdale 2011). In addition, Yca1p has also exhibited non-death roles affecting cell cycle progression and growth. It has been shown that Yca1p accelerates the G1/S transition and slows down the G2/M transition. Yca1p can also contribute to clearance of insoluble protein aggregates during the natural lifespan of yeast, thus improving its longevity and fitness (Lee et al. 2008, 2010). In the case of exposure to very high concentrations of pro-apoptotic factors, yeast cells can undergo the other type of cell death: accidental cell death such as necrosis. This type of cell death was observed in the case of high doses of H_2O_2 (Madeo et al. 1999), acetic acid (Ludovico et al. 2001) and heavy metals (Liang and Zhou 2007). Necrosis is identified by some characteristic features, such as plasma membrane permeabilisation and disintegration of subcellular structures (Eisenberg et al. 2010). It is also worth mentioning that in the light of recent studies, necrosis should not be considered only as accidental and uncontrollable cell death. Yeast cells may also undergo a so-called regulated necrosis which is executed by a genetically encoded molecular machinery. This type of cell death can be inhibited, which indicates that it results from activation of a molecular mechanism (Carmona-Gutierrez et al. 2018).

The relationships that exist between cell cycle progression and induction of cell death pathways are very complex. However, adequate regulation of these processes, especially under stress conditions, is necessary for survival of individual cells but also essential for the survival of the population.

1.9 Response to Oxidative Stress: Yeast and Other Fungi

The fungal kingdom encompasses an enormous diversity of taxa which can live as saprophytes, pathogens/parasites and symbionts. This diversity can arise from the differences in the life cycle strategies, the morphologies ranging from unicellular to multicellular forms, the type of metabolism and also the environmental niches which they occupy. In general, the ways of ROS production, antioxidant machinery and the stages of response to oxidative stress are similar and conserved between eukaryotic cells, including fungal organisms. However, some differences may be observed, inter alia in intracellular ROS level, cellular function of ROS, the presence of specific antioxidant enzymes or low molecular antioxidants. Such examples are enzymes connected with generation of ROS in fungi, namely, NADPH oxidases. In the *S. cerevisiae* yeast, one NADPH oxidase (Yno1p) has been found (Rinnerthaler et al. 2012); in turn the filamentous fungi have three classes of classical fungal NADPH oxidases: Nox A, Nox B and Nox C (Breitenbach et al. 2015). Information about the mechanism of the oxidative stress response is based on a limited number of yeast species such as *Saccharomyces cerevisiae*, *Schizosaccharomyces pombe* or *Candida albicans*. However, most of the studies have been carried out on the *S. cerevisiae* yeast. Particular yeast species share common signalling responsive pathways to stress, and some proteins involved in these pathways are evolutionarily conserved. Generally, in the diverse yeast species, there are three conserved signalling pathways to sense oxidative stress and regulate gene expression: (1) the SAPK cascade (the stress-responsive MAPK cascade), (2) the multistep phosphorelay and (3) AP-1 (Activating Protein-1) (Toone et al. 2001; Ikner and Shiozaki 2005; Morigasaki et al. 2008). Some differences between these species arise from adaptation to the specific ecological niche, biological activity, life strategies or sugar metabolism (Gonzalez-Parraga et al. 2008; Brion et al. 2016). In *Sc. pombe*, the SAPK pathway is sensitive to a variety of agents that induce oxidative stress, such as

H_2O_2, paraquat or UV irradiation (Degols et al. 1996; Degols and Russell 1997), and oxidative stress response is integrated in a general stress response (Chen et al. 2003). In turn, in the *S. cerevisiae* yeast, the role of SAPK in oxidative stress response is not necessarily evident (Bilsland et al. 2004), while for *C. albicans* a Hog1-mediated MAP kinase pathway plays an important role in resistance to oxidative stress, and *HOG1* gene is homologous to that of the *S. cerevisiae* and *Sc. pombe* yeast (Alonso-Monge et al. 2003). The second pathway—the multistep phosphorelay—is similar in *S. cerevisiae* and *Sc. pombe* and is required for the cellular response to H_2O_2 (Singh 2000). In turn, in the case of the third pathway, the difference is that some of the oxidative stress response genes in *S. cerevisiae* are regulated by Yap1p but not by Skn7p, while in *Sc. pombe* such regulation involves various signalling molecules, depending on the range of the oxidative stress (Quinn et al. 2002).

The relationship between the type of metabolic pathways of sugar oxidation and the oxidative stress response regulation was documented through the example of *S. cerevisiae* (Magherini et al. 2009) and *Kluyveromyces lactis* (Overkamp et al. 2002; Tarrio et al. 2006). Under aerobic condition, *K. lactis* carries out respiratory metabolism, in contrast to the *S. cerevisiae* yeast which carries out fermentative metabolism (Gonzalez-Siso et al. 2000). Hence, the mechanisms involved in the reoxidation of the NADPH, e.g. redox metabolism mechanism, provide the basis of molecular difference between these two yeast species (Tarrio et al. 2006). The genome sequence analysis shows that genes related to ROS detoxification and glutathione synthesis such as genes coding superoxide dismutases, catalases and peroxidases, glutathione/glutaredoxin and thioredoxin systems are well conserved in yeast but differ as far as transcriptional regulation is concerned (Gonzalez-Siso et al. 2009). An example might be hypoxia conditions, which in *S. cerevisiae* lead to an increased mRNA level of several genes of the stress response, while in *K. lactis* that level is low (Blanco et al. 2007). In turn, under aerobic conditions, transcription induction of *CTA1* and *CTT1* genes encoding catalases occurs in *S. cerevisiae* (Martínez-Pastor et al. 1996), yet such induction is not observed in the case of *K. lactis* (Blanco et al. 2007). These differences might be a result of differences in respiro-fermentative metabolism.

A specific group of fungi with regard to the type of metabolism is represented by methylotrophs. They are organisms able to obtain energy from molecules that have no C–C bond. Eukaryotic methylotrophs include a number of yeast species capable to grow on methanol as a carbon and energy source, e.g. *Candida*, *Pichia*, *Ogataea* and *Torulopsis* (Van Der Klei et al. 2006; Yurimoto et al. 2011). Methanol metabolism in this yeast is strictly connected with oxidative reactions and massive proliferation of peroxisomes, which contain several methanol-metabolising enzymes (Yurimoto et al. 2011). Methanol oxidation catalysed by alcohol oxidase (AO) results in the formation of formaldehyde and hydrogen peroxide. Therefore, peroxisomes contain catalase (CAT), which prevents leakage of H_2O_2 produced by AO into the cytosol. In comparison to the *S. cerevisiae* yeast, the activity of CAT is essential to methylotrophic yeast, which was confirmed by studies showing that deletion of *CAT* gene results in a decrease or inability of growth of methylotrophic yeast on methanol-containing media (Verduyn et al. 1988; Horiguchi et al. 2001a;

Nakagawa et al. 2010). Peroxisomes of methylotrophs may also contain glutathione peroxidases, which were discovered in *Candida boidinii*, in which peroxisomal peripheral membrane protein (Pmp20p) was identified as a protein with glutathione peroxidase activity, reacting with H_2O_2 and alkyl hydroperoxides (Sakai et al. 1998; Horiguchi et al. 2001b). Therefore, peroxisomes of methylotrophs contain relevant level of GSH (Horiguchi et al. 2001b), which is used not only by glutathione peroxidase but is mainly involved in conversion of the formaldehyde, which is the second toxic compound generated by AO activity (Yurimoto et al. 2011).

An example of oxidative stress response adjustment to environmental habitats may be a fungus growing on wood, known as a wood-decay fungus (mainly from *Basidiomycota*), which uses radical reactions and ROS to degrade lignocellulose. The degradation of lignin, characteristic of white rot fungi, is closely connected with oxidative action of H_2O_2 used by extracellular peroxidases such as lignin peroxidases and manganese peroxidases (Breitenbach et al. 2015). Important enzymes allowing for lignin degradation are laccases, which do not use H_2O_2 but create phenolic radicals that can break up the C–C bonds in the lignin (Bugg et al. 2011; Breitenbach et al. 2015). In lignin degradation, it is also possible to use the strategy based on Fenton reaction and generation of ˙OH, which was observed for brown rot fungi (Gessler et al. 2007; Breitenbach et al. 2015).

An important question is the oxidative stress response of pathogenic fungal species, which include in particular *Candida* but also *Cryptococcus* or *Pneumocystis*. The ability of these species to tolerate high levels of ROS is crucial for their pathogenicity. The reaction between the host organism and the pathogenic fungi is multifaceted. On the one hand, pathogenic fungi may generate ROS to infect the host organism, while on the other hand macrophages, neutrophils and phagocytic cells of host organism may generate ROS (and also RNS) and disturb the redox homeostasis of fungi (Bogdan et al. 2000). Hence, in comparison to *S. cerevisiae*, enhanced antioxidant defence mechanisms are observed in pathogenic fungi, which help the fungal pathogens survive (Missall et al. 2004; Gessler et al. 2007; Brown et al. 2009; Staerck et al. 2017). For example, *Candida albicans* have six *SOD* genes; *Aspergillus fumigatus* has three catalases (one conidial, CatAp; two mycelial, Cat1p and Cat2p) (Missall et al. 2004). Moreover, pathogenic fungi have several non-enzymatic antioxidants, which are not characteristic for the *S. cerevisiae* yeast. Among them, melanins, carotenoids and polyols are present in the cells of pathogenic fungi. Melanins are located in the cell wall of *Cryptococcus neoformans*, *Wangiella dermatitidis*, *A. fumigatus* and *Magnaporthe grisea*. A high level of carotenoids was found in *Rhodotorula mucilaginosa*, while mannitol (ROS scavenging polyol) is characteristic for *C. neoformans* (Chaturvedi et al. 1996; Hamilton and Gomez 2002; Missall et al. 2004; Gessler et al. 2007). What is more, in the case of *C. albicans*, the key regulators of stress signalling pathway are evolutionarily conserved and similar to those of *S. cerevisiae*, but some of these regulators might display distinct roles (Ramsdale et al. 2008; Brown et al. 2009). The Cap1p redox-sensitive cysteine residues become oxidised under oxidative stress, which leads to nuclear accumulation of Cap1p and activation of genes involved in (1) detoxification of oxidative stress, *SOD1* and *CAT1*; (2) glutathione synthesis, *GSH1*; and (3) redox homeostasis and oxidative damage repair, *GLR1* and *TRX1*. The role of Cap1p was proven by its inactivation, which attenuates induction of these genes and thus increases sensitivity of *C. albicans* to oxidative stress (Enjalbert et al. 2006; Znaidi et al. 2009).

1.10 Conclusions

All aerobic organisms are continuously exposed to reactive oxygen species, which by-products of cellular metabolism but may also be derived from environmental sources. Therefore, the most important thing is to maintain the balance between production of reactive oxygen species and antioxidant defences, as perturbation of such balance may lead to molecular damage and thereby give rise to stress. Consequently, the ability to activate stress response is necessary for yeast and other fungi to survive and adapt at to dynamic changes in their environmental niches. Cellular response and survival under stress conditions require rapid reprograming of the transcriptional activity towards restoration of the redox homeostasis, molecular damage repair and cell survival. Therefore, the mechanism of stress response can be based on few important principles applicable to all fungi: (1) the ability to detect stress signals, (2) the ability to transduce these signals to regulate the cellular processes and (3) triggering the response which includes processes to counteract or detoxify stress induction factors and then repair or removal of the molecular damage caused by stress, which is necessary for cell adaptation and survival in stress conditions. Effectiveness of these activities may be crucial for triggering of the programmed cell death pathway as the last element of stress response.

Acknowledgements Numerous conclusions presented in this chapter were based on the results of studies supported by the Polish National Science Centre Grant numbers: 2013/09/B/NZ3/01352 and 2017/01/X/NZ1/00153.

References

Aguilaniu H, Gustafsson L, Rigoulet M, Nystrom T (2001) Protein oxidation in G0 cells of *Saccharomyces cerevisiae* depends on the state rather than rate of respiration and is enhanced in pos9 but not yap1 mutants. J Biol Chem 276:35396–35404
Aiken CT, Kaake RM, Wang X, Huang L (2011) Oxidative stress-mediated regulation of proteasome complexes. Mol Cell Proteomics 10:R110.006924
Almeida T, Marques M, Mojzita D, Amorim MA, Silva RD, Almeida B, Rodrigues P, Ludovico P, Hohmann S, Moradas-Ferreira P, Corte-Real M, Costa V (2008) Isc1p plays a key role in hydrogen peroxide resistance and chronological lifespan through modulation of iron levels and apoptosis. Mol Biol Cell 19:865–876
Alonso-Monge R, Navarro-Garcia F, Roman E, Negredo AI, Eisman B, Nombela C, Pla J (2003) The Hog1 mitogen-activated protein kinase is essential in the oxidative stress response and chlamydospore formation in *Candida albicans*. Eukaryot Cell 2:351–361
Andersson V, Hanzén S, Liu B, Molin M, Nyström T (2013) Enhancing protein disaggregation restores proteasome activity in aged cells. Aging (Albany NY) 5:802–812

Auesukaree C (2017) Molecular mechanisms of the yeast adaptive response and tolerance to stresses encountered during ethanol fermentation. J Biosci Bioeng 124:133–142

Aung-Htut MT, Ayer A, Breitenbach M, Dawes IW (2012) Oxidative stresses and ageing. Subcell Biochem 57:13–54

Avery AM, Avery SV (2001) *Saccharomyces cerevisiae* expresses three phospholipid hydroperoxide glutathione peroxidases. J Biol Chem 276:33730–33735

Ayer A, Sanwald J, Pillay BA, Meyer AJ, Perrone GG, Dawes IW (2013) Distinct redox regulation in sub-cellular compartments in response to various stress conditions in *Saccharomyces cerevisiae*. PLoS One 8:e65240

Azevedo D, Tacnet F, Delaunay A, Rodrigues-Pousada C, Toledano MB (2003) Two redox centers within Yap1 for H_2O_2 and thiol-reactive chemicals signaling. Free Radic Biol Med 35:889–900

Barker MG, Brimage LJ, Smart KA (1999) Effect of Cu, Zn superoxide dismutase disruption mutation on replicative senescence in *Saccharomyces cerevisiae*. FEMS Microbiol Lett 177:199–204

Bartosz G (2009) Reactive oxygen species: destroyers or messengers? Biochem Pharmacol 77:1303–1315

Bhatia-Kiššová I, Camougrand N (2010) Mitophagy in yeast: actors and physiological roles. FEMS Yeast Res 10:1023–1034

Bigarella CL, Liang R, Ghaffari S (2014) Stem cells and the impact of ROS signaling. Development 141:4206–4218

Biliński T, Krawiec Z, Liczmański A, Litwińska J (1985a) Is hydroxyl radical generated by the Fenton reaction in vivo? Biochem Biophys Res Commun 130:533–539

Biliński T, Litwińska J, Błaszczyński M (1985b) Selective killing of respiratory sufficient yeast cells by paraquat. Acta Microbiol Pol 34:15–17

Bilsland E, Molin C, Swaminathan S, Ramne A, Sunnerhagen P (2004) Rck1 and Rck2 MAPKAP kinases and the HOG pathway are required for oxidative stress resistance. Mol Microbiol 53:1743–1756

Blanco M, Nunez L, Tarrio N, Canto E, Becerra M, Gonzalez-Siso MI, Cerdan ME (2007) An approach to the hypoxic and oxidative stress responses in *Kluyveromyces lactis* by analysis of mRNA levels. FEMS Yeast Res 7:702–714

Bogdan C, Rollinghoff M, Diefenbach A (2000) Reactive oxygen and reactive nitrogen intermediates in innate and specific immunity. Curr Opin Immunol 12:64–76

Bourbouloux A, Shahi P, Chakladar A, Delrot S, Bachhawat AK (2000) Hgt1p, a high affinity glutathione transporter from the yeast *Saccharomyces cerevisiae*. J Biol Chem 275:13259–13265

Braconi D, Bernardini G, Fiorani M, Azzolini C, Marzocchi B, Proietti F, Collodel G, Santucci A (2010) Oxidative damage induced by herbicides is mediated by thiol oxidation and hydroperoxides production. Free Radic Res 44:891–906

Breitenbach M, Rinnerthaler M, Hartl J, Stincone A, Vowinckel J, Breitenbach-Koller H, Ralser M (2014) Mitochondria in ageing: there is metabolism beyond the ROS. FEMS Yeast Res 14:198–212

Breitenbach M, Weber M, Rinnerthaler M, Karl T, Breitenbach-Koller L (2015) Oxidative stress in fungi: its function in signal transduction, interaction with plant hosts, and lignocellulose degradation. Biomol Ther 5:318–342

Brennan RJ, Schiestl RH (1996) Cadmium is an inducer of oxidative stress in yeast. Mutat Res 356:171–178

Brion C, Pflieger D, Souali-Crespo S, Friedrich A, Schacherer J (2016) Differences in environmental stress response among yeasts is consistent with species-specific lifestyles. Mol Biol Cell 27:1694–1705

Brown AJ, Haynes K, Quinn J (2009) Nitrosative and oxidative stress responses in fungal pathogenicity. Curr Opin Microbiol 12:384–391

Brown AJP, Cowen LE, Di Pietro A, Quinn J (2017) Stress adaptation. Microbiol Spectr 5. https://doi.org/10.1128/microbiolspec.FUNK-0048-2016

Bugg TD, Ahmad M, Hardiman EM, Rahmanpour R (2011) Pathways for degradation of lignin in bacteria and fungi. Nat Prod Rep 28:1883–1896

Burhans WC, Heintz NH (2009) The cell cycle is a redox cycle: linking phase-specific targets to cell fate. Free Radic Biol Med 47:1282–1293

Busti S, Mapelli V, Tripodi F, Sanvito R, Magni F, Coccetti P, Rocchetti M, Nielsen J, Alberghina L, Vanoni M (2016) Respiratory metabolism and calorie restriction relieve persistent endoplasmic reticulum stress induced by calcium shortage in yeast. Sci Rep 6:27942

Buttner S, Eisenberg T, Herker E, Carmona-Gutierrez D, Kroemer G, Madeo F (2006) Why yeast cells can undergo apoptosis: death in times of peace, love, and war. J Cell Biol 175:521–525

Buttner S, Eisenberg T, Carmona-Gutierrez D, Ruli D, Knauer H, Ruckenstuhl C, Sigrist C, Wissing S, Kollroser M, Frohlich KU, Sigrist S, Madeo F (2007) Endonuclease G regulates budding yeast life and death. Mol Cell 25:233–246

Cabiscol E, Piulats E, Echave P, Herrero E, Ros J (2000) Oxidative stress promotes specific protein damage in *Saccharomyces cerevisiae*. J Biol Chem 275:27393–27398

Cadet J, Wagner JR (2013) DNA base damage by reactive oxygen species, oxidizing agents, and UV radiation. Cold Spring Harb Perspect Biol 5

Cai Z, Yan LJ (2013) Protein oxidative modifications: beneficial roles in disease and health. J Biochem Pharmacol Res 1:15–26

Carmona-Gutierrez D, Eisenberg T, Buttner S, Meisinger C, Kroemer G, Madeo F (2010) Apoptosis in yeast: triggers, pathways, subroutines. Cell Death Differ 17:763–773

Carmona-Gutierrez D, Bauer MA, Zimmermann A et al (2018) Guidelines and recommendations on yeast cell death nomenclature. Microb Cell 5:4–31

Castello PR, David PS, Mcclure T, Crook Z, Poyton RO (2006) Mitochondrial cytochrome oxidase produces nitric oxide under hypoxic conditions: implications for oxygen sensing and hypoxic signaling in eukaryotes. Cell Metab 3:277–287

Castro FA, Mariani D, Panek AD, Eleutherio EC, Pereira MD (2008) Cytotoxicity mechanism of two naphthoquinones (menadione and plumbagin) in *Saccharomyces cerevisiae*. PLoS One 3: e3999

Causton HC, Ren B, Koh SS, Harbison CT, Kanin E, Jennings EG, Lee TI, True HL, Lander ES, Young RA (2001) Remodeling of yeast genome expression in response to environmental changes. Mol Biol Cell 12:323–337

Cebollero E, Reggiori F (2009) Regulation of autophagy in yeast *Saccharomyces cerevisiae*. Biochim Biophys Acta 1793:1413–1421

Charizanis C, Juhnke H, Krems B, Entian KD (1999) The mitochondrial cytochrome c peroxidase Ccp1 of *Saccharomyces cerevisiae* is involved in conveying an oxidative stress signal to the transcription factor Pos9 (Skn7). Mol Gen Genet 262:437–447

Charoenbhakdi S, Dokpikul T, Burphan T, Techo T, Auesukaree C (2016) Vacuolar H+-ATPase protects *Saccharomyces cerevisiae* cells against ethanol-induced oxidative and cell wall stresses. Appl Environ Microbiol 82:3121–3130

Chaturvedi V, Wong B, Newman SL (1996) Oxidative killing of *Cryptococcus neoformans* by human neutrophils. Evidence that fungal mannitol protects by scavenging reactive oxygen intermediates. J Immunol 156:3836–3840

Chen D, Toone WM, Mata J, Lyne R, Burns G, Kivinen K, Brazma A, Jones N, Bahler J (2003) Global transcriptional responses of fission yeast to environmental stress. Mol Biol Cell 14:214–229

Chiu J, Dawes IW (2012) Redox control of cell proliferation. Trends Cell Biol 22:592–601

Chiu J, Tactacan CM, Tan SX, Lin RC, Wouters MA, Dawes IW (2011) Cell cycle sensing of oxidative stress in *Saccharomyces cerevisiae* by oxidation of a specific cysteine residue in the transcription factor Swi6p. J Biol Chem 286:5204–5214

Chondrogianni N, Sakellari M, Lefaki M, Papaevgeniou N, Gonos ES (2014) Proteasome activation delays aging in vitro and in vivo. Free Radic Biol Med 71:303–320

Crešnar B, Petrič S (2011) Cytochrome P450 enzymes in the fungal kingdom. Biochim Biophys Acta 1814:29–35

Culotta VC, Yang M, O'halloran TV (2006) Activation of superoxide dismutases: putting the metal to the pedal. Biochim Biophys Acta 1763:747–758

Dahl JU, Gray MJ, Jakob U (2015) Protein quality control under oxidative stress conditions. J Mol Biol 427:1549–1563

Degols G, Russell P (1997) Discrete roles of the Spc1 kinase and the Atf1 transcription factor in the UV response of *Schizosaccharomyces pombe*. Mol Cell Biol 17:3356–3363

Degols G, Shiozaki K, Russell P (1996) Activation and regulation of the Spc1 stress-activated protein kinase in *Schizosaccharomyces pombe*. Mol Cell Biol 16:2870–2877

Delaunay A, Isnard AD, Toledano MB (2000) H_2O_2 sensing through oxidation of the Yap1 transcription factor. EMBO J 19:5157–5166

Delaunay A, Pflieger D, Barrault M-B, Vinh J, Toledano MB (2002) A thiol peroxidase is an H_2O_2 receptor and redox-transducer in gene activation. Cell 111:471–481

Demasi M, Netto LE, Silva GM, Hand A, De Oliveira CL, Bicev RN, Gozzo F, Barros MH, Leme JM, Ohara E (2013) Redox regulation of the proteasome via S-glutathionylation. Redox Biol 2:44–51

Dennis KJ, Shibamoto T (1989) Production of malonaldehyde from squalene, a major skin surface lipid, during UV-irradiation. Photochem Photobiol 49:711–716

Deponte M (2013) Glutathione catalysis and the reaction mechanisms of glutathione-dependent enzymes. Biochim Biophys Acta 1830:3217–3266

Drakulic T, Temple MD, Guido R, Jarolim S, Breitenbach M, Attfield PV, Dawes IW (2005) Involvement of oxidative stress response genes in redox homeostasis, the level of reactive oxygen species, and ageing in *Saccharomyces cerevisiae*. FEMS Yeast Res 5:1215–1228

Du L, Yu Y, Chen J, Liu Y, Xia Y, Chen Q, Liu X (2007) Arsenic induces caspase- and mitochondria-mediated apoptosis in *Saccharomyces cerevisiae*. FEMS Yeast Res 7:860–865

Eisenberg T, Buttner S, Kroemer G, Madeo F (2007) The mitochondrial pathway in yeast apoptosis. Apoptosis 12:1011–1023

Eisenberg T, Carmona-Gutierrez D, Buttner S, Tavernarakis N, Madeo F (2010) Necrosis in yeast. Apoptosis 15:257–268

Enjalbert B, Smith DA, Cornell MJ, Alam I, Nicholls S, Brown AJ, Quinn J (2006) Role of the Hog1 stress-activated protein kinase in the global transcriptional response to stress in the fungal pathogen *Candida albicans*. Mol Biol Cell 17:1018–1032

Fabrizio P, Battistella L, Vardavas R, Gattazzo C, Liou LL, Diaspro A, Dossen JW, Gralla EB, Longo VD (2004) Superoxide is a mediator of an altruistic aging program in *Saccharomyces cerevisiae*. J Cell Biol 166:1055–1067

Fang J, Beattie DS (2003) External alternative NADH dehydrogenase of *Saccharomyces cerevisiae*: a potential source of superoxide. Free Radic Biol Med 34:478–488

Farré JC, Krick R, Subramani S, Thumm M (2009) Turnover of organelles by autophagy in yeast. Curr Opin Cell Biol 21:522–530

Farrugia G, Balzan R (2012) Oxidative stress and programmed cell death in yeast. Front Oncol 2:64

Finley D, Ulrich HD, Sommer T, Kaiser P (2012) The ubiquitin-proteasome system of *Saccharomyces cerevisiae*. Genetics 192:319–360

Flattery-O'brien JA, Dawes IW (1998) Hydrogen peroxide causes RAD9-dependent cell cycle arrest in G2 in *Saccharomyces cerevisiae* whereas menadione causes G1 arrest independent of RAD9 function. J Biol Chem 273:8564–8571

Fong CS, Temple MD, Alic N, Chiu J, Durchdewald M, Thorpe GW, Higgins VJ, Dawes IW (2008) Oxidant-induced cell-cycle delay in *Saccharomyces cerevisiae*: the involvement of the SWI6 transcription factor. FEMS Yeast Res 8:386–399

Fourquet S, Huang ME, D'autreaux B, Toledano MB (2008) The dual functions of thiol-based peroxidases in H_2O_2 scavenging and signaling. Antioxid Redox Signal 10:1565–1576

Garrido EO, Grant CM (2002) Role of thioredoxins in the response of *Saccharomyces cerevisiae* to oxidative stress induced by hydroperoxides. Mol Microbiol 43:993–1003

Gessler NN, Aver'yanov AA, Belozerskaya TA (2007) Reactive oxygen species in regulation of fungal development. Biochemistry (Mosc) 72:1091–1109

Gonzalez-Parraga P, Sanchez-Fresneda R, Martinez-Esparza M, Arguelles JC (2008) Stress responses in yeasts: what rules apply? Arch Microbiol 189:293–296

Gonzalez-Siso MI, Freire-Picos MA, Ramil E, Gonzalez-Dominguez M, Rodriguez Torres A, Cerdan ME (2000) Respirofermentative metabolism in *Kluyveromyces lactis*: insights and perspectives. Enzym Microb Technol 26:699–705

Gonzalez-Siso MI, Garcia-Leiro A, Tarrio N, Cerdan ME (2009) Sugar metabolism, redox balance and oxidative stress response in the respiratory yeast *Kluyveromyces lactis*. Microb Cell Factories 8:46

Gostimskaya I, Grant CM (2016) Yeast mitochondrial glutathione is an essential antioxidant with mitochondrial thioredoxin providing a back-up system. Free Radic Biol Med 94:55–65

Gourlay CW, Ayscough KR (2006) Actin-induced hyperactivation of the Ras signaling pathway leads to apoptosis in *Saccharomyces cerevisiae*. Mol Cell Biol 26:6487–6501

Gralla EB, Valentine JS (1991) Null mutants of *Saccharomyces cerevisiae* Cu,Zn superoxide dismutase: characterization and spontaneous mutation rates. J Bacteriol 173:5918–5920

Grant CM (2001) Role of the glutathione/glutaredoxin and thioredoxin systems in yeast growth and response to stress conditions. Mol Microbiol 39:533–541

Grant CM, Collinson LP, Roe JH, Dawes IW (1996) Yeast glutathione reductase is required for protection against oxidative stress and is a target gene for yAP-1 transcriptional regulation. Mol Microbiol 21:171–179

Grant CM, Perrone G, Dawes IW (1998) Glutathione and catalase provide overlapping defenses for protection against hydrogen peroxide in the yeast *Saccharomyces cerevisiae*. Biochem Biophys Res Commun 253:893–898

Groll M, Ditzel L, Löwe J, Stock D, Bochtler M, Bartunik HD, Huber R (1997) Structure of 20S proteasome from yeast at 2.4 A resolution. Nature 386:463–471

Gruhlke MCH, Schlembach I, Leontiev R, Uebachs A, Gollwitzer PUG, Weiss A, Delaunay A, Toledano M, Slusarenko AJ (2017) Yap1p, the central regulator of the *S. cerevisiae* oxidative stress response, is activated by allicin, a natural oxidant and defence substance of garlic. Free Radic Biol Med 108:793–802

Hamilton AJ, Gomez BL (2002) Melanins in fungal pathogens. J Med Microbiol 51:189–191

Hecht KA, O'donnell AF, Brodsky JL (2014) The proteolytic landscape of the yeast vacuole. Cell Logist 4:e28023

Herrero E, Ros J, Bellí G, Cabiscol E (2008) Redox control and oxidative stress in yeast cells. Biochim Biophys Acta Gen Subj 1780:1217–1235

Herrero E, Bellí G, Casa C (2010) Structural and functional diversity of glutaredoxins in yeast. Curr Protein Pept Sci 11:659–668

Hershko A, Ciechanover A (1998) The ubiquitin system. Annu Rev Biochem 67:425–479

Horiguchi H, Yurimoto H, Goh T, Nakagawa T, Kato N, Sakai Y (2001a) Peroxisomal catalase in the methylotrophic yeast *Candida boidinii*: transport efficiency and metabolic significance. J Bacteriol 183:6372–6383

Horiguchi H, Yurimoto H, Kato N, Sakai Y (2001b) Antioxidant system within yeast peroxisome. Biochemical and physiological characterization of CbPmp20 in the methylotrophic yeast *Candida boidinii*. J Biol Chem 276:14279–14288

Ikner A, Shiozaki K (2005) Yeast signaling pathways in the oxidative stress response. Mutat Res Fundam Mol Mech Mutagen 569:13–27

Inoue Y, Matsuda T, Sugiyama K, Izawa S, Kimura A (1999) Genetic analysis of glutathione peroxidase in oxidative stress response of *Saccharomyces cerevisiae*. J Biol Chem 274:27002–27009

Isoyama T, Murayama A, Nomoto A, Kuge S (2001) Nuclear import of the yeast AP-1-like transcription factor Yap1p is mediated by transport receptor Pse1p, and this import step is not affected by oxidative stress. J Biol Chem 276:21863–21869

Izawa S, Inoue Y, Kimura A (1996) Importance of catalase in the adaptive response to hydrogen peroxide: analysis of acatalasaemic *Saccharomyces cerevisiae*. Biochem J 320(Pt 1):61–67

Izawa S, Maeda K, Miki T, Mano J, Inoue Y, Kimura A (1998) Importance of glucose-6-phosphate dehydrogenase in the adaptive response to hydrogen peroxide in *Saccharomyces cerevisiae*. Biochem J 330(Pt 2):811–817

Izawa S, Maeda K, Sugiyama K-I, Mano JI, Inoue Y, Kimura A (1999) Thioredoxin deficiency causes the constitutive activation of Yap1, an AP-1-like transcription factor in *Saccharomyces cerevisiae*. J Biol Chem 274:28459–28465

Izquierdo A, Casas C, Mühlenhoff U, Lillig CH, Herrero E (2008) *Saccharomyces cerevisiae* Grx6 and Grx7 are monothiol glutaredoxins associated with the early secretory pathway. Eukaryot Cell 7:1415–1426

Jiang H, English AM (2006) Phenotypic analysis of the ccp1Delta and ccp1Delta-ccp1W191F mutant strains of *Saccharomyces cerevisiae* indicates that cytochrome c peroxidase functions in oxidative-stress signaling. J Inorg Biochem 100:1996–2008

Jiang YY, Kong DX, Qin T, Zhang HY (2010) How does oxygen rise drive evolution? Clues from oxygen-dependent biosynthesis of nuclear receptor ligands. Biochem Biophys Res Commun 391:1158–1160

Jung T, Höhn A, Grune T (2014) The proteasome and the degradation of oxidized proteins: Part II – protein oxidation and proteasomal degradation. Redox Biol 2:99–104

Kanki T, Klionsky DJ (2008) Mitophagy in yeast occurs through a selective mechanism. J Biol Chem 283:32386–32393

Kanki T, Wang K, Cao Y, Baba M, Klionsky DJ (2009) Atg32 is a mitochondrial protein that confers selectivity during mitophagy. Dev Cell 17:98–109

Kathiresan M, English AM (2016) Targeted proteomics identify metabolism-dependent interactors of yeast cytochrome c peroxidase: implications in stress response and heme trafficking. Metallomics 8:434–443

Kaya A, Lobanov AV, Gerashchenko MV, Koren A, Fomenko DE, Koc A, Gladyshev VN (2014) Thiol peroxidase deficiency leads to increased mutational load and decreased fitness in *Saccharomyces cerevisiae*. Genetics 198:905–917

Ketela T, Brown JL, Stewart RC, Bussey H (1998) Yeast Skn7p activity is modulated by the Sln1p-Ypd1p osmosensor and contributes to regulation of the HOG pathway. Mol Gen Genet MGG 259:372–378

Kisselev AF, Callard A, Goldberg AL (2006) Importance of the different proteolytic sites of the proteasome and the efficacy of inhibitors varies with the protein substrate. J Biol Chem 281:8582–8590

Kissová I, Deffieu M, Manon S, Camougrand N (2004) Uth1p is involved in the autophagic degradation of mitochondria. J Biol Chem 279:39068–39074

Kissová I, Deffieu M, Samokhvalov V, Velours G, Bessoule JJ, Manon S, Camougrand N (2006) Lipid oxidation and autophagy in yeast. Free Radic Biol Med 41:1655–1661

Klionsky DJ, Codogno P, Cuervo AM, Deretic V, Elazar Z, Fueyo-Margareto J, Gewirtz DA, Kroemer G, Levine B, Mizushima N, Rubinsztein DC, Thumm M, Tooze SA (2010) A comprehensive glossary of autophagy-related molecules and processes. Autophagy 6:438–448

Klöppel C, Michels C, Zimmer J, Herrmann JM, Riemer J (2010) In yeast redistribution of Sod1 to the mitochondrial intermembrane space provides protection against respiration derived oxidative stress. Biochem Biophys Res Commun 403:114–119

Korolchuk VI, Menzies FM, Rubinsztein DC (2010) Mechanisms of cross-talk between the ubiquitin-proteasome and autophagy-lysosome systems. FEBS Lett 584:1393–1398

Koziol S, Zagulski M, Bilinski T, Bartosz G (2005) Antioxidants protect the yeast *Saccharomyces cerevisiae* against hypertonic stress. Free Radic Res 39:365–371

Kozmin S, Slezak G, Reynaud-Angelin A, Elie C, De Rycke Y, Boiteux S, Sage E (2005) UVA radiation is highly mutagenic in cells that are unable to repair 7,8-dihydro-8-oxoguanine in *Saccharomyces cerevisiae*. Proc Natl Acad Sci U S A 102:13538–13543

Kraft C, Reggiori F, Peter M (2009) Selective types of autophagy in yeast. Biochim Biophys Acta 1793:1404–1412

Kraikivski P, Chen KC, Laomettachit T, Murali TM, Tyson JJ (2015) From START to FINISH: computational analysis of cell cycle control in budding yeast. NPJ Syst Biol Appl 1:15016

Kuge S, Jones N, Nomoto A (1997) Regulation of yAP-1 nuclear localization in response to oxidative stress. EMBO J 16:1710–1720

Kuge S, Arita M, Murayama A, Maeta K, Izawa S, Inoue Y, Nomoto A (2001) Regulation of the yeast Yap1p nuclear export signal is mediated by redox signal-induced reversible disulfide bond formation. Mol Cell Biol 21:6139–6150

Kwolek-Mirek M, Zadrag-Tecza R (2014) Comparison of methods used for assessing the viability and vitality of yeast cells. FEMS Yeast Res 14:1068–1079

Kwolek-Mirek M, Bednarska S, Bartosz G, Bilinski T (2009) Acrolein toxicity involves oxidative stress caused by glutathione depletion in the yeast Saccharomyces cerevisiae. Cell Biol Toxicol 25:363–378

Kwolek-Mirek M, Bartosz G, Spickett CM (2011) Sensitivity of antioxidant-deficient yeast to hypochlorite and chlorite. Yeast 28:595–609

Kwolek-Mirek M, Zadrag-Tecza R, Bednarska S, Bartosz G (2015) Acrolein-induced oxidative stress and cell death exhibiting features of apoptosis in the yeast Saccharomyces cerevisiae deficient in SOD1. Cell Biochem Biophys 71:1525–1536

Leadsham JE, Sanders G, Giannaki S, Bastow EL, Hutton R, Naeimi WR, Breitenbach M, Gourlay CW (2013) Loss of cytochrome c oxidase promotes RAS-dependent ROS production from the ER resident NADPH oxidase, Yno1p, in yeast. Cell Metab 18:279–286

Lee J, Godon C, Lagniel G, Spector D, Garin J, Labarre J, Toledano MB (1999) Yap1 and Skn7 control two specialized oxidative stress response regulons in yeast. J Biol Chem 274:16040–16046

Lee JC, Straffon MJ, Jang TY, Higgins VJ, Grant CM, Dawes IW (2001) The essential and ancillary role of glutathione in Saccharomyces cerevisiae analysed using a grande gsh1 disruptant strain. FEMS Yeast Res 1:57–65

Lee YJ, Hoe KL, Maeng PJ (2007) Yeast cells lacking the CIT1-encoded mitochondrial citrate synthase are hypersusceptible to heat- or aging-induced apoptosis. Mol Biol Cell 18:3556–3567

Lee RE, Puente LG, Kaern M, Megeney LA (2008) A non-death role of the yeast metacaspase: Yca1p alters cell cycle dynamics. PLoS One 3:e2956

Lee RE, Brunette S, Puente LG, Megeney LA (2010) Metacaspase Yca1 is required for clearance of insoluble protein aggregates. Proc Natl Acad Sci U S A 107:13348–13353

Lemasters JJ (2005) Selective mitochondrial autophagy, or mitophagy, as a targeted defense against oxidative stress, mitochondrial dysfunction, and aging. Rejuv Res 8:3–5

Levine RL, Moskovitz J, Stadtman ER (2000) Oxidation of methionine in proteins: roles in antioxidant defense and cellular regulation. IUBMB Life 50:301–307

Li SC, Kane PM (2009) The yeast lysosome-like vacuole: endpoint and crossroads. Biochim Biophys Acta 1793:650–663

Liang Q, Zhou B (2007) Copper and manganese induce yeast apoptosis via different pathways. Mol Biol Cell 18:4741–4749

Longo VD, Gralla EB, Valentine JS (1996) Superoxide dismutase activity is essential for stationary phase survival in Saccharomyces cerevisiae. Mitochondrial production of toxic oxygen species in vivo. J Biol Chem 271:12275–12280

López-Mirabal HR, Winther JR (2008) Redox characteristics of the eukaryotic cytosol. Biochim Biophys Acta 1783:629–640

López-Mirabal HR, Thorsen M, Kielland-Brandt MC, Toledano MB, Winther JR (2007) Cytoplasmic glutathione redox status determines survival upon exposure to the thiol-oxidant 4,4-'-dipyridyl disulfide. FEMS Yeast Res 7:391–403

Lu K, Psakhye I, Jentsch S (2014) A new class of ubiquitin-Atg8 receptors involved in selective autophagy and polyQ protein clearance. Autophagy 10:2381–2382

Ludovico P, Sousa MJ, Silva MT, Leão C, Côrte-Real M (2001) Saccharomyces cerevisiae commits to a programmed cell death process in response to acetic acid. Microbiology 147:2409–2415

Ludovico P, Rodrigues F, Almeida A, Silva MT, Barrientos A, Corte-Real M (2002) Cytochrome c release and mitochondria involvement in programmed cell death induced by acetic acid in *Saccharomyces cerevisiae*. Mol Biol Cell 13:2598–2606

Madeo F, Frohlich E, Frohlich KU (1997) A yeast mutant showing diagnostic markers of early and late apoptosis. J Cell Biol 139:729–734

Madeo F, Frohlich E, Ligr M, Grey M, Sigrist SJ, Wolf DH, Frohlich KU (1999) Oxygen stress: a regulator of apoptosis in yeast. J Cell Biol 145:757–767

Madeo F, Herker E, Maldener C, Wissing S, Lachelt S, Herlan M, Fehr M, Lauber K, Sigrist SJ, Wesselborg S, Frohlich KU (2002) A caspase-related protease regulates apoptosis in yeast. Mol Cell 9:911–917

Madeo F, Herker E, Wissing S, Jungwirth H, Eisenberg T, Frohlich KU (2004) Apoptosis in yeast. Curr Opin Microbiol 7:655–660

Madeo F, Carmona-Gutierrez D, Ring J, Buttner S, Eisenberg T, Kroemer G (2009) Caspase-dependent and caspase-independent cell death pathways in yeast. Biochem Biophys Res Commun 382:227–231

Magherini F, Carpentieri A, Amoresano A, Gamberi T, De Filippo C, Rizzetto L, Biagini M, Pucci P, Modesti A (2009) Different carbon sources affect lifespan and protein redox state during *Saccharomyces cerevisiae* chronological ageing. Cell Mol Life Sci 66:933–947

Marques M, Mojzita D, Amorim MA, Almeida T, Hohmann S, Moradas-Ferreira P, Costa V (2006) The Pep4p vacuolar proteinase contributes to the turnover of oxidized proteins but PEP4 overexpression is not sufficient to increase chronological lifespan in *Saccharomyces cerevisiae*. Microbiology 152:3595–3605

Marshall RS, Mcloughlin F, Vierstra RD (2016) Autophagic turnover of inactive 26S proteasomes in yeast is directed by the ubiquitin receptor Cue5 and the Hsp42 chaperone. Cell Rep 16:1717–1732

Martínez-Pastor MT, Marchler G, Schüller C, Marchler-Bauer A, Ruis H, Estruch F (1996) The *Saccharomyces cerevisiae* zinc finger proteins Msn2p and Msn4p are required for transcriptional induction through the stress response element (STRE). EMBO J 15:2227–2235

Maslanka R, Kwolek-Mirek M, Zadrag-Tecza R (2017) Consequences of calorie restriction and calorie excess for the physiological parameters of the yeast *Saccharomyces cerevisiae* cells. FEMS Yeast Res 17

Mason JT, Kim S-K, Knaff DB, Wood MJ (2006) Thermodynamic basis for redox regulation of the Yap1 signal transduction pathway. Biochemistry 45:13409–13417

Matsuzawa A (2017) Thioredoxin and redox signaling: roles of the thioredoxin system in control of cell fate. Arch Biochem Biophys 617:101–105

Missall TA, Lodge JK, Mcewen JE (2004) Mechanisms of resistance to oxidative and nitrosative stress: implications for fungal survival in mammalian hosts. Eukaryot Cell 3:835–846

Miyake T, Shibamoto T (1999) Formation of malonaldehyde and acetaldehyde from the oxidation of 2'-deoxyribonucleosides. J Agric Food Chem 47:2782–2785

Moraitis C, Curran BP (2004) Reactive oxygen species may influence the heat shock response and stress tolerance in the yeast *Saccharomyces cerevisiae*. Yeast 21:313–323

Morano KA, Grant CM, Moye-Rowley WS (2012) The response to heat shock and oxidative stress in *Saccharomyces cerevisiae*. Genetics 190:1157–1195

Morigasaki S, Shimada K, Ikner A, Yanagida M, Shiozaki K (2008) Glycolytic enzyme GAPDH promotes peroxide stress signaling through multistep phosphorelay to a MAPK cascade. Mol Cell 30:108–113

Mulford KE, Fassler JS (2011) Association of the Skn7 and Yap1 transcription factors in the *Saccharomyces cerevisiae* oxidative stress response. Eukaryot Cell 10:761–769

Muller EG (1991) Thioredoxin deficiency in yeast prolongs S phase and shortens the G1 interval of the cell cycle. J Biol Chem 266:9194–9202

Nakagawa T, Yoshida K, Takeuchi A, Ito T, Fujimura S, Matsufuji Y, Tomizuka N, Yurimoto H, Sakai Y, Hayakawa T (2010) The peroxisomal catalase gene in the methylotrophic yeast Pichia methanolica. Biosci Biotechnol Biochem 74:1733–1735

Nguyên D-T, Alarco A-M, Raymond M (2001) Multiple Yap1p-binding sites mediate induction of the yeast major facilitator *FLR1* gene in response to drugs, oxidants, and alkylating agents. J Biol Chem 276:1138–1145

Nishimoto T, Watanabe T, Furuta M, Kataoka M, Kishida M (2016) Roles of catalase and trehalose in the protection from hydrogen peroxide toxicity in *Saccharomyces cerevisiae*. Biocontrol Sci 21:179–182

Osório NS, Carvalho A, Almeida AJ, Padilla-Lopez S, Leão C, Laranjinha J, Ludovico P, Pearce DA, Rodrigues F (2007) Nitric oxide signaling is disrupted in the yeast model for Batten disease. Mol Biol Cell 18:2755–2767

Østergaard H, Tachibana C, Winther JR (2004) Monitoring disulfide bond formation in the eukaryotic cytosol. J Cell Biol 166:337–345

Ouyang X, Tran QT, Goodwin S, Wible RS, Sutter CH, Sutter TR (2011) Yap1 activation by H_2O_2 or thiol-reactive chemicals elicits distinct adaptive gene responses. Free Radic Biol Med 50:1–13

Overkamp KM, Bakker BM, Steensma HY, Van Dijken JP, Pronk JT (2002) Two mechanisms for oxidation of cytosolic NADPH by *Kluyveromyces lactis* mitochondria. Yeast 19:813–824

Park SG, Cha MK, Jeong W, Kim IH (2000) Distinct physiological functions of thiol peroxidase isoenzymes in *Saccharomyces cerevisiae*. J Biol Chem 275:5723–5732

Pedrajas JR, Kosmidou E, Miranda-Vizuete A, Gustafsson JA, Wright AP, Spyrou G (1999) Identification and functional characterization of a novel mitochondrial thioredoxin system in *Saccharomyces cerevisiae*. J Biol Chem 274:6366–6373

Pedrajas JR, Porras P, Martínez-Galisteo E, Padilla CA, Miranda-Vizuete A, Bárcena JA (2002) Two isoforms of *Saccharomyces cerevisiae* glutaredoxin 2 are expressed in vivo and localize to different subcellular compartments. Biochem J 364:617–623

Penninckx MJ (2002) An overview on glutathione in Saccharomyces versus non-conventional yeasts. FEMS Yeast Res 2:295–305

Perez-Gallardo RV, Briones LS, Diaz-Perez AL, Gutierrez S, Rodriguez-Zavala JS, Campos-Garcia J (2013) Reactive oxygen species production induced by ethanol in *Saccharomyces cerevisiae* increases because of a dysfunctional mitochondrial iron-sulfur cluster assembly system. FEMS Yeast Res 13:804–819

Perrone GG, Tan SX, Dawes IW (2008) Reactive oxygen species and yeast apoptosis. Biochim Biophys Acta 1783:1354–1368

Petrova VY, Drescher D, Kujumdzieva AV, Schmitt MJ (2004) Dual targeting of yeast catalase A to peroxisomes and mitochondria. Biochem J 380:393–400

Pinto AV, Deodato EL, Cardoso JS, Oliveira EF, Machado SL, Toma HK, Leitao AC, De Padula M (2010) Enzymatic recognition of DNA damage induced by UVB-photosensitized titanium dioxide and biological consequences in *Saccharomyces cerevisiae*: evidence for oxidatively DNA damage generation. Mutat Res 688:3–11

Pollard TD, Earnshaw WC, Lippincott-Schwartz J, Johnson G (2017) Cell biology, 3rd edn. Elsevier, Philadelphia, PA

Pujol-Carrion N, Belli G, Herrero E, Nogues A, De La Torre-Ruiz MA (2006) Glutaredoxins Grx3 and Grx4 regulate nuclear localisation of Aft1 and the oxidative stress response in *Saccharomyces cerevisiae*. J Cell Sci 119:4554–4564

Quinn J, Findlay VJ, Dawson K, Millar JB, Jones N, Morgan BA, Toone WM (2002) Distinct regulatory proteins control the graded transcriptional response to increasing H_2O_2 levels in fission yeast *Schizosaccharomyces pombe*. Mol Biol Cell 13:805–816

Raitt DC, Johnson AL, Erkine AM, Makino K, Morgan B, Gross DS, Johnston LH, Craig E (2000) The Skn7 response regulator of *Saccharomyces cerevisiae* interacts with Hsf1 in vivo and is required for the induction of heat shock genes by oxidative stress. Mol Biol Cell 11:2335–2347

Ralser M, Wamelink MM, Kowald A, Gerisch B, Heeren G, Struys EA, Klipp E, Jakobs C, Breitenbach M, Lehrach H, Krobitsch S (2007) Dynamic rerouting of the carbohydrate flux is key to counteracting oxidative stress. J Biol 6:10

Ramsdale M, Selway L, Stead D, Walker J, Yin Z, Nicholls SM, Crowe J, Sheils EM, Brown AJ (2008) MNL1 regulates weak acid-induced stress responses of the fungal pathogen *Candida albicans*. Mol Biol Cell 19:4393–4403

Reddi AR, Culotta VC (2013) SOD1 integrates signals from oxygen and glucose to repress respiration. Cell 152:224–235

Reinhard CT, Planavsky NJ, Olson SL, Lyons TW, Erwin DH (2016) Earth's oxygen cycle and the evolution of animal life. Proc Natl Acad Sci U S A 113:8933–8938

Reverter-Branchat G, Cabiscol E, Tamarit J, Ros J (2004) Oxidative damage to specific proteins in replicative and chronological-aged *Saccharomyces cerevisiae*: common targets and prevention by calorie restriction. J Biol Chem 279:31983–31989

Rinnerthaler M, Büttner S, Laun P et al (2012) Yno1p/Aim14p, a NADPH-oxidase ortholog, controls extramitochondrial reactive oxygen species generation, apoptosis, and actin cable formation in yeast. Proc Natl Acad Sci U S A 109:8658–8663

Rodríguez-Manzaneque MT, Tamarit J, Bellí G, Ros J, Herrero E (2002) Grx5 is a mitochondrial glutaredoxin required for the activity of iron/sulfur enzymes. Mol Biol Cell 13:1109–1121

Russell SJ, Steger KA, Johnston SA (1999) Subcellular localization, stoichiometry, and protein levels of 26 S proteasome subunits in yeast. J Biol Chem 274:21943–21952

Sakai Y, Yurimoto H, Matsuo H, Kato N (1998) Regulation of peroxisomal proteins and organelle proliferation by multiple carbon sources in the methylotrophic yeast, *Candida boidinii*. Yeast 14:1175–1187

Sapienza K, Bannister W, Balzan R (2008) Mitochondrial involvement in aspirin-induced apoptosis in yeast. Microbiology 154:2740–2747

Schafer FQ, Buettner GR (2001) Redox environment of the cell as viewed through the redox state of the glutathione disulfide/glutathione couple. Free Radic Biol Med 30:1191–1212

Scherz-Shouval R, Shvets E, Fass E, Shorer H, Gil L, Elazar Z (2007) Reactive oxygen species are essential for autophagy and specifically regulate the activity of Atg4. EMBO J 26:1749–1760

Schieber M, Chandel NS (2014) ROS function in redox signaling and oxidative stress. Curr Biol 24: R453–R462

Schroeder EA, Shadel GS (2014) Crosstalk between mitochondrial stress signals regulates yeast chronological lifespan. Mech Ageing Dev 135:41–49

Semchyshyn HM, Miedzobrodzki J, Bayliak MM, Lozinska LM, Homza BV (2014) Fructose compared with glucose is more a potent glycoxidation agent in vitro, but not under carbohydrate-induced stress in vivo: potential role of antioxidant and antiglycation enzymes. Carbohydr Res 384:61–69

Severin FF, Hyman AA (2002) Pheromone induces programmed cell death in *S. cerevisiae*. Curr Biol 12:R233–R235

Shenton D, Grant CM (2003) Protein S-thiolation targets glycolysis and protein synthesis in response to oxidative stress in the yeast *Saccharomyces cerevisiae*. Biochem J 374:513–519

Silva RD, Sotoca R, Johansson B, Ludovico P, Sansonetty F, Silva MT, Peinado JM, Corte-Real M (2005) Hyperosmotic stress induces metacaspase- and mitochondria-dependent apoptosis in *Saccharomyces cerevisiae*. Mol Microbiol 58:824–834

Singh KK (2000) The *Saccharomyces cerevisiae* Sln1p-Ssk1p two-component system mediates response to oxidative stress and in an oxidant-specific fashion. Free Radic Biol Med 29:1043–1050

Srinivasan C, Liba A, Imlay JA, Valentine JS, Gralla EB (2000) Yeast lacking superoxide dismutase(s) show elevated levels of "free iron" as measured by whole cell electron paramagnetic resonance. J Biol Chem 275:29187–29192

Staerck C, Gastebois A, Vandeputte P, Calenda A, Larcher G, Gillmann L, Papon N, Bouchara JP, Fleury MJJ (2017) Microbial antioxidant defense enzymes. Microb Pathog 110:56–65

Strich R (2015) Programmed cell death initiation and execution in budding yeast. Genetics 200:1003–1014

Sturtz LA, Diekert K, Jensen LT, Lill R, Culotta VC (2001) A fraction of yeast Cu,Zn-superoxide dismutase and its metallochaperone, CCS, localize to the intermembrane space of mitochondria.

A physiological role for SOD1 in guarding against mitochondrial oxidative damage. J Biol Chem 276:38084–38089

Sun Z, Fay DS, Marini F, Foiani M, Stern DF (1996) Spk1/Rad53 is regulated by Mec1-dependent protein phosphorylation in DNA replication and damage checkpoint pathways. Genes Dev 10:395–406

Suzuki SW, Onodera J, Ohsumi Y (2011) Starvation induced cell death in autophagy-defective yeast mutants is caused by mitochondria dysfunction. PLoS One 6:e17412

Swiecilo A (2016) Cross-stress resistance in *Saccharomyces cerevisiae* yeast--new insight into an old phenomenon. Cell Stress Chaperones 21:187–200

Tal R, Winter G, Ecker N, Klionsky DJ, Abeliovich H (2007) Aup1p, a yeast mitochondrial protein phosphatase homolog, is required for efficient stationary phase mitophagy and cell survival. J Biol Chem 282:5617–5624

Tanaka T, Izawa S, Inoue Y (2005) GPX2, encoding a phospholipid hydroperoxide glutathione peroxidase homologue, codes for an atypical 2-Cys peroxiredoxin in *Saccharomyces cerevisiae*. J Biol Chem 280:42078–42087

Tarrio N, Becerra M, Cerdan ME, Gonzalez Siso MI (2006) Reoxidation of cytosolic NADPH in *Kluyveromyces lactis*. FEMS Yeast Res 6:371–380

Taverne YJ, Merkus D, Bogers AJ, Halliwell B, Duncker DJ, Lyons TW (2018) Reactive oxygen species: radical factors in the evolution of animal life: a molecular timescale from Earth's earliest history to the rise of complex life. BioEssays 40

Temple MD, Perrone GG, Dawes IW (2005) Complex cellular responses to reactive oxygen species. Trends Cell Biol 15:319–326

Theopold U (2009) Developmental biology: a bad boy comes good. Nature 461:486–487

Toledano MB, Kumar C, Le Moan N, Spector D, Tacnet F (2007) The system biology of thiol redox system in *Escherichia coli* and yeast: differential functions in oxidative stress, iron metabolism and DNA synthesis. FEBS Lett 581:3598–3607

Toone WM, Morgan BA, Jones N (2001) Redox control of AP-1-like factors in yeast and beyond. Oncogene 20:2336–2346

Trotter EW, Grant CM (2002) Thioredoxins are required for protection against a reductive stress in the yeast *Saccharomyces cerevisiae*. Mol Microbiol 46:869–878

Trotter EW, Rand JD, Vickerstaff J, Grant CM (2008) The yeast Tsa1 peroxiredoxin is a ribosome-associated antioxidant. Biochem J 412:73–80

Tsang CK, Liu Y, Thomas J, Zhang Y, Zheng XF (2014) Superoxide dismutase 1 acts as a nuclear transcription factor to regulate oxidative stress resistance. Nat Commun 5:3446

Tsukada M, Ohsumi Y (1993) Isolation and characterization of autophagy-defective mutants of *Saccharomyces cerevisiae*. FEBS Lett 333:169–174

Vachova L, Palkova Z (2005) Physiological regulation of yeast cell death in multicellular colonies is triggered by ammonia. J Cell Biol 169:711–717

Van Der Klei IJ, Yurimoto H, Sakai Y, Veenhuis M (2006) The significance of peroxisomes in methanol metabolism in methylotrophic yeast. Biochim Biophys Acta 1763:1453–1462

Veal EA, Ross SJ, Malakasi P, Peacock E, Morgan BA (2003) Ybp1 is required for the hydrogen peroxide-induced oxidation of the Yap1 transcription factor. J Biol Chem 278:30896–30904

Verduyn C, Giuseppin ML, Scheffers WA, Van Dijken JP (1988) Hydrogen peroxide metabolism in yeasts. Appl Environ Microbiol 54:2086–2090

Vialard JE, Gilbert CS, Green CM, Lowndes NF (1998) The budding yeast Rad9 checkpoint protein is subjected to Mec1/Tel1-dependent hyperphosphorylation and interacts with Rad53 after DNA damage. EMBO J 17:5679–5688

Vignols F, Bréhélin C, Surdin-Kerjan Y, Thomas D, Meyer Y (2005) A yeast two-hybrid knockout strain to explore thioredoxin-interacting proteins in vivo. Proc Natl Acad Sci U S A 102:16729–16734

Wadskog I, Maldener C, Proksch A, Madeo F, Adler L (2004) Yeast lacking the SRO7/SOP1-encoded tumor suppressor homologue show increased susceptibility to apoptosis-like cell death on exposure to NaCl stress. Mol Biol Cell 15:1436–1444

Wallace MA, Liou LL, Martins J, Clement MH, Bailey S, Longo VD, Valentine JS, Gralla EB (2004) Superoxide inhibits 4Fe-4S cluster enzymes involved in amino acid biosynthesis. Cross-compartment protection by CuZn-superoxide dismutase. J Biol Chem 279:32055–32062

Wang X, Yen J, Kaiser P, Huang L (2010) Regulation of the 26S proteasome complex during oxidative stress. Sci Signal 3:ra88

Wawryn J, Krzepiłko A, Myszka A, Biliński T (1999) Deficiency in superoxide dismutases shortens life span of yeast cells. Acta Biochim Pol 46:249–253

Weinberger M, Mesquita A, Caroll T, Marks L, Yang H, Zhang Z, Ludovico P, Burhans WC (2010) Growth signaling promotes chronological aging in budding yeast by inducing superoxide anions that inhibit quiescence. Aging (Albany NY) 2:709–726

Westermann B (2008) Molecular machinery of mitochondrial fusion and fission. J Biol Chem 283:13501–13505

Wieser R, Adam G, Wagner A, Schüller C, Marchler G, Ruis H, Krawiec Z, Bilinski T (1991) Heat shock factor-independent heat control of transcription of the CTT1 gene encoding the cytosolic catalase T of Saccharomyces cerevisiae. J Biol Chem 266:12406–12411

Wilkinson D, Ramsdale M (2011) Proteases and caspase-like activity in the yeast Saccharomyces cerevisiae. Biochem Soc Trans 39:1502–1508

Wolf DH, Hilt W (2004) The proteasome: a proteolytic nanomachine of cell regulation and waste disposal. Biochim Biophys Acta 1695:19–31

Wood MJ, Andrade EC, Storz G (2003) The redox domain of the Yap1p transcription factor contains two disulfide bonds. Biochemistry 42:11982–11991

Xiao W, Wang RS, Handy DE, Loscalzo J (2018) NAD(H) and NADP(H) redox couples and cellular energy metabolism. Antioxid Redox Signal 28:251–272

Yan C, Lee LH, Davis LI (1998) Crm1p mediates regulated nuclear export of a yeast AP-1-like transcription factor. EMBO J 17:7416–7429

Yurimoto H, Oku M, Sakai Y (2011) Yeast methylotrophy: metabolism, gene regulation and peroxisome homeostasis. Int J Microbiol 2011:101298

Zadrag-Tecza R, Kwolek-Mirek M, Bartosz G, Bilinski T (2009) Cell volume as a factor limiting the replicative lifespan of the yeast Saccharomyces cerevisiae. Biogerontology 10:481–488

Zadziński R, Maszewski J, Bartosz G (1996) Transport of glutathione S-conjugates in the yeasts Saccharomyces cerevisiae. Cell Biol Int 20:325–330

Zhang Y, Qi H, Taylor R, Xu W, Liu LF, Jin S (2007) The role of autophagy in mitochondria maintenance: characterization of mitochondrial functions in autophagy-deficient S. cerevisiae strains. Autophagy 3:337–346

Zhang C, Li Z, Zhang X, Yuan L, Dai H, Xiao W (2016) Transcriptomic profiling of chemical exposure reveals roles of Yap1 in protecting yeast cells from oxidative and other types of stresses. Yeast 33:5–19

Znaidi S, Barker KS, Weber S, Alarco AM, Liu TT, Boucher G, Rogers PD, Raymond M (2009) Identification of the Candida albicans Cap1p regulon. Eukaryot Cell 8:806–820

Response Mechanisms to Chemical and Physical Stresses in Yeast and Filamentous Fungi

2

Marek Skoneczny and Adrianna Skoneczna

Abstract

Free-living unicellular eukaryotes, including yeast and filamentous fungi, remain in close contact with the surrounding environment throughout their existence. Such a lifestyle facilitates the absorption of life-supporting nutrients but inevitably exposes these organisms to the hostility of the elements. Consequently, they experience various chemical and physical stresses, often collectively dubbed as abiotic stress, and mounting appropriate protective measures against the aggressive exterior is the essence of their survival. In this chapter, the stresses of extreme pH, low temperature, extreme hydrostatic pressure, hyperosmosis, and dehydration will be discussed. The modes of toxicity of these stresses will be described, as well as the methods used by cells to tolerate them. In accordance with the central theme of this book, the machineries of sensing and responses to those stresses will be emphasized. On the surface they seem diverse; however, at the basal level, the damage to cellular functions is often similar, as reflected by the resemblance of the cellular regulatory responses. Moreover, cellular protection mechanisms against various physical and chemical stresses are also strikingly convergent. These similarities will be highlighted in this chapter. The importance of studying fungal response mechanisms to chemical and physical stresses for industry and medicine will also be succinctly delineated.

2.1 Introduction

Throughout their life, yeast and filamentous fungi cells remain in close contact with the environment, isolated from it only with the thin and porous cell wall made of polysaccharides. However, some of these organisms can thrive in seemingly

M. Skoneczny (✉) · A. Skoneczna
Institute of Biochemistry and Biophysics, Polish Academy of Sciences, Warszawa, Poland
e-mail: kicia@ibb.waw.pl

© Springer Nature Switzerland AG 2018
M. Skoneczny (ed.), *Stress Response Mechanisms in Fungi*,
https://doi.org/10.1007/978-3-030-00683-9_2

uninhabitable ecological niches, which is possible owing to the presence of various defense mechanisms that neutralize or minimize the harmful effects of the harsh surroundings.

The stresses that will be discussed in this chapter—extreme pH, low temperature, extreme hydrostatic pressure, hyperosmotic, and dehydration stresses—are collectively called environmental or abiotic stresses (Thammavongs et al. 2008), although the latter name is more often used in the context of plants rather than yeast and fungi. In principle, these names accurately depict their nature. However, low pH stress is also a tool that is used as a defense against yeast and filamentous fungi by other organisms; therefore, it is biotic rather than abiotic. These stresses seem diverse and, superficially, have little in common; thus, this chapter might be perceived as a collection of miscellanies. Surprisingly though, this is quite often not true. As the following subsections will reveal, various and apparently dissimilar stresses are indeed closely related to each other in several aspects. The links between dehydration and hyperosmotic stresses may not be surprising. More intriguing is the close resemblance of hyperbaric and low-temperature stresses. In addition, there are general-purpose regulatory pathways that control the cellular response to several external insults with common mechanisms to circumvent their deleterious effects and acquire tolerance to them. A good example is the high osmolarity glycerol (HOG) signaling pathway: one of the hubs that transmit environmental stress signals, originally described as a pathway communicating hyperosmotic stress but now known to respond to multiple stresses, sometimes together with other cellular mitogen-activated protein kinase (MAPK) pathways.

While extremophiles are often the organisms of choice in studies on environmental stress response mechanisms (Turk and Gostinčar 2018), fungal responses to chemical and physical stresses (and to other stresses discussed in other chapters of this book) are, by far, the most comprehensively examined with *Saccharomyces cerevisiae* as a model organism. Therefore, the reviewed data mostly regard this yeast species. The data collected for other fungi are added when available, and they reveal the high level of conservation of these mechanisms.

Response of yeast and fungi to chemical and physical stresses, besides general, pure scientific interest, receives a lot of attention for practical reasons. Most of the environmental stresses belonging to this category are encountered by industrially important yeast and fungi species in the plethora of biotechnological processes. These issues will also be addressed in this chapter.

The response mechanisms to some of the stresses that can be assigned to the chemical or physical stress category are described in other chapters of this book and will not be discussed here. The response to oxidative stress is the subject of Chap. 1; the responses to UV, γ-irradiation, and various chemical compounds reactive to and damaging DNA are covered in Chap. 3; and the response to thermal stress is mentioned in Chap. 5.

2.2 Fungal Responses to pH Extremities

Most reactions occurring inside the cells of any organism have their optimum near neutral pH. *S. cerevisiae* cells grown exponentially on glucose have a neutral pH in the cytosol and in mitochondria (Orij et al. 2009). However, many organisms can invade niches with strongly alkaline or acidic pH; therefore, they must possess the means to isolate themselves from harsh exterior conditions and preserve the favorable conditions of the cell interior. This applies especially to unicellular organisms such as yeast and fungi. As emphasized in the introductory chapter of this book, these organisms are renowned for their capacity to survive under extreme conditions and to exploit seemingly unabsorbable nutrients. To perform this efficiently, they must be resistant to the severity of the conditions experienced (Amils et al. 2011; Wiegel 2011). Nevertheless, even the less exotic yeast and fungi species commonly known and used as research models usually tolerate a broad range of pH values, as low as 2 and as high as 10, in their environment (Zvyagilskaya et al. 2001; Peñalva and Arst 2004; Cornet and Gaillardin 2014). However, there are differences among them, especially in the alkaline part of the scale. For instance, *S. cerevisiae* does not grow above pH 8 (Ariño 2010), and *Schizosaccharomyces pombe* does not tolerate pH values above 6.7, which is not in fact alkaline (Higuchi et al. 2018). These differences reflect the presence or absence in their cells of specific tolerance mechanisms that will be described in this section. The pH value denotes the negative logarithm of the concentration of H^+ ions; thus, a low pH translates into an excess of protons, whereas a high pH means their deficit. Consequently, fungal cells are affected by these disparate conditions in opposite ways, and the mechanisms of tolerance they employ are dissimilar. However, the response mechanisms to low and high pH, at least those that are better characterized, are, in part, common.

Notably, the mechanism of toxicity to fungi cells of low pH, i.e., of high concentration of protons in the surrounding medium, is more complicated and multifaceted than one would expect. It appears that fungi cells respond differently to acidic conditions provoked by different categories of compounds. When low pH stress was applied by adjusting the pH of the growth medium with a strong inorganic acid such as HCl (Kawahata et al. 2006) or H_2SO_4 (Chen et al. 2009; de Lucena et al. 2012), i.e., acidic pH stress per se, the response of yeast cells was remarkably different than when the stress conditions were created using organic acids, such as acetic or lactic acids, more likely to be found in the natural environment of fungi (Kawahata et al. 2006). So, unsurprisingly, the mechanisms developed by yeast and filamentous fungi to endure and grow a tolerance to these stresses are diverse, and some of them are targeted against organic acids (de Melo et al. 2010). Some of these acids are called weak organic acids (WOA) and are used in the food industry as antispoilage agents; this topic will be discussed in a separate subsection.

2.2.1   Acidic pH Stress

Most yeast and fungi species are not extremophiles, and only a few of them can be found in inorganic acidic niches, which are typically populated by photo- and chemosynthetic prokaryotes (Amils et al. 2011). However, low pH conditions are very common in the natural environment of yeast and fungi. Many yeast species ferment sugars and release carbon dioxide that, although volatile, transiently lowers the pH of the surrounding medium. Must fermented by wine yeast strains may contain up to 0.2% of acetic acid in addition to the main glycolysis end product, ethanol (Vilela-Moura et al. 2011). Some wine spoilage yeast and bacteria coexisting with wine yeast in the must may raise its acetic acid concentration up to 0.7% (Vilela-Moura et al. 2011). In the wild, yeast and fungi typically share their niches with other organisms, including bacteria that release organic acids, such as acetic, propionic, or lactic as the end products of fermentation. Last but not least, various fruits, the common growth substrate of yeast and fungi, may contain a high concentration of organic acids. For all these reasons, it is of prime importance for the evolutionary success of these organisms to have robust mechanisms of tolerance to low external pH. So it is explicable why lowering the pH of the growth medium to as low as 3.0 has no apparent influence on the *S. cerevisiae* growth capacity (Chen et al. 2009; Dong et al. 2017). Moreover, such a low pH value is not even perceived as stress because, as the genome-wide study revealed, the shift of *S. cerevisiae* cells from medium buffered at pH 6 to that at pH 3 **downregulates** stress-responsive genes (Chen et al. 2009). The authors conclude that maintaining the pH of the growth medium at 6.0 is more stressful for this organism than its gradual acidification.

Interestingly, lowering the pH further, below 2.5, rapidly kills the cells (de Melo et al. 2010; de Lucena et al. 2012; Dong et al. 2017). This abrupt change in cell viability should not be surprising because, at pH 2.5, the H^+ ion concentration is threefold higher than at pH 3.0. Likewise, 1 mM HCl in the medium is well tolerated by budding yeast cells, but 10 mM caused their rapid death (Malakar et al. 2006). Apparently, an external concentration of 1 mM H^+, but not much more, can be controlled by them.

2.2.2   Mechanisms of Tolerance to the Acidic pH

A neutral pH of the cell interior is maintained by the plasma membrane H^+ pump ATPase, which is encoded by the *PMA1* gene in *S. cerevisiae* (Portillo 2000) and by the vacuolar H+ ATPase complex (Kawahata et al. 2006), although the latter is more commonly associated with alkaline pH tolerance (Kuroki et al. 2002) (see below). This process is energetically costly. Thus, in addition to the high activity of the plasma membrane H^+ pump ATPase, a high rate of ATP production by the mitochondrial respiratory chain and oxidative phosphorylation is necessary (Fletcher et al. 2015). The toxicity of low pH is due to an excessively steep H^+ gradient across the plasma membrane that forces the cell to expend energy to fight against this gradient. Thus, when energy reserves are exhausted, this leads to acidification of the cell interior outside the optimal range for enzymatic reactions that destroy cell metabolism. In extreme cases, a low internal pH leads to protein misfolding and denaturation and unavoidable cell death. However, is the pumping out of protons the only way to avoid acidification of the cell interior and its destructive consequences?

Numerous studies were dedicated to deciphering the mechanisms of yeast and filamentous fungi tolerance to low pH. Gene expression approaches have revealed that yeast cells respond to low external pH in multiple ways. Phenotypic analyses performed on individual strains deleted for various genes or on the whole collection of *S. cerevisiae* knockout strains, as well as microevolution studies, revealed the importance of various pathways for the tolerance of an acidic medium. As mentioned previously, an acidic pH value higher than 3.0 is not perceived by *S. cerevisiae* as stressful; therefore, in these studies, the applied pH stress was in the range of 2.0–2.8. The budding yeast response to pH 2.5 adjusted with sulfuric acid is mainly regulated by the cell wall integrity (CWI) pathway. Moreover, the high osmolarity glycerol (HOG) and calcineurin signaling pathways are also involved (de Lucena et al. 2012). In some microarray studies, a pH of 2.5 adjusted with sulfuric acid caused the induction of the general stress response (GSR), mostly the heat shock response (de Melo et al. 2010). Individual proteins involved in cell wall assembly, such as Gas1p, a β-1,3-glucanosyltransferase anchored to the cell surface via glycosylphosphatidylinositol (GPI) domain, are also important for low pH tolerance (Matsushika et al. 2017). The lipid composition of cellular membranes is also crucial because the evolved *S. cerevisiae* strain that acquired higher tolerance to pH 2.8 of the medium, adjusted with HCl, had altered sterol composition (Fletcher et al. 2017). Indeed, a low pH must affect cellular lipid structures because 10 mM HCl (~pH 2) caused morphological anomalies of the plasma membrane in this organism (Malakar et al. 2006). Lipid homeostasis is also important for *Candida glabrata* adaptation to pH 2.0 adjusted with the same acid (Lin et al. 2017), and manipulation of the lipid composition resulted in increased tolerance of this yeast species to acid stress (Qi et al. 2017). *S. cerevisiae* grown in the medium acidified with lactic or hydrochloric acid displayed vacuole fragmentation, and lactic acid caused severe perturbations in amino acid homeostasis (Suzuki et al. 2012). Acidic pH stress results in Aft1p transcription factor-dependent induction of genes whose products are involved in iron utilization and homeostasis and upregulates proteins involved in cell wall biogenesis. The resistance to acidic pH is affected by the absence of genes involved in diverse cellular processes with a strong representation of various aspects of intracellular vesicle transport (Kawahata et al. 2006). Despite the abundance of these data, no consistent image of yeast tolerance to acidic pH has emerged except the well-established role of the essential plasma membrane H+ ATPase Pma1p in maintaining the neutral pH of the cell interior.

2.2.3 Alkaline pH Stress

While budding yeast cells tolerate well the acidic part of the pH spectrum, showing no signs of growth slowdown until the pH of the medium drops below 3.0, they display far lower tolerance to alkaline pH. This is perhaps explicable by alkaline pH being mostly found in specific environmental niches, not populated by this organism. It was noticed independently in various laboratories that increasing the pH of the growth medium by one unit from neutrality—i.e., to pH 8.0—results in complete growth arrest of most laboratory *S. cerevisiae* strains (Ariño 2010; Canadell et al. 2015). Furthermore, the fission yeast *Schizosaccharomyces pombe* cannot grow above pH 6.7 (Higuchi et al. 2018). However, other organisms can tolerate alkalinity up to pH 10, such as *Aspergillus* sp. (Peñalva and Arst 2004; Cornet and Gaillardin 2014), or pH 11, such as *Yarrowia lipolytica* (Zvyagilskaya et al. 2001), although they are not alkalophiles.

The function of plasma membrane active transporters is driven by the permanent gradient of protons between the cell exterior and interior that is maintained in *S. cerevisiae* by Pma1p H+-ATPase and in other fungi by its respective orthologs (Portillo 2000). This gradient is necessary for the efficient uptake of various vital compounds (van der Rest et al. 1995), and, while acidic pH influences it slightly, high pH destroys it. Cells cannot neutralize the surrounding alkaline medium; thus, upon exposure to alkaline pH, they are starved for nutrients, such as phosphate (Serra-Cardona et al. 2014) and glucose (Casamayor et al. 2012). Moreover, high external pH decreases the ionization of essential transition metals, including iron and copper, which affects their uptake, resulting in starvation for these cations (Serrano et al. 2004).

2.2.4 Resistance to Alkaline pH

In *S. cerevisiae*, the tolerance to alkaline pH and also to salt is conferred by Ena1p, a plasma membrane Na+ ATPase, which counteracts internal alkalinization by the efflux of Na+ cations (Haro et al. 1991; Yenush 2016). Vacuolar H+-ATPase (V-ATPase) is also required for high pH tolerance in yeast, and the role of vacuoles in stabilizing the intracellular pH was demonstrated (Brett et al. 2011; Rane et al. 2014).

To compensate for the compromised glucose uptake under alkaline pH stress, *S. cerevisiae* cells mobilize glycogen (Casamayor et al. 2012). *N. crassa* cells exposed to the same stress conditions lower the accumulation of the reserve carbohydrates glycogen and trehalose (Virgilio et al. 2017).

Other storage molecules, polyphosphates, which are present in the cells of all organisms, including yeast and fungi, were suggested long ago to play a role in high pH tolerance. An increased pH of the cytosol activates the hydrolysis of polyphosphates, which benefits the cell in two ways by compensating for the shortage of phosphate and restoring the correct pH of the cell interior. This phenomenon was shown not only for halotolerant alga *Dunaliella salina* (Pick et al. 1990) but also for *S. cerevisiae* (Castro et al. 1995) and *Yarrowia lipolytica* (Zvyagilskaya et al. 2001). Since then, our knowledge concerning the roles of these ubiquitous compounds in living cells, including their protective function against various stresses, has expanded considerably (Gray and Jakob 2015); however, their means of conferring tolerance to alkaline stress in eukaryotes, particularly in yeast and fungi, are not fully understood.

Since alkaline pH results in the starvation of *S. cerevisiae* cells, it triggers the morphological switch to the invasive pseudohyphal growth of haploids and sporulation in diploids (Lamb and Mitchell 2003; Piccirillo et al. 2010).

2.2.5   Fungal Response Mechanisms to Changes in Ambient pH

The major system regulating the budding yeast response to ambient pH changes is the Rim101p/Nrg1p pathway (Lamb et al. 2001; Platara et al. 2006) (see Fig. 2.1). In filamentous fungi, such as *Aspergillus nidulans*, *Neurospora crassa*, or *Trichoderma reesei* (and most likely also in other fungi species), orthologous signaling pathways exist, in which pivotal roles are played by Rim101p orthologs PacC (Tilburn et al. 1995), PAC-3 (Virgilio et al. 2017), and PAC1 (Häkkinen et al. 2015) transcription

Fig. 2.1 Current view of the fungal cells signaling of ambient pH stress showing high level of conservancy of this pathway between yeast and filamentous fungi. All *S. cerevisiae* Rim proteins have their Pal equivalents in *A. nidulans*. In budding yeast Dfg1p is also somehow involved in sensing ambient pH. See the text for more details. *PM* plasma membrane (Adapted from Cornet and Gaillardin 2014)

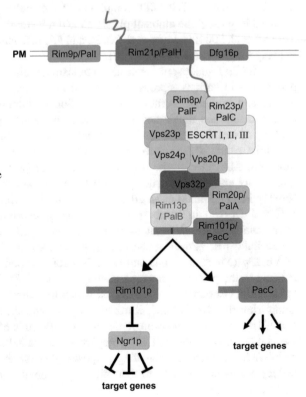

factors (TFs), respectively. Interestingly, in *A. nidulans*, the PacC-dependent pathway regulates both alkaline-specific and acid-specific genes, activating the former and repressing the latter ones and preparing cells for both high and low ambient pH stress (Tilburn et al. 1995; Häkkinen et al. 2015). On the other hand, in *S. cerevisiae*, the Rim101p/Nrg1p pathway is responsible for adaptation only to alkaline pH, and no evidence exists for a Rim101p-mediated response to low pH stress in this organism (Mira et al. 2009; Serra-Cardona et al. 2015). Instead, it seems to be required for the full response to WOA stress (Mira et al. 2009) (see Sect. 2.2.9). Although Rim101p and PacC TFs are orthologous proteins, their settings in the pathway transmitting the ambient pH signal in the respective organisms are different. While Rim101p activates alkaline-responsive genes by repressing Nrg1p, a downstream repressor of genes conferring high pH tolerance (Lamb and Mitchell 2003), PacC in *A. nidulans* activates its cognate alkaline-responsive genes directly (Tilburn et al. 1995). The Slg1p/Wsc1p-Pkc1p-Slt2p MAPK pathway that is involved in cell wall integrity signaling and one of five known MAP kinase cascades (Gustin et al. 1998) also plays an important role in the adaptive response of *S. cerevisiae* to alkali stress (Serrano et al. 2006), indicating that external alkaline pH affects the cell wall of this organism.

The mechanism of ambient pH perception involves at least two sensors transmitting the signal to their respective pathways. It was shown for *S. cerevisiae* and *A. nidulans* that the ESCORT complexes of the multivesicular body (MVB) are involved in sensing the ambient pH signal and transmitting it to the Rim101p/PacC TFs (Xu et al. 2004; Galindo et al. 2012), and this mechanism is probably conserved in many fungi species (Ost et al. 2015).

The ESCRT (endosomal sorting complexes required for transport) machinery participates in various topologically related processes involving membrane deformation and budding, including the formation of intraluminal vesicles in MVB, plasma membrane repair and microvesicle formation, abscission in cytokinesis, budding of certain viruses, and nuclear membrane resealing (see the recent review by Christ et al. 2017). However, ESCRT components also have nonendosomal functions. Activation of the Rim (in *S. cerevisiae*)/Pal (in *A. nidulans*) pathway is initiated by extracellular sensing of pH by the 7-transmembrane domain receptor, Rim21p/PalH (Galindo et al. 2012), which, together with Rim8p/PalF and Rim9p/PalI and Dfg16p in some species such as *S. cerevisiae*, forms a pH-sensing membrane complex. It remains to be elucidated how these proteins detect the extracellular pH. However, the next step involves the recruitment by Rim8/PalF of Vps23p (Xu et al. 2004) and other ESCRT components to form an assembly tethered to the plasma membrane (Galindo et al. 2012). This assembly constitutes the scaffold (Herrador et al. 2010), on which the Rim proteolysis complex forms, comprising the Rim23/PalC, Rim20/PalA, and Rim13/PalB proteases, to activate Rim101p/PacC TF by its proteolytic cleavage (Peñalva et al. 2014). It appears that the components of the ESCRT complexes are "hijacked" from the endocytosis pathway for ambient pH signaling purposes (Peñalva et al. 2014), and functional endocytosis is not required to transmit the pH signal to the target genes (Lucena-Agell et al. 2015).

Another protein implicated as the sensor of high pH stress is Wsc1p/Slg1p because *S. cerevisiae slg1Δ* cells are hypersensitive to this stress. Wsc1p/Slg1p, together with two other closely related proteins, Wsc2p and Wsc3p, are plasma membrane stress sensors of the cell wall integrity MAPK signaling cascade. All three share the same architecture with a single transmembrane domain, with the C-terminus facing the cytosol and interacting with the downstream proteins and the highly mannosylated N-terminus extending into the cell wall. They are believed to be mechanosensors or nanosprings that detect cell wall perturbations (Kock et al. 2015). Interestingly, the absence of only one of them, Slg1p, results in alkaline pH hypersensitivity, although all three function in the same MAPK signaling cascade and supposedly have redundant functions. Apparently they have not (Rodicio and Heinisch 2010). Notably, Hkr1p and Msb2p mucins of similar architecture, also believed to be mechanosensors (Brewster and Gustin 2014), detect hyperosmotic stress in HOG MAPK signaling cascade (Tatebayashi et al. 2007) (see Sect. 2.4).

The toxic effects of alkaline pH are diverse and so are the regulatory systems triggered by this stress, affecting the expression of hundreds of genes when *S. cerevisiae* cells are shifted from moderately acidic conditions (pH 5.5) to the growth limiting pH 8.0 (Ariño 2010). In budding yeast and in other fungi species, many of those genes respond to high or low pH in a Rim101/PacC-independent fashion; therefore, they must be controlled by other ambient pH response systems (Häkkinen et al. 2015) in addition to those mentioned above.

The calcium-activated calcineurin-Crz1p pathway, which transmits the signals of diverse stimuli, such as exposure to mating pheromones, high salt or osmolarity, endoplasmic reticulum stress, and others (Thewes 2014), is also involved in the response to high pH stress (Viladevall et al. 2004; Roque et al. 2016). Both Rim101p and Crz1p were shown to participate in the alkaline pH response of *Candida albicans* (Wang et al. 2011).

Alkaline pH also compromises the uptake of phosphate, and the depletion of this compound elicits the response of Pho85p cyclin-dependent kinase and Pho4p TF, which are required for high pH tolerance. Phosphate limitation is signaled by Rim101p whose nuclear accumulation is regulated by Pho85p (Nishizawa et al. 2010).

As mentioned above, the alkaline pH stress results in the glucose starvation of *S. cerevisiae* cells, as reflected by the activation of the Snf1p kinase AMP-activated protein kinase (AMPK) pathway in response to nutrient restriction (Hong and Carlson 2007). Notably, approximately 75% of the genes induced by alkaline pH respond also to glucose limitation (Casamayor et al. 2012). In response to alkaline pH stress, *S. cerevisiae* Snf1p phosphorylates Mig2p TF, resulting in the activation of many genes that are repressed in the presence of glucose in the medium (Serra-Cardona et al. 2014). By the same mechanism, alkaline pH also inactivates the glucose-responsive cAMP/PKA protein kinase A pathway, resulting in the induction of the Msn2/4-dependent general stress response. Approximately half of the several hundred genes induced by alkaline pH are dependent on this TF (Casado et al. 2011). Curiously, no alkaline pH response pathways orthologous to those described for *S. cerevisiae* and *A. nidulans* were found in *S. pombe*, corresponding to the aforementioned inability of this organism to grow when the pH of the medium is greater than 6.7 (Higuchi et al. 2018).

The crucial elements of fungi resistance to alkaline pH are ion-pumping ATPases. One of them is the plasma membrane Na^+-ATPase, involved in Na^+ and Li^+ efflux, which in budding yeast is encoded by the *ENA1* gene. Its activity is also crucial for salt tolerance (Yenush 2016). Its high pH stress-induced expression is controlled by three of the regulatory pathways mentioned above: those involving Rim101p, Crz1p, and Snf1p TFs (Platara et al. 2006; Petrezsélyová et al. 2016). *VMA2* and *VMA4* genes encoding the subunits of V-ATPase, another protein important for alkaline pH tolerance, are upregulated in response to high pH stress in a Rim101p-dependent manner (Pérez-Sampietro and Herrero 2014). The fungal ortholog of *VMA4* from *Aspergillus oryzae*, *vmaA*, is also induced by alkaline pH (Kuroki et al. 2002).

Interestingly, the increased level of polyunsaturated fatty acids in *S. cerevisiae* cells enhanced their tolerance to alkaline pH that is dependent on Rim101p/Nrg1p regulators (Yazawa et al. 2009).

As mentioned previously, the Rim101p pathway is considered in the context of the *S. cerevisiae* cell response to alkaline pH stress, whereas its orthologous PacC pathway in *A. nidulans* is responsive to both high and low ambient pH. Nevertheless, a substantial number of budding yeast genes, whose absence results in hypersensitivity to acidic pH, is involved in vesicular trafficking (Kawahata et al. 2006), suggesting that, in *S. cerevisiae*, similar to *A. nidulans*, the ambient pH signaling pathway involving ESCORT components regulates the response to both acidic and alkaline pH stress.

2.2.6   Mechanisms of the Response and Tolerance to Weak Organic Acids (WOAs)

Among the organic acids, there is one distinct group attracting much attention that encompasses acetic, propionic, sorbic, and benzoic acids, which are used as food preservatives. They are often referred to as weak organic acids (WOA), which can be misleading because the strength of their acidity does not distinguish them from other organic acids.

Despite the differences in molecular weight and structure, the dissociation constants—i.e., the pK_a values—of most organic acids are very similar, falling within the range between 4.2 and 4.9 (Mira et al. 2010c), which is relatively high compared with the strong inorganic acids ($pK_a < 1$). Because of this high pK_a value, all these organic acids should be described as weak; however, the toxicity of individual organic acids for yeast cells varies widely. While lactic or citric acid becomes toxic to *S. cerevisiae* cells when present in the medium at a 1–2 M concentration, acetic acid is toxic to this organism at 130 mM and sorbic acid is toxic at a concentration as low as 3 mM (Stratford et al. 2013b). This indicates that, although yeast and fungi are well prepared to resist low external pH, as described in the previous subsection, some organic acids can breach that resistance barrier.

2.2.7 Mechanisms of WOA Toxicity

It is now well established that the mechanisms of detection, response to, and protection against some organic acids are not identical to those of the acidic pH itself. Although ultimately the cell is hit by weak organic acids via the decrease in the intracellular pH, the mechanism of their toxicity is quite clever, which could be expected for an efficient antispoilage agent.

The differences in the mode of toxicity of various WOAs result from the differences in their chemical properties. However, they have one feature in common. Because of their high pK_a, they are mostly in the undissociated state in a moderately acidic environment (pH ~4.0). In that state, they can passively diffuse through the plasma membrane with an efficiency proportional to their lipophilicity.

The diffusion of WOA into fungi cells is quite rapid, within 1 min of exposure (Ullah et al. 2013). Once inside the neutral cell interior, WOAs become dissociated. This results in constant disequilibrium of undissociated WOAs between the cell interior and exterior, leading to the accumulation of WOAs inside cells to concentrations far exceeding their level in the growth medium (Stratford et al. 2013b). Acidification of the cytoplasm can be overcome by pumping out H+ at the expense of ATP by Pma1p, but this results in further WOA accumulation. It is postulated that, instead of wasting energy for a futile fight with cytosol acidification, cells exposed to WOA stress slow down their growth and metabolism and enter a dormant state until the environmental conditions become more favorable (Ullah et al. 2013) or until they themselves adapt to the WOA stress (Fernandes et al. 2005). It is also possible that this growth arrest is not the regulated process, and cytosol acidification simply disrupts cell metabolism so that energy production is inefficient.

It appears that yeast cell responses to various WOAs are different (Mira et al. 2010c), and the mechanisms of their toxicity are also, at least to some extent, dissimilar. In particular, acetic acid seems to be different from propionic, sorbic, and benzoic acids (Ullah et al. 2012; Stratford et al. 2013a). The growth inhibitory concentration of acetic acid is much higher than that of the other three, and their lipophilicity, especially that of sorbic and benzoic acids, makes them much more toxic than acetic acid (Piper et al. 2001). It was shown that the minimal inhibitory concentration of aliphatic acids of various chain lengths was inversely proportional to their lipophilicity measured as the partition coefficient between octane and water (Stratford et al. 2013b).

Because sorbic and benzoic acids are toxic at a much lower level, its intracellular concentration—i.e., cytosol acidification—is also lower. Thus, the mechanism of their toxicity might be different. Indeed, they have been shown to affect membranes and membrane trafficking and to induce oxidative stress due to the increased production of free radicals in mitochondria (Piper 1999). It is not unthinkable that they do so by affecting the mitochondrial membranes, thus damaging the respiratory chain.

Another feature that distinguishes acetic acid from larger WOAs is the manner in which they enter cells. While more lipophilic propionic, sorbic, and benzoic acids do so by passive diffusion, it was suggested that Fps1p aquaglyceroporin, the plasma membrane channel, provides the entrance for undissociated acetic acid into the cytosol (Mollapour and Piper 2007).

2.2.8 Resistance to WOA

In addition to maintaining a neutral pH by the action of Pma1p H+ ATPase (Ullah et al. 2012), budding yeast cells counteract WOA toxicity by pumping them out of the cell via the plasma membrane ABC transporter Pdr12p (Piper et al. 1998). Budding yeast cells adapt and acquire tolerance to these acids by strong induction of the *PDR12* gene. Without stress, the Pdr12p protein is barely detectable but becomes abundant following sorbic or benzoic acid treatment. The adaptation to these compounds requires several hours of growth arrest. However, when adapted, the cells no longer enter the dormant state when transferred again to sorbic or benzoic acid-containing medium (Piper et al. 2001).

Highly WOA-resistant yeast species, such as *Zygosaccharomyces* spp., but not *S. cerevisiae*, can also oxidize sorbic and benzoic acids via a mitochondrial monooxygenase (Mollapour and Piper 2001). Acetic and propionic acids can be utilized as a carbon source by *S. cerevisiae*, but not in the presence of a repressible carbon source, such as glucose, whereas acetic acid can be coassimilated with glucose by *Zygosaccharomyces* spp. (Guerreiro et al. 2012).

Pma1p and Pdr12p are not the only proteins identified thus far that participate in weak organic acid resistance. Numerous genome-wide studies have revealed many proteins involved in weak organic acid tolerance that belong to diverse functional categories: multidrug resistance transporters other than Pdr12p, proteins involved in the metabolism of lipids, amino acids, and carbohydrates, in cell wall function, in ribosome biogenesis, in vesicular trafficking, and in protein folding (reviewed in Mira et al. 2010c). Only a few of these genes and their products were studied individually. It was demonstrated that Yro2p and Mrh1p, two paralogs related to the HSP30 family residing in the plasma membrane, are required for acetic acid tolerance (Takabatake et al. 2015). However, their molecular function and their mechanism of action are unknown. Plasma membrane polyamine transporters belonging to the DHA1 family of drug H(+) antiporters—Tpo1p, Tpo2p, and Tpo3p—have also been shown to confer tolerance to WOAs. Interestingly, while Tpo1p confers resistance to benzoic acid and its upregulation is under the control of the Gcn4p and Stp1p TFs (Godinho et al. 2017), Tpo2p and Tpo3p confer tolerance to less lipophilic, shorter chain acetic and propionic acids, and their upregulation is controlled by Haa1p TF (Fernandes et al. 2005). These differences also support the concept of the multitude of WOA tolerance mechanisms in the possession of fungal cells, tailored to the individual compounds.

The defense mechanisms are multifaceted and are not limited to metabolic changes. It is conceivable that the diffusion of WOAs into cells can be restrained by modification of the plasma membrane lipid composition. Indeed, both *Zygosaccharomyces bailii* and *S. cerevisiae* grown in the presence of acetic acid demonstrated an increase in saturated glycerophospholipids and complex sphingolipids. Moreover, regardless of the presence of WOAs, the complex sphingolipids were more abundant in *Z. bailii*, more tolerant to these compounds than in less tolerant *S. cerevisiae*, indicating the importance of a higher level of lipid saturation and sphingolipid content in the plasma membrane for acetic acid tolerance (Lindberg et al. 2013). Intriguingly, in another study, *S. cerevisiae* cells exposed to acetic acid showed decreased levels of ceramides (Guerreiro et al. 2017). Despite this inconsistency, the plasma membrane lipid composition is an important determinant of yeast cell tolerance to acetic acid and other WOAs. Alteration of the cell wall structure and composition may also lower its permeability to WOAs. While no direct evidence of such a causative link exists, several proteins involved in the *S. cerevisiae* cell wall biogenesis, such as Spi1p, a member of the glycosylphosphatidylinositol-anchored cell wall protein family (Simões et al. 2006), or Ygp1p, a cell wall-related secretory glycoprotein (Mira et al. 2010b), were shown to confer resistance to WOA.

It should be noted that in addition to all mechanisms of tolerance to WOAs involving their active efflux, prevention of their influx, and neutralization of their toxicity, at least some WOAs can simply be metabolized by fungi. While *S. cerevisiae* has a glucose repression mechanism that turns off the metabolism of any other carbon sources when glucose is available (Kayikci and Nielsen 2015), *Z. bailii* can metabolize acetic acid in parallel with glucose (Rodrigues et al. 2012). Thus, assimilation of WOAs can be an alternative mechanism of tolerance to this stress for yeast and fungi whose carbon source assimilation is not glucose-repressible.

2.2.9 Response Mechanisms to WOA Stress

WOA stress is commonly experienced by yeast and filamentous fungi in nature. Therefore, it may not be surprising that their responses to this stress are elaborate, developed during a long-lasting arms race between them and plants trying to protect their fruits and other organs against fungal infection. Employing the *S. cerevisiae* collection of knockout strains allowed identification of 650 genes that appeared as important to maintain acetic acid tolerance, most of them not previously associated with this process (Mira et al. 2010b). The resulting gene set was enriched with genes having diverse functional assignments, including transcription, internal pH homeostasis, carbohydrate metabolism, cell wall assembly, biogenesis of mitochondria, ribosomes and vacuoles, and the sensing, signaling, and uptake of various nutrients, particularly iron, potassium, glucose, and amino acids.

WOA stress elicits various signaling and regulatory systems (reviewed in Mira et al. 2010c). Two main regulatory systems exist that respond to WOA. One is dependent on Haa1p TF and regulates the response to the less lipophilic WOAs acetic and propionic acids (Fernandes et al. 2005). Whole-transcriptome analyses of yeast cells revealed that more than 200 *S. cerevisiae* genes are induced under acetic acid stress and that 80% of them respond to this stress in an Haa1p-dependent manner (Mira et al. 2010a). They encode, among others, multidrug resistance transporters, proteins involved in carbohydrate and lipid metabolism and proteins associated with nucleic acid processing. In the spoilage yeast *Z. bailii*, the Haa1 regulon encompasses similar genes and functional groups, demonstrating the conservation of the response to acetic acid among yeast species (Palma et al. 2017).

The much smaller group of genes is dependent on War1p TF, which regulates the response to the more lipophilic WOAs sorbic, benzoic, octanoic, and propionic acids. War1p regulates the response to WOA stress and upregulates, among others, the synthesis of Pma1p H^+ ATPase and the Pdr12p plasma membrane ATP-binding cassette transporter (Kren et al. 2003). It was suggested that War1p TF may be activated through the direct binding of the weak acid anion (Piper 2011).

More recently, the target of rapamycin complex 2-Ypk1 signaling pathway was shown to be activated in *S. cerevisiae* in response to acetic acid stress and to confer tolerance to this compound by the induction of sphingolipid biosynthesis (Guerreiro et al. 2016). Rim101p, a TF that is better known as a regulator of the alkaline pH stress response involved in the response to alkaline stress (see Sect. 2.2.5), is also required for the full response to WOA stress (Mira et al. 2009).

In addition, the general stress response Msn2/4 TF is involved in response to WOAs (Schüller et al. 2004), and its activation by dephosphorylation is probably mediated by the stress-responding Whi2p-Psr1p phosphatase complex (Chen et al. 2016).

The HOG signaling pathway is well known for sensing and relaying the osmostress signal (see Sect. 2.4) but plays a more universal role in sensing environmental stresses in *S. cerevisiae* cells. In addition to osmostress and some other stresses discussed in this chapter, this pathway also participates in the response to acetic acid stress (Mollapour and Piper 2006). Of the two known HOG subpathways, only the one dependent on the Sln1p membrane sensor is functional in sensing acetic acid stress. In *S. cerevisiae*, Hog1p protein is necessary for acetic acid tolerance, but its protective function does not involve increased glycerol production unlike Hog1p-mediated osmostress resistance (Mollapour and Piper 2006). Instead, to prevent the influx of acetic acid via aquaglyceroporin Fps1p, Hog1p phosphorylates it, resulting in its removal from the plasma membrane (Mollapour and Piper 2007).

Acetic acid and, to some extent, propionic and sorbic acids, but not lactic acid, cause endoplasmic reticulum (ER) stress and trigger unfolded protein response (UPR). The Ire1p-Hac1p UPR signaling pathway (see Chap. 5) is required for the maximum tolerance to acetic acid (Kawazoe et al. 2017).

2.2.10 Response Mechanisms to WOA Stress in Other Yeast and Fungi

Various species of the genus *Aspergillus* have differing sensitivities to acetic and sorbic acids but similar to those of *S. cerevisiae*. The rule of dependence of this sensitivity on the extracellular pH and, hence, the fraction of acid that is in the undissociated state also applies, as well as the correlation between the toxicity and hydrophobicity, amounting to an approximately 20-fold lower concentration of sorbic acid than acetic acid that is necessary to inhibit the growth of *Aspergillus* spp. (Alcano et al. 2016).

Whole-transcriptome analyses of yeast cells under acetic acid-induced low pH stress revealed that 80% of genes in *S. cerevisiae* and 75% of *C. glabrata* genes responding to this stress are controlled by ScHaa1p and CgHaa1p TFs (Mira et al. 2010a; Bernardo et al. 2017), respectively. Despite this superficial similarity, the respective gene sets overlap only partially (Bernardo et al. 2017), accounting for the difference in environmental niches occupied by these two yeast species.

In *Candida glabrata*, the response and resistance to sorbic acid stress depend on the plasma membrane (PM) transporter Pdr12p and is partially dependent on the HOG pathway. The *hog1Δ* strain is oversensitive to sorbic acid, and, following sorbic acid exposure, Hog1 becomes more abundant and phosphorylated, directing it to the nucleus (Jandric et al. 2013).

Valuable insights into the nature of low pH and WOA stress came from studies on the persistent spoilage yeast *Zygosaccharomyces bailii*. This organism is a real trouble to the food industry because it is threefold more resistant than *S. cerevisiae* and other yeast species to various stresses, not only to WOAs but also to high osmotic pressure and ethanol concentration. It is resistant to food preservatives at concentrations that exceed the highest allowable limits of these substances in food products. On the other hand, being resistant to a harsh environment, it can become useful in the production of sustainable ethanol from lignocellulose hydrolysates, which contain high amounts of acetic acid (Koppram et al. 2014). A recent review by Kuanyshev et al. described *Z. bailii* peculiarities, some of which are mentioned here (Kuanyshev et al. 2017). It turned out that so high resistance is attained by the heterogeneity of *Z. bailii* cell population with only a small fraction of resistant cells that dominate the culture under WOA stress conditions. This increased tolerance is maintained only under constant stress exposure and is reset during overnight growth in the standard medium (Stratford et al. 2013b). Notably, the resistant cell subpopulation grows more slowly and has a more acidic internal pH, which affects their metabolism but lowers the uptake of WOAs, contributing to higher tolerance to these compounds. Thus, the slower growth is a trade-off for higher resistance but is a disadvantage under nonstressful conditions. Such a strategy of resistance to high levels of WOAs may not be unique to organisms such as *Z. bailii*. The variability of the resistance of individual cells to acetic acid within the clonal population was recently demonstrated also for *S. cerevisiae* (Fernández-Niño et al. 2015). Notably, the threefold higher tolerance of *Z. bailii* relative to *S. cerevisiae* is distinctive for the WOA category. *Z. bailii* and *S. cerevisiae* were equally tolerant to other organic compounds of various types. Within the WOA category, there was a strong inverse correlation between the lipophilicity of the acid and its minimum inhibitory concentration (MIC), indicating that the rate of diffusion through PM determines the toxicity of each WOA (Stratford et al. 2013b). The authors of the discussed paper tested 30 WOAs, many of them occurring naturally in various plants. For all of them, *Z. bailii* was approximately 3x more tolerant than *S. cerevisiae* (Stratford et al. 2013b). This vaguely suggests that more organic acids naturally occurring in various plant species may in fact play a role as antifungal agents.

2.2.11 Other Antispoilage WOAs

Most studies regarding the toxicity of WOA for yeast and fungi limit their interest to acetic, propionic, sorbic, and benzoic acids, which are used as antispoilage agents in the food industry. However, many other WOAs existing in nature may also have antispoilage potential (Stratford et al. 2013b). One of them is butyric acid, which, although useless as a food preservative because of the obnoxious smell, may play a role as a ginkgo tree fruit protectant. Intriguingly, together with acetic, propionic, and lactic acids, butyric acid is the product of the human bacterial microbiome, and perhaps it is an important agent in the battle between bacteria and pathogenic fungi for dominance within the various niches of the human body (Cottier et al. 2015).

2.2.12 WOA and Programmed Cell Death

It has been known for many years that exposure of *S. cerevisiae* cells to 20–80 mM *(0.2–0.5%)* acetic acid induces their programmed cell death (PCD) (Ludovico et al. 2001). Since the time of this discovery, acetic acid is commonly used as a tool in PCD studies; however, the exact chain of events leading from acetic acid stress to apoptotic death is not well understood. Several whole genomic and proteomic studies on acetic acid-induced PCD have revealed multiple signaling pathways involved in triggering it, including TOR and the general amino acid control systems (Almeida et al. 2009), and demonstrated that virtually every aspect of cellular metabolism is affected en route to apoptosis: mitochondrial function, protein synthesis, modifications and degradation, vesicular traffic, amino acid transport and biosynthesis, oxidative stress response, cell growth and differentiation, histone deacetylation, carbohydrate and lipid metabolism, and stress response (Sousa et al. 2013; Longo et al. 2015; Dong et al. 2017). Despite this, or maybe just because of the multitude of processes involved, the clear scheme of acetic acid-induced PCD has not yet been delineated (Palma et al. 2018). On the other hand, as has been pointed out above, acidification of the cytosol has profound and devastating consequences for many cellular functions. Cytosol acidification was implicated as an early event triggering apoptosis in mammalian cells (Matsuyama et al. 2000). Thus, it may be the sole and sufficient primary stimulus leading also the yeast cell to PCD. Quite recently, it was demonstrated that propionic acid, similar to acetic acid, kills *C. albicans* cells via PCD mechanism (Yun and Lee 2016). Interestingly, the propolis that is made by bees from plant sap and beeswax induces PCD of budding yeast cells (de Castro et al. 2011), and it was recently shown to contain significant amounts of phenolic acids (Galeotti et al. 2018). Thus, the array of antifungal WOAs present in nature is definitely much wider than the most studied acetic, propionic, sorbic, and benzoic acids. They all may have the same mechanism of toxicity toward fungal cells, and it is conceivable that plants in their battle against fungal infections invented the use of WOA to exploit the preexisting PCD mechanism that otherwise is beneficial to fungi under certain circumstances.

2.3   Low-Temperature Stress

Low temperature is a common environmental stress for yeast and filamentous fungi living in a temperate or colder climate. These organisms are periodically exposed to such challenges in sync with the day/night rhythm. However, in contrast to the very well-characterized response of high-temperature stress (see Chap. 5), cold stress is less commonly studied on fungal models.

Numerous genome-wide studies have revealed that the cell response to this stress and therefore the stress itself is not uniform. It can be distinguished as follows: (1) moderately low temperature (18–10 °C); (2) low but above the water freezing point (10–0 °C), which slows down cell metabolism and, as we soon learn, also compromises important cellular processes; and (3) the stress of freezing the cells (below 0 °C), thereby suddenly stopping all their activities and exposing them to frost-inflicted structural damage. For this reason the resistance to freeze should be considered as the ability to resume the vital functions after thawing. Moreover, nonfreezing low temperature may trigger a diversified response depending on the actual temperature value, to which cells are exposed.

2.3.1   Harmful Effects of Low-Temperature Stress

A low temperature above the water freezing point establishes a distinct set of challenges that affect cell survival: (1) lowered membrane fluidity (Russell 2008; Ernst et al. 2016), with all its consequences of distorted transmembrane transactions (Los and Murata 2004) and increased passive permeability (Hoekstra et al. 2001); (2) decelerated metabolic flow, therefore, the insults requiring energy expenditure are more difficult to battle with; (3) prolonged time of protein synthesis and folding; (4) general decrease in enzymatic activity (Gerday et al. 1997); and (5) lower elasticity of proteins that affects all processes relying on recognition of spatial features of molecules, including fundamental cellular processes, such as DNA replication, RNA transcription, and protein synthesis (Jones and Inouye 1996; Farewell and Neidhardt 1998). Moreover, the uptake of inorganic phosphate is a factor limiting yeast growth at low temperatures (Vicent et al. 2015).

In the context of the harmful effects of low temperature on yeast cells, it is worth mentioning the phenomenon known for many years that *S. cerevisiae* strains auxotrophic for tryptophan are cold sensitive (Chen et al. 1994). While it is not the effect of low temperature alone because prototrophic strains are unaffected, it is a consequence of distorted membrane fluidity by cold stress. It was also shown that heightening membrane fluidity by the introduction of polyunsaturated fatty acids affects the uptake of tryptophan at low temperatures (Rodríguez-Vargas et al. 2007).

2.3.2 Resistance to Low-Temperature Stress

Several countermeasures are undertaken by fungi cells when they are exposed to cold stress. Glycerol is a well-known osmoprotectant accumulated in response to external hyperosmotic stress, but it is also important for *S. cerevisiae* survival of freeze stress (Izawa et al. 2004). Its intracellular concentration increases through the action of plasma membrane glycerol/H+ symporter Stl1p, which pumps this compound inside the cell (Tulha et al. 2010) and/or by enhancing its synthesis during cold adaptation (Oliveira et al. 2014). Glycerol has a protective role against freeze stress, although it seems dispensable for the survival of the low but nonfreezing temperature stress (Panadero et al. 2006), which supports the thesis that the cold stress response, in fact, prepares the cell for freeze stress that, in nature, usually follows the decrease in ambient temperature to near 0 °C.

Another compound important for low-temperature tolerance is trehalose. It accumulates in budding yeast cells exposed to low, near-freezing temperatures (10–0 °C), but not following the cold shock with temperatures above 10 °C. The accumulation of trehalose also supported the survival of yeast cells kept at 0 °C and subjected to freezing (Kandror et al. 2004). Unlike the desiccation stress (see Sect. 2.5), no mechanisms of trehalose protection against cold stress were proposed. The trehalose levels are increased under multiple stress conditions, and for a long time, it was regarded as a general stress protectant. Recently, its importance for tolerance to many stressors was questioned (Petitjean et al. 2015; Gibney et al. 2015). Nevertheless, it seems indispensable to endure freezing (Mahmud et al. 2010) and desiccation (Tapia and Koshland 2014; Tapia et al. 2015) (see Sect. 2.5). Proline also appears to have cryoprotective capabilities in *S. cerevisiae*, similar to those of glycerol and trehalose (Takagi et al. 2000), and it also supports the growth of *S. cerevisiae* cells at low temperatures (Vicent et al. 2015).

In line with the reported upregulation of genes encoding chaperone proteins in response to low-temperature conditions, a subset of these proteins conferred the increased tolerance of budding yeast to freeze-thaw stress (Kandror et al. 2004; Naicker et al. 2012). One of them, Hsp12p, a small heat shock protein and a hydrophilin that functions as a cell wall plasticizer (Karreman et al. 2007), was strongly induced by severe cold stress (4 °C) (Murata et al. 2006). The contribution of other hydrophilins not only to desiccation tolerance but also to freeze tolerance was demonstrated as well (Dang and Hincha 2011).

The decrease in the fluidity of lipid membranes is a hallmark of low-temperature stress, and this phenomenon affects fundamental cellular functions. It can be reversed by the increase in the proportion of unsaturated fatty acids in the membrane lipids (Nakagawa et al. 2002; Martin et al. 2007).

The proportion and the variety of sterols do not change in adaptation to cold (Russell 2008). Ergosterol is a crucial component of *S. cerevisiae* lipid membranes and, as mentioned in the next subchapter, it is important for the survival of desiccation stress (Dupont et al. 2012). While there are no data on the importance of this sterol for cold or freeze stress, it appears that the adequate level of ergosterol in the plasma membrane is necessary for the proper targeting of the tryptophan permease Tat2p (Umebayashi and Nakano 2003), and ergosterol biosynthesis improves low-temperature growth that is dependent on tryptophan uptake (Vicent et al. 2015).

The accumulation of all these compounds and macromolecules at low temperature constitutes the inducible response to cold stress of yeast species living in a moderate or warm climate. However, psychrophilic and psychrotolerant yeast species inhabiting the cold regions of our planet have constant adaptations to low temperature. In addition to the well-established increased content of mono- and polyunsaturated fatty acids in their membranes, it was suggested that they secrete antifreeze proteins similar to those found in higher organisms (Alcaíno et al. 2015).

Stationary phase, starved, or dormant *S. cerevisiae* cells are more resistant to freezing than log phase cells; moreover, increased tolerance to freezing can be induced by preexposure to other stresses, suggesting that cells prepared to endure various environmental stresses are better equipped to withstand freezing as well (Park et al. 1997). Clearly, adaptations to low temperature, although they develop when the ambient temperature is above the freezing point and cells can still perform their functions, are tailored to allow survival under subzero, freezing temperatures.

2.3.3   Sensor of Low-Temperature Stress

The mechanisms of detection and transmission of the signal of cold stress to the target genes in *S. cerevisiae* cells are depicted in Fig. 2.2.

Low temperature influences the whole cell so unlike, for instance, the external hyperosmosis, the sensor of cold stress does not need to be located on the cell surface and in fact it is built-in into the ER membrane. Two paralogous and mostly redundant proteins, Mga2p and Spt23p, can form homodimers anchored to the ER membrane, and their conformation depends on the membrane fluidity. When the fluidity decreases, the structure of the dimer changes and the proteins become susceptible to ubiquitination and proteolytic cleavage. This event releases the soluble, globular part of one of the molecules of the dimer, called p90, constituting the functional TF, that, upon entering the nucleus, induces a subset of genes, including *OLE1* encoding $\Delta(9)$ fatty acid desaturase (Nakagawa et al. 2002; Covino et al. 2016; Ballweg and Ernst 2017).

Interestingly, low-temperature stress signaling is performed also by the HOG pathway (Hayashi and Maeda 2006; Panadero et al. 2006). This pathway is well known for sensing and relaying the osmostress signal (see Sect. 2.4 and Fig. 2.3) but plays a more universal role in sensing environmental stresses in *S. cerevisiae* cells. In addition to osmostress and low-temperature stress, it participates in the response to acetic acid stress (Mollapour and Piper 2006) and oxidative stress (Bilsland et al. 2004). Of the two known subpathways that converge at the MAP kinase kinase Pbs2p (see Sect. 2.4), only the one beginning with the membrane sensor protein

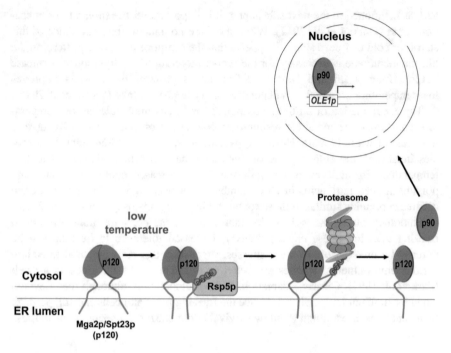

Fig. 2.2 Mechanisms of detection and transmission of the cold stress signal in budding yeast. See the text for details (Adapted from Ballweg and Ernst 2017)

Sln1p is functional in sensing low-temperature stress. Cell exposure to dimethyl sulfoxide, an organic solvent known to increase lipid membrane rigidity, thus mimicking the effect of low temperature on these structures (Sangwan et al. 2002), also activates Sln1p (Panadero et al. 2006). Thus, it seems that both Mga2p/Spt23p and Sln1p detect the same effect of low temperature, the decreased membrane fluidity, although the detection methods are different. Notably, the decrease in the membrane lipid fluidity was shown also under hyperosmotic stress (Laroche et al. 2001).

Remarkably, the absence of Hog1p, the pivotal protein of the HOG pathway, does not sensitize budding yeast cells to lower temperature (12 °C) but makes them unable to prepare themselves to freeze stress (Panadero et al. 2006), indicating the already mentioned diversion of the response mechanisms to moderately decreased temperature (18–10 °C) and to near-freezing temperature (10–0 °C) that prepares cells to endure incoming freezing conditions. These two stresses may incite the same sensor, Sln1p, which likely acts as a mechanosensor to detect the change in the diffusion rate within the cells.

CWI and target of rapamycin complex 1 signaling pathways were also implicated in the response to cold stress because their inactivation results in a cold-sensitivity phenotype, yet no further data on this subject are available (Córcoles-Sáez et al. 2012).

2.3.4   Response Mechanisms to Low-Temperature Stress

The *S. cerevisiae* cell response to low temperature depends on stress severity. A moderately lower temperature (18–12 °C) slows down cellular processes but also causes upregulation of genes involved in protein translation. This effect is even stronger in its close relative *Saccharomyces kudriavzevii*, which is better adapted for growth at low temperatures, and in *S. cerevisiae* × *S. kudriavzevii* hybrids (Combina et al. 2012; Tronchoni et al. 2014). On the other hand, protein synthesis genes are strongly repressed in *S. cerevisiae* exposed to a near-freezing temperature (4 °C) (Murata et al. 2006). These results demonstrate that cells have different programs for cold stress that slows down but do not prevent growth and for conditions that make the growth impossible.

The *S. cerevisiae* cell response to low-temperature stress can also be dissected, based on stress duration, into two or even three phases (Sahara et al. 2002; Schade et al. 2004). The early response (up to 2 h) induced the expression of proteins involved in transcription, RNA processing enzymes (Schade et al. 2004), and fatty acid desaturase Ole1p that can modify membrane fluidity (Nakagawa et al. 2002; Martin et al. 2007) and other phospholipid biosynthesis enzymes, such as Ino1p and Opi3p (Murata et al. 2006). Transient induction of the *OLE1* gene was shown also for cold-exposed *Dekkera bruxellensis* (Galafassi et al. 2015). The late response (between 12 and 60 h) induces accumulation of trehalose (Kandror et al. 2004) and glycerol (Panadero et al. 2006), some chaperone proteins (Hsp12p, Hsp26p, Hsp42p, and Hsp104p), and oxidative stress protection enzymes (Sahara et al. 2002; Schade et al. 2004; Murata et al. 2006). These and other proteins of the long duration cold stress, such as chaperone proteins and trehalose synthesis enzymes, are upregulated in the general stress response, Msn2/4-dependent manner, preparing cells to survive freeze-thaw stress (Kandror et al. 2004).

The accumulation of glycogen, but not trehalose, and the increased proportion of unsaturated fatty acids in cellular lipids were also demonstrated for *Dekkera bruxellensis*. Similarly to the *S. cerevisiae*, HOG-MAP kinase signaling pathway was also activated by cold stress in this organism (Galafassi et al. 2015), indicating the conserved role of this kinase in signaling low-temperature stress.

Low-temperature-instigated growth limitation of tryptophan auxotrophs is dependent on Gcn2p, a protein kinase that downregulates protein synthesis in response to an unfavorable growth environment. The upsurge of P-bodies at low temperatures is necessary for cold tolerance, and their proper formation is affected by tryptophan limitation (Ballester-Tomás et al. 2017).

2.3.5   High Hydrostatic Pressure: An Intriguing Resemblance to Low-Temperature Stress

High hydrostatic pressure (HHP) stress is not frequently studied with yeast or fungi as a model organism. Perhaps, it is believed that this stress is not encountered by these organisms in nature, so symptoms of specific, biologically meaningful responses are not expected. There are anthropogenic niches in which yeast *S. cerevisiae* are subjected to overpressure—for instance, during secondary fermentation of sparkling wine production—and the influence of such overpressure on yeast and fungi model organisms in the context of industrial processes is investigated. The extent of overpressure used in those studies may be up to 0.6 MPa (6 atm) (Coelho et al. 2004), which is similar to the internal turgor pressure of filamentous fungi (Lew et al. 2004) and yeast cells (Schaber et al. 2010). However, no yeast and fungi species, at least not *S. cerevisiae*, normally live under extremely high hyperbaric pressure in a range of 25 to more than 100 MPa, i.e., as high as the pressure at the bottom of the deepest ocean trenches. However, this is the range of pressure magnitudes that was applied in most of the experiments described in the papers cited in this subsection. The pressure of 150 MPa causes profound changes in the substructures of *S. cerevisiae* cells, leading to their death within 30 min (Kawarai et al. 2006) and indicating the lack of any tolerance mechanisms to HHP stress of such magnitude in this organism.

HHP affects all components of living cells. Under extremely high pressure, the structure of protein molecules is modified (Roche et al. 2012). Hydration of nucleic acids is also affected by HHP (Giel-Pietraszuk and Barciszewski 2012). Therefore, this topic is more interesting for structural biologists and beyond the scope of this chapter. Nevertheless, one aspect of this stress deserves mentioning: the influence of HHP on the structure of lipid membranes. HHP causes the phase transition of lipid membranes, making them less fluid and more rigid, closely resembling the cellular effects of low temperature (Matsuki et al. 2013). Thus, it is reasonable to assume that the effects of HHP observed in budding yeast largely mimic the effect of low temperature. Indeed, increasing membrane fluidity by increasing the proportion of unsaturated lipids results in a higher survival rate of yeast cells under HHP (de Freitas et al. 2012). There is also a substantial overlap between the group of budding yeast knockout strains that are sensitive to high pressure and those that are sensitive to low temperature (Abe and Minegishi 2008). It was also demonstrated that cold-sensitive tryptophan auxotrophic *S. cerevisiae* strains are oversensitive to HHP stress and that overexpression of the *TAT2* gene encoding the high-affinity tryptophan permease or supplementation of the growth medium with an excess of tryptophan reverses the growth defects of tryptophan auxotrophs in HHP, just like in low temperature (Abe and Horikoshi 2000). Thus, it seems clear that the explanation of the link between these two stresses lies in the similar changes in the fluidity of lipid membranes caused by low temperature and by high pressure.

Exposure to moderately high hydrostatic pressure (25–50 MPa) that does not immediately kill cells results in the accumulation of reactive oxygen species and induction of oxidative stress protectants in budding yeast cells (Bravim et al. 2016). It results in the increase in the levels of chaperone proteins (Miura et al. 2006) and triggers the general stress response (Domitrovic et al. 2006; Bravim et al. 2012). This most likely reflects the general damage to all cell components and deterioration of cell metabolism provoked by HHP (Kawarai et al. 2006). Transcriptome analyses revealed the induction of genes involved in membrane and cell wall biosynthesis, as well as in protein degradation (Iwahashi 2015).

2.4   Osmotic Stress

Among the environmental extremities mentioned in this chapter, quite common are the biotopes with the concentration of various salts far exceeding their concentration inside living cells. Despite their hostility, they are densely populated by organisms belonging to all systematic kingdoms. Even the exemplary Dead Sea with its 34% salinity is not, in fact, dead (Grant 2004; Jacob et al. 2017). Yeast and filamentous fungi can also be found in such niches, but most of them, including the budding yeast, the most commonly used in studies on hyperosmotic stress, are not halophilic organisms. Nevertheless, they too find themselves in the conditions of a high external osmotic pressure. Living inside and on the surface of fruits, they are usually exposed to highly concentrated sugars. Moreover, they may suffer rapid changes in the surrounding osmoticum incited by alternating dry and wet weather, so they should have the appropriate countermeasures to defend themselves against it.

The studies on osmotic stress and adaptation using yeast as a model organism are very extensive and were conducted for several decades (Brown and Simpson 1972). This stress is probably the best known of all stresses discussed in this chapter (Hohmann 2002, 2009, 2015; Grant 2004; Ma and Li 2013; Brewster and Gustin 2014; de Nadal and Posas 2015; Saxena and Sitaraman 2016).

2.4.1   Resistance to Osmotic Stress

Water movement through the lipid membranes is driven by the osmotic gradient and will escape from the cells placed in a hyperosmotic environment. This causes its shrinking, which increases molecular crowding and slows down all metabolic processes (Miermont et al. 2013). Cell shrinkage is also a mechanical stress that can cause damage to the plasma membrane and cell wall. Further loss of water leads to dehydration stress, which will be discussed in Sect. 2.5. These deleterious effects can be alleviated by increasing the osmotic strength of the cell interior. To accomplish that, the xerophilic organisms can accumulate diverse osmolytes (compatible solutes) inside their cells, such as polyols, sugars, betaines, thetines, amino acids, modified and unmodified, and ectoines (Grant 2004). Prokaryotes are especially ingenious in this respect, whereas the methods of adaptation to the increased osmotic strength of the surrounding medium found in *S. cerevisiae* and in other fungi species are less diverse and involve the accumulation of polyols, mostly glycerol but also arabitol, erythritol, sorbitol, and mannitol (Mager and Siderius 2002; Kumar and Gummadi 2009; Galafassi et al. 2013; de Lima Alves et al. 2015; Turk and Gostinčar 2018). This is accomplished either by their increased synthesis (Mager and Siderius 2002) or their uptake from the environment. In *S. cerevisiae*, active uptake of glycerol is realized by the glycerol/H^+ symporter Stl1p (Ferreira et al. 2005). When the osmotic strength outside the cell is reduced, the cell can dispose of these molecules via the aquaglyceroporin Fps1p, a plasma membrane channel allowing the efflux of excessive glycerol and other solutes (Tamás et al. 1999; Ahmadpour et al. 2014). Perhaps, this same excessive glycerol is uptaken again as an osmolyte when the conditions become hyperosmotic again or is used as a carbon source when fermentable sugars are exhausted from the currently inhabited niche. Proteins of adequate function exist in other yeast species, such as *Zygosaccharomyces rouxii* (Dušková et al. 2015), and many of them can uptake glycerol under hyperosmotic stress (Lages et al. 1999).

When hyperosmotic stress is produced by excessive salinity, in addition to osmolyte accumulation, cells can actively pump out metal ions. P-type ATPase Ena1p and the sodium-proton antiporter Nha1p are responsible for this task and confer resistance not only to high salinity but also to alkaline pH stress (Yenush 2016) (see Sect. 2.2).

2.4.2 Response Mechanisms to Osmotic Stress

Hyperosmotic stress causes cell shrinkage and increases molecular crowding, which slows down all metabolic processes. Notably, these include also the transmission of stress signals to target genes (Miermont et al. 2013), including the signal of hyperosmosis (Babazadeh et al. 2013), indicating that this signal must be transmitted sufficiently early to preempt the disabling effect of a water deficit.

The response to osmotic stress is regulated by the well-known signaling pathway, HOG, which is one of five MAP kinase signaling cascades known in the budding yeast that convey various signals and regulate diverse cellular processes (Gustin et al. 1998). In addition to osmotic stress signaling performed by HOG, they are involved in the cell-wall integrity pathway, the spore wall assembly pathway, filamentous/invasive growth pathway, and pheromone response pathway (reviewed in Chen and Thorner 2007). They share some of their components yet control separate cellular processes (reviewed in Saito 2010).

The HOG pathway was already mentioned several times in this chapter, documenting that it plays a more diversified role in signaling environmental stresses, beyond that in hyperosmosis one. In *S. cerevisiae* cells, in addition to informing about the change in external solute concentration, it conveys the signals of other stresses, such as low-temperature stress (Hayashi and Maeda 2006; Panadero et al. 2006), acetic acid stress (Mollapour and Piper 2006), oxidative stress (Bilsland et al. 2004), and, together with the cell wall integrity (CWI) pathway, also mechanostress via its Sho1p branch (Delarue et al. 2017) (see Fig. 2.3). The multifunctionality of the HOG pathway is seen not only in *S. cerevisiae* but also in other yeast species, such as the highly osmotolerant and halotolerant yeast *Debaryomyces hansenii* (Sharma et al. 2005), and in filamentous fungi, *Aspergillus nidulans* (Furukawa et al. 2005) and *Aspergillus fumigatus*, in which this pathway contributes to the response to high-temperature stress (Ji et al. 2012). However, the most extensively studied and, historically, the earliest recognized is the involvement of the HOG pathway in the regulation of the osmotic stress response, which will be described in more detail in this subsection.

Other yeast species (Kumar and Gummadi 2009; Galafassi et al. 2013; Enjalbert et al. 2006) and filamentous fungus *Aspergillus nidulans* (Duran et al. 2010) were also used as models in studies on the osmotic stress response, and their results have revealed strong similarities of the HOG MAP kinase signaling pathway organization and functions across fungal evolutionary tree. However, the *S. cerevisiae* HOG pathway looks considerably more elaborate (see Fig. 2.3), perhaps reflecting that the response to osmostress was studied the most extensively in this organism.

2.4.3 Description of the HOG Signaling Pathway

The *S. cerevisiae* HOG pathway comprises two subpathways called the Sln1 branch and Sho1 branch after the names of sensor proteins residing in the plasma membrane (Fig. 2.3). These two subpathways converge at the MAPKK step of the kinase cascade, performed by Pbs2p. The exact mechanism of hyperosmotic stress detection by these sensors is not known. One might expect that mechanical deformation because of this stress could be detected by the mechanosensor embedded in the plasma membrane. Indeed, mucins Msb2p and Hkr1p, which are membrane osmosensors in a Sho1p branch of HOG, could be such mechanosensors. These transmembrane proteins with a heavily glycosylated domain extending into the cell wall could sense the detachment of the plasma membrane from the cell wall during cell shrinkage. Notably, these proteins also function as sensors in another MAPK signaling pathway, triggering the filamentous growth in budding yeast (reviewed in Brewster and Gustin 2014). Moreover, Msb2p and Sho1p sense the compression signal and trigger the adaptation response to increase cell survival under compressive stress (Delarue et al. 2017), supporting the hypothesis that the signals sensed by the Hkr1p/Msb2p/Sho1p branch of the HOG pathway are mechanical in nature. Sho1p functions as an osmosensor and a scaffold for a multiprotein signaling complex comprising the plasma membrane Opy2p and Hkr1p and cytoplasmic adaptor protein Ste50p that is essential for Hog1p activation (Tatebayashi et al. 2015). It was also proposed that the existence of multiple sensors in the Sho1 branch functions as a logical AND gate, avoiding illegitimate activation of this branch of HOG when only one sensor is triggered (Tatebayashi et al. 2015).

The Sln1p branch of the HOG pathway begins with the sensor protein histidine kinase Sln1p. While the exact method by which hyperosmotic stress is sensed by this protein is not known, it is postulated to detect molecular crowding, i.e., increased compaction of molecules upon water loss (Hohmann 2015). Because activation of the Sln1p branch of the HOG pathway occurs via the decrease in Sln1p histidine kinase activity, it is easy to imagine the mechanism of stress sensing because, as mentioned earlier, molecular crowding slows down all metabolic processes. Notably, the Sln1p-dependent subpathway is one of the pathways transmitting the signal of low-temperature stress (Panadero et al. 2006) (see Sect. 2.3). This justifies speculation that the change in the internal milieu that is detected by Sln1p is the same under either stress. Both low temperature and water deficit-inflicted molecular crowding result in the slowdown of diffusion within the cell and related effects, such as lower elasticity of macromolecules. Thus, it is reasonable that the same pathway operates in the signaling of both these stresses. It was

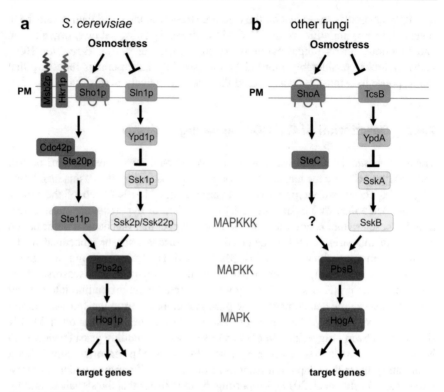

Fig. 2.3 HOG pathway in *S. cerevisiae* and other model fungi species. (**a**) In *S. cerevisiae*, both Sln1p- and Sho1p-dependent subpathways of HOG converge at the MAPKK step of the kinase cascade, performed by Pbs2p. The Sln1p sensor is a histidine kinase, which downregulates its branch of the HOG pathway. Hyperosmotic shock inactivates Sln1p, which leads, via Ypd1p and Ssk1p phosphorelays, to activation of the MAPKKKs Ssk2p and Ssk22p comprising the first step in the HOG pathway kinase cascade. Ssk2p and Ssk22p phosphorylate and activate MAPKK Pbs2p that, in turn, phosphorylates and activates the MAPK Hog1p. The components of the Sho1 branch form a dynamic multiprotein complex associated with the plasma membrane using the Sho1p as a scaffold protein. Hkr1p and Msb2p mucins function as hyperosmotic stress sensors in this branch. The detection of stress probably induces changes in the interaction between the proteins within the complex. This ultimately leads to the phosphorylation and activation of the MAPKKK Ste11p by Ste20p-Cla4p protein kinases, which connect this branch to the kinase cascade common for both branches. Active Hog1p, the last step of the cascade, accumulates in the nucleus where it initiates the transcription of its target genes, including those encoding proteins involved in glycerol synthesis and uptake (see the text for more details) (de Nadal and Posas 2015). (**b**) Our knowledge on HOG in other fungal species is far less complete; however, from what is known about this pathway, it seems to be highly conserved among fungi (Adapted from Ma and Li 2013; Brewster and Gustin 2014; Hohmann 2015; Suescún-Bolívar and Thomé 2015)

shown that even moderate hyperosmotic stress leads to a slowdown of molecular processes (Mika and Poolman 2011), agreeing with the postulated function of Sln1p as a sensor of molecular crowding. Incidentally, the increased osmotic pressure, similar to low temperature, also decreases the fluidity of the membranes (Laroche et al. 2001).

And that leads to another presumption. Activation of a more sensitive Sln1p branch perhaps results in the upregulation of proteins not only involved in maintaining the proper cell turgor but also assuring optimal cellular functions when the diffusion rate and protein elasticity are diminished. On the other hand, the less sensitive Sho1p branch is triggered in the alarming situation of hyperosmotic stress-provoked mechanical damage to the plasma membrane and other cellular components.

Interestingly, there is a remarkable difference between the two branches in their mode of response to hyperosmotic stress. The Sln1p branch response is gradual, proportional to the strength of the hyperosmotic stress, whereas the Sho1p branch functions like an on-off switch, eliciting a full response above a certain threshold (Maeda et al. 1995). These data may indicate that these two branches detect different types of stimuli and are compatible with the suggestion that Msb2p and Hkr1p mucins in the Sho1 branch detect the mechanical stress resulting from plasma membrane detachment from the cell wall, which may be an abrupt event, whereas Sln1p detects molecular crowding that will increase gradually along with the water loss.

There is cross-talk between the Sho1p branch of the HOG pathway and MAPK pathway, triggering the invasive growth in response to nutrient starvation and the pheromone response. In the filamentous fungus *A. nidulans*, both branches of the HOG pathway exist, but only the branch orthologous to Sln1p branch in *S. cerevisiae* is well characterized. Morover, the Sho branch seemingly transmits other stress stimuli by sensing contact with surfaces, which is crucial for filamentous growth (Furukawa et al. 2005). In line with these data, the same branch in yeast *Cryptococcus neoformans* was implicated as important for the virulence of this pathogenic yeast (Malachowski et al. 2016). Recent studies on *Neurospora crassa* have suggested that the HOG pathway may sense the nutritional signal of the free sugar abundance in the environment (Huberman et al. 2017).

Most of the studies on the HOG signaling pathway are focused on its role in the adjustment of intracellular glycerol as an osmoprotectant. Hog1p controls the expression of glycerol biosynthetic genes *GPD1* and *GPP2* [via Hot1p TF (Rep et al. 2000)] as well as Stl1p, a plasma membrane protein responsible for active glycerol uptake (Ferreira et al. 2005). Fps1p, an aquaglyceroporin constituting the glycerol export channel, closes under hyperosmotic stress to prevent glycerol from leaking out from the cell (Ahmadpour et al. 2014). Therefore it could be expected that the activity of Fps1p is also controlled by Hog1p. Nevertheless, the data on that issue are inconsistent (Beese et al. 2009; Babazadeh et al. 2014), suggesting that it may be under the control of other HOG-independent signaling pathways. On the other hand, Hog1p probably downregulates the expression of the aquaporin Aqy2p, thereby restricting water loss (Furukawa et al. 2009).

It appears that the intracellular level of glycerol in *S. cerevisiae* is not controlled exclusively by HOG because cells with a dysfunctional pathway can accumulate some glycerol in response to hyperosmotic stress. This is explicable by the ability of other MAPK signaling pathways to take over the missing HOG functionality, especially that the plasma membrane sensors of hyperosmotic stress can communicate with them (see above). Nevertheless, cells with an inactive HOG pathway are highly sensitive to hyperosmotic stress (Hohmann 2015).

The functions of the HOG pathway go beyond the regulation of the intracellular level of glycerol. In fact, Hog1p targets are multiple because the deletion of the *HOG1* gene affects the expression of numerous genes. They include a range of TFs that control the general response to environmental stress, such as Msn2/4 (Rep et al. 2000) and Sko1p (Marotta et al. 2013), resulting in the induction of genes encoding stress-protecting proteins. In response to incoming environmental stress, the HOG pathway triggers the cell cycle delay and influences every step of mRNA biogenesis—the initiation and elongation of transcription, mRNA processing, and export from the nucleus, as well as its stability and translation (reviewed in Solé et al. 2015; de Nadal and Posas 2015). Thus, it appears that HOG is one of the pivotal stress response regulatory hubs.

Although the importance of the HOG signaling pathway for budding yeast cell survival under hyperosmotic stress is solidly and extensively documented, the picture of their response to this stress does not seem to be complete. There must be some other ways to acquire tolerance to hyperosmosis, independent of HOG (Westfall et al. 2008; Saxena and Sitaraman 2016). It was shown for several yeast species that a high concentration of monovalent cations activated, in addition to the HOG pathway, AMPK (Ssp2 in fission yeast *S. pombe* and Snf1p in *S. cerevisiae*), a sensor and regulator of cellular energy status (Hong and Carlson 2007; Schutt and Moseley 2017). Notably, in fission yeast, but not in budding yeast, the same activation was induced with hyperosmotic sorbitol. Osmotic stress provoked with excessive monovalent cations caused a transient decrease in the cellular ATP levels (Schutt and Moseley 2017), most likely due to the energy expense on the active efflux of these cations by ATPase sodium pump, Ena1p. The transient energy deficit explains AMPK stimulation.

2.4.4   Hypoosmotic Stress

For various living organisms, both hypoosmotic and hyperosmotic conditions may be stressful, yet yeast and fungi cells are surrounded by a rigid chitin cell wall so it might be expected that they easily withstand the excessive internal osmotic pressure when exposed to the conditions of low solute concentration, without engaging any specific protective measures. If at all, under extreme cases, when transferred to a strongly diluted medium, they will experience nutrient depletion. It is common knowledge that *S. cerevisiae* cells starved for nutrients by incubation in deionized water preserve their viability for days (e.g., Granot and Snyder 1991). However, as mentioned earlier, yeast and fungi in their natural environment are constantly exposed to dry-wet cycles and experience both hyper- and hypoosmotic stresses, so they frequently must adjust the osmolarity of their internal milieu to that of the external medium. Nevertheless, hypoosmotic stress has been far less thoroughly investigated. It seems that the low solute concentration is not by itself stressful to yeast cells. However, when they respond to hyperosmotic conditions by increasing the internal glycerol or other osmolytes, rapid transfer to the diluted medium may be damaging to the cell.

In *S. cerevisiae*, the appropriate reduction of internal osmotic strength is accomplished by the aquaglyceroporin Fps1p, which allows the efflux of excessive glycerol and other osmolytes (Tamás et al. 1999), whose role in fine-tuning the intracellular osmoticum has already been discussed (Beese et al. 2009; Babazadeh et al. 2014).

An alternative protective mechanism against hypoosmotic stress exists in *Schizosaccharomyces pombe*. Two calcium channel proteins, Msy1p and Msy2p, reside in the endoplasmic reticulum membrane of this species, and they were shown to regulate intracellular Ca^{2+} and the cell volume, supporting the survival of hypoosmotic shock-exposed *S. pombe* cells (Nakayama et al. 2012). Interestingly, no apparent homolog of these proteins was found in *S. cerevisiae*, although the calcium ions may play a role in hypoosmotic shock signaling in this yeast species (Batiza et al. 1996).

As mentioned above, decreased osmolality of the cell interior activates Sln1p histidine kinase and turns off the HOG signaling pathway, but this change signals the absence of hyperosmotic stress rather than the appearance of hypoosmotic insult. Indeed, the budding yeast cell response to switching from hyper- to hypoosmotic conditions was dominated by withdrawal from the changes induced by the hyperosmotic stress (Gasch et al. 2000). In addition, the cell integrity MAPK pathway seems to be involved in sensing hypoosmotic stress (Davenport et al. 1995).

A high concentration of solutes in the surrounding medium lowers water activity, similar to dehydration (Grant 2004). Therefore, one might expect that hyperosmotic and desiccation stresses have much in common in terms of the damage they cause and cellular responses they induce, especially that, in nature, dehydration is preceded by hyperosmosis, during which the cell can prepare itself to survive the life-suspending dehydration phase. However, as we will learn from the next subchapter, the responses to desiccation and hyperosmotic stresses are not identical. In particular, inactivation of the HOG signaling pathway does not result in the hypersensitivity of *S. cerevisiae* cells to dehydration.

2.5 Desiccation Stress

Water is the essence of life on Earth and, as discussed in the previous section, even a moderate decrease in water activity results in profound changes of the cell interior that slow down all biological processes and may lead to cell envelope damage and cell death. However, there are organisms from all kingdoms that can inhabit various environmental niches of extremely low water activity because of the dry climate, a high concentration of salts and other solutes, or both (Grant 2004). Moreover, many species, not necessarily xerophiles, had ventured into the dry areas of our planet and have exposed themselves to the constant danger of losing water and not being able to perform life-supporting chemical reactions. Therefore, developing mechanisms to protect them against dehydration is crucial. These mechanisms are not yet fully understood, although they have been studied for several decades now and the major agents assuring tolerance to desiccation have been known for a long time (Crowe et al. 1992). It is in profound contrast to the response mechanisms to osmotic stress that are perhaps the most thoroughly elucidated of all environmental stress response mechanisms (see Sect. 2.4). While a water deficit inside living cells slows down all their processes, the absence of water arrests them so it seems obvious that the mechanisms of the response to desiccation stress must be triggered when the cell senses the danger of approaching dehydration **before** it occurs, or the cell must be constantly prepared for this stress. As we will soon learn, the cells employ both these strategies.

Water is not only a solvent, i.e. the donor of protons or hydroxyl groups and medium for chemical reactions. It is estimated that approximately 15% of water molecules inside living cells is involved in the interactions with cellular macromolecules, and the preservation of their native state depends on these interactions (Persson and Halle 2008).

The capacity to survive dehydration is not common in nature; the group of organisms that have this ability is small and includes several yeast species, bacteria, tardigrades, rotifers, brine shrimps, plant seeds, mosses, and nematodes (Crowe et al. 1992). Fortunately for the studies on the desiccation stress the budding yeast, a common research model organism belongs to this group as well (Dupont et al. 2014). This may seem surprising for yeast biologists who usually grow them in liquid media or on plates kept in a humid environment. However, in nature, yeast often live on the surface of plants or animals and are in constant contact with open air. Their foraging ground is rotten fruit, which can quickly become dry during sunny weather and rehydrate when it is rainy. This brings to our attention the very clue of the dehydration tolerance, which is the ability to resume vital functions suspended by temporary water removal because the removal of water from the cells can be as dangerous to them as their rehydration.

2.5.1 Resistance to Dehydration

Desiccation causes an arrest of all living functions, so the harmful effects of this stress are manifested during the rehydration phase when the decision about the cell fate is made. Water forms hydration envelopes around all cellular molecules, so removal of water results in their denaturation, misfolding, and aggregation (Ball 2008). Upon rehydration, the proper envelopes may not be sufficiently restored and the cell may not resume its functions. Cellular lipid membranes are also vulnerable to cycles of alternating dryness and rehydration periods (Crowe et al. 1984). Moreover, dehydration exposes cells to air and, hence, to oxidative stress (Garre et al. 2010).

The basic concepts of desiccation tolerance mechanisms of organisms from all kingdoms were formulated decades ago (Crowe et al. 1992). The role of trehalose and other sugars, particularly the process of their vitrification, in the stabilization of vital cellular components— lipid membranes and proteins—in a state that allows them to resume their functions following rehydration, was established (Crowe et al. 1998; Potts 2001).

Two characteristics that are crucial for dehydration tolerance clearly stand out: (1) exponentially grown *S. cerevisiae* cells are several orders of magnitude more sensitive to desiccation stress than postdiauxic cells, and (2) functional respiration is crucial for survival of this stress (Ratnakumar and Tunnacliffe 2006; Calahan et al. 2011; Ratnakumar et al. 2011). Notably, these two characteristics are related to each other. It is known that budding yeast cells growing fast on glucose contain much lower levels of stress-protecting proteins even though they may meet some environmental stresses also under these circumstances. It turns out that respiratory incompetent cells have much lower levels of these proteins as well, because, once they use up glucose, they are energy starved and hence cannot synthesize them. Thus, perhaps the most important adaptation to desiccation stress (and to many other stresses for that matter) is the switch that occurs at the diauxic shift, from fast fermentative growth fueled with a preferred carbon source assimilated with the use of streamlined metabolism to slower respiratory growth usually followed by the dormant state when nutrients are depleted (Singh et al. 2005). Because in the absence of water no life-supporting reactions in the cells are possible, rapid dehydration prevents the inducible response to desiccation stress. Therefore, cells must be prepared for this stress before it occurs and while nutrients are still available.

The survival of yeast cells under desiccation stress is assured by the accumulation of certain compounds. One of them is trehalose, which was already mentioned in this chapter in the context of other stresses. It accomplishes its protective function by the process called vitrification, i.e., turning into an amorphous state that can replace absent water (Crowe et al. 1998). This assures the protection of proteins against denaturation when there are insufficient water molecules to hydrate them. Trehalose and sucrose can also replace water in stabilizing hydrophilic heads of membrane lipids, thereby preventing irreparable and deadly membrane breakage (Crowe et al. 1992). Despite the accumulated knowledge, the exact mechanism of protecting cells against dehydration by trehalose has not been clarified (Crowe 2015).

The survival of yeast cells under desiccation stress is supported by hydrophilins, Sip18p (Rodríguez-Porrata et al. 2012) and Stf2p (López-Martínez et al. 2012). Hydrophilins belong to the special class of proteins called intrinsically disordered proteins (IDP) (Boothby and Pielak 2017). As their name suggests, they do not have stable tertiary structures but constantly cycle between many equally probable conformation states. This feature is the consequence of a high proportion of glycine residues in their polypeptide chain. Some of them, including hydrophilins, also have numerous hydrophilic amino acids in their sequence. Despite accumulating evidence of their abilities to protect cells against desiccation stress, the exact mechanism is not clarified. Remarkably, the proposed mechanisms are similar to those of trehalose—via vitrification, water replacement, molecular shielding, membrane stabilization, and preservation of cellular organization and structure—and also by water retention because of their high hydrophilicity or by scavenging of reactive oxygen species (Boothby and Pielak 2017). IDPs are present in various desiccation-resistant organisms that do not accumulate trehalose (Boothby et al. 2017). Apparently, high dehydration resistance of *S. cerevisiae* depends both on trehalose and on 12 proteins belonging to the family of hydrophilins (Dang and Hincha 2011).

Dehydration exposes cells to air and, hence, to oxidative stress, and it was demonstrated that agents protecting against oxidative stress are upregulated under dehydration stress (Zambuto et al. 2017) and that glutathione or catalase deficiency affects the budding yeast survival of dehydration stress (Espindola et al. 2003; França et al. 2005). It was also shown that some agents protecting against desiccation such as trehalose or hydrophilins act by protecting against oxidative stress, thereby preventing the cells from entering apoptosis (Rodríguez-Porrata et al. 2012; López-Martínez et al. 2012). However, they may also decrease this stress by preserving the redox enzymes in the native state and preventing their dysfunctional free radical generation.

Autophagy is important for the survival of desiccation stress, and the postulated mechanism is the removal, after rehydration, of molecules and cellular structures damaged during the desiccation period, at the same time reclaiming the substrates for new syntheses (Ratnakumar et al. 2011).

The composition and structural rearrangement of the plasma membrane is also important for *S. cerevisiae* resistance to dehydration, as well as the timing of dehydration, with slow dehydration as a prognostic of a higher survival rate (Dupont et al. 2010). Yeast lipid membranes contain an unusually high level of ergosterol, which influences its mechanical properties. Perhaps, this single characteristic accounts for much of the *S. cerevisiae* cells high resistance to desiccation because its substitution with other sterols has severe consequences (Dupont et al. 2012). In addition to improving membrane rigidity, ergosterol protects against oxidative stress (Dupont et al. 2012). Notably, the ergosterol content, although divergent in certain *S. cerevisiae* strains, is much higher in cells of aerated cultures (Strydom et al. 1982). The demonstration of the correlation between the yeast cell ergosterol content and their resistance to desiccation would corroborate its importance for survival under these conditions.

Additionally, the elasticity of the cell wall, which prevents the plasmolysis and deformation damage of the PM, is crucial for successful rehydration (Dupont et al. 2010).

2.5.2 Response Mechanisms to Dehydration

Water deficit increases the external solute concentrations, so dehydration stress inevitably begins with hyperosmotic stress. However, the response mechanisms to these stresses show little overlap. Among others, the synthesis of osmoprotectants, such as glycerol or trehalose, is increased. Dehydration stress is also associated with the increased tendency for protein misfolding and aggregation (see Chap. 5). It was demonstrated that the increased tolerance to dehydration stress of yeast winery strains is correlated with higher levels of the enzymatic proteins involved in the synthesis of glycerol (Gpd1p), trehalose (Tps1p, Tps2p), protein chaperones

(Hsp12p, Hsp104p, and HSP70 family Ssa proteins), and oxidative protectants (Cttlp, Sodlp) (Capece et al. 2016). Increased expression of enzymes protecting against the oxidative stress suggests that desiccation leads also to the increased vulnerability to this stress (Zambuto et al. 2017). This is to be expected because dried cells are more accessible to oxygen.

Interestingly, the absence of individual genes encoding proteins predicted to be important for dehydration stress tolerance, such as hydrophilins, the components of HOG osmostress signaling pathway, or those conferring protection against high salinity stress, has a negligible or minor effect on dehydration stress tolerance (Calahan et al. 2011). The same is true for the trehalose-synthesizing enzymes (Ratnakumar and Tunnacliffe 2006). Therefore, it is assumed that the resistance to this stress is a compound trait, dependent on multiple gene functions (López-Martínez et al. 2015). Moreover, the absence of some components of the desiccation stress protection system may be compensated by the increased abundance of other ones. For instance, the *S. cerevisiae hsp12Δ* strain lacking Hsp12p heat shock-induced hydrophilin has more trehalose in its cells, leading to the paradoxical higher resistance of this strain to dehydration than the parental wild type (Shamrock and Lindsey 2008).

Slow dehydration enables the cell to respond to this stress, and the first signal of environmental hostility is the increased concentration of the solutes that triggers the HOG response pathway and accumulation of glycerol in the cell. However, this phenomenon does not seem to be a part of the true response to desiccation because the dysfunctional HOG response pathway does not make *S. cerevisiae* cells more sensitive to desiccation stress (Ratnakumar et al. 2011). However, the response to desiccation stress was shown to be under the control of the TOR and RAS pathways (Singh et al. 2005; Welch et al. 2013).

2.6 Practical Aspects of Studying Response Mechanisms to Chemical and Physical Stresses

All stresses described in this chapter have important practical, industrial, and, sometimes, medical relevance. In the industry, the strains resistant to all stresses are desirable because, in the commercial applications, they are exposed to much more severe conditions than those in the laboratory.

The issue of tolerance to acidic pH and WOA is of high interest for the industry of renewable ethanol fuel production. The main substrate in this process is lignocellulose hydrolysate, which is acidic and contains a substantial concentration of acetic acid. Better understanding of the acidic pH response and tolerance mechanisms will help to develop strains optimized for this process (Brodeur et al. 2011).

Strain optimization will be beneficial also for wine and bakery production. Some alcoholic beverages, such as lager beer, are fermented at low temperature. Low temperature decreases the risk of bacterial contamination of the fermentation tank. It is also believed that a lower temperature of fermentation influences the flavor of wines, shifting it toward a fresher, fruitier nose, although the results of rigorous tests were inconsistent (Molina et al. 2007; Deed et al. 2017). For these reasons, strain improvement for wine- and beer-making at low temperature is desirable. Their increased tolerance to acidic pH will also help them to better withstand decontamination with diluted sulfuric acid when recovered after fermentation prior to their reinoculation (de Melo et al. 2010). Filamentous fungi are used for the large-scale production of organic acids such as citric or lactic acid, so their resistance to a higher concentration of these substances would be beneficial for the manufacturing process.

Commercial yeast strains used to inoculate dough or starter cultures in bakery, brewery, winery, and distillery processes are sold in a dried form, so their activity depends on the accuracy of their preparation and ability to survive dehydration (Mille et al. 2005; Zambuto et al. 2017). A common practice of the distribution of frozen dough in current mass food production also necessitates the improvement of bakers' yeast strain survival of freezing (Tan et al. 2014).

Sorbic and benzoic acids are commonly used as food preservatives, and better knowledge on spoilage fungi resistance to these compounds is important to the food industry.

Growth-inhibitory and growth-killing abilities of HHP recently led to concepts to employ extreme pressure (50–150 MPa) for prolonged storage of unprocessed food at room temperature, which would preserve all of its nutritional values and texture (Moreira et al. 2015; Freitas et al. 2016; Lemos et al. 2017) and would avoid the addition of any food preservatives.

Yeast species from polar habitats adapted for food industry purposes could be the source of essential polyunsaturated fatty acids, an interesting alternative to fish oil, devoid of an unwanted "fishy" smell inherent to the latter (Alcaíno et al. 2015). Hydrolytic enzymes secreted by psychrophilic yeast, such as amylases, cellulases, invertases, proteases, and lipases, of high activity at low temperatures could also have a high industrial potential (Alcaíno et al. 2015).

Dry yeasts also have other, less obvious applications, such as bioremediation of the environment (Rapoport et al. 2016).

Pathogenic yeast and fungi invade the environment of the human body whose pH is 7.4, which is slightly alkaline (Davis 2009). They are tolerant to the broad range of pH in their surroundings, and the alkaline pH induces adaptations to increase their pathogenicity. For instance, in dimorphic yeast, such as *C. albicans*, alkaline pH induces the morphological switch from saprophytic yeast to invasive filamentous growth (reviewed in Cornet and Gaillardin 2014). Since the alkaline pH signaling pathways found in yeast and fungi, including the pathogenic ones, are specific to them, they can potentially be used as targets in the development of new antifungal drugs with minimal side effects to the human host (Cornet and Gaillardin 2014). It was also shown that the ability of *C. glabrata* to colonize vagina epithelium at low pH is strongly increased by the expression of Haal TF (Bernardo et al. 2017). Therefore, full understanding of the responses of these organisms to ambient pH (and to other stresses as well) is of great medical importance.

These and other aspects of yeast and filamentous fungi resistance to various stresses in industrial, agricultural, and medical contexts are described in more detail in Chaps. 6 and 7 of this book.

2.7 Conclusions and Perspectives

Our knowledge on the mechanisms of response and tolerance to environmental stresses has expanded considerably in recent years. It was possible to a great extent thanks to the extreme convenience of yeast and filamentous fungi as experimental models. This chapter provides only a concise digest of this topic and that was our intention. Despite the diversity of the environmental stresses, certain common themes become visible, such as the participation of the HOG signaling pathway as a sensor and transducer of several of the stresses discussed in this chapter.

Despite the accumulation of the wealth of data regarding the stress response mechanisms, many of their aspects are still not clarified. The immediate cell exterior, the periplasm, is an important compartment that is the most exposed to, for instance, extreme pH stress but is not protected against this stress in any way. The plasma membrane ATPases pumping H^+ and Na^+ ions function further below and protect only the cell interior. Are there any mechanisms protecting the periplasmic space against low and high pH or other elements? This is a very interesting and largely unanswered question, although the periplasmic proteome was studied for many years and we know that it contains various stress-protecting proteins (Nombela et al. 2006).

Another interesting and probably challenging question regards the adaptations to low temperature and extreme hydrostatic pressure, as both stresses are constantly experienced by organisms living in the deepest oceanic trenches where the water temperature usually is barely above 0 °C. They must cope with the conditions extremely unfavorable for their membrane fluidity. The presence of polyunsaturated fatty acids seems to be their adaptation to these circumstances (Bergé and Barnathan 2005). Along the same lines, HHP together with high temperatures and low pH near the hydrothermal vents on the bottom of the ocean imposes a strong challenge for the creatures living there as well, and their specific adaptations to these conditions include peculiar membrane lipids composition (van de Vossenberg et al. 1998).

The issue that surfaced several times in this chapter is the role of trehalose as a stress protectant. Its importance for the survival of heat and desiccation stresses was suggested long ago (Hottiger et al. 1987), and, for many years, this sugar was believed to be the universal guard increasing the survival of yeast cells exposed to various stresses. In addition to heat (Verghese et al. 2012) and desiccation, it was demonstrated to be an osmoprotectant (Iturriaga et al. 2009), and its presence in much higher amounts in postdiauxic *S. cerevisiae* cells, known to be highly resistant to various environmental stresses, implied its general protective abilities. Unexpectedly, this view was challenged by the results showing that trehalose does not protect against any of the stresses that were thought to be alleviated by this disaccharide (Gibney et al. 2015; Petitjean et al. 2015), except the desiccation stress (Tapia et al. 2015) during which trehalose functions as a low-molecular-weight chaperone (Tapia and Koshland 2014). However, earlier studies have demonstrated that it was dispensable for drought resistance (Ratnakumar and Tunnacliffe 2006). Of course, this controversy can easily be explained by the multitude of stress-protecting compounds that are present in the cell in addition to trehalose—glycerol and other polyols, proline, hydrophilins, chaperone proteins, and other compounds mentioned in this chapter. These substances may have, at least partially, overlapping and redundant functions, so the absence of just one of them may not manifest with the easily discernible phenotype. Similarly, the deletion of individual genes encoding hydrophilins has a minor or negligible influence on budding yeast cell survival (Dang and Hincha 2011). Thus, the participation of individual classes of compounds in the protection against the environmental stresses is still not fully elucidated.

Another unanswered question regards the mechanisms of detection of some of the stresses described in this chapter. Sensors of low temperature or hyperosmotic stress are mechanosensors detecting the change in plasma membrane rigidity, shape, or its detachment from the cell wall. However, how is the ambient pH sensed? The participation of the ESCRT components in ambient pH signaling suggests that, here too, the membrane deformation is somehow involved. We may expect more exciting discoveries to be made in this area in the future.

Acknowledgments This work was supported by Polish National Science Center grant 2016/21/B/ NZ3/03641.

References

Abe F, Horikoshi K (2000) Tryptophan permease gene TAT2 confers high-pressure growth in *Saccharomyces cerevisiae*. Mol Cell Biol 20:8093–8102

Abe F, Minegishi H (2008) Global screening of genes essential for growth in high-pressure and cold environments: searching for basic adaptive strategies using a yeast deletion library. Genetics 178:851–872. https://doi.org/10.1534/genetics.107.083063

Ahmadpour D, Geijer C, Tamás MJ, Lindkvist-Petersson K, Hohmann S (2014) Yeast reveals unexpected roles and regulatory features of aquaporins and aquaglyceroporins. Biochim Biophys Acta 1840:1482–1491. https://doi.org/10.1016/j.bbagen.2013.09.027

Alcaíno J, Cifuentes V, Baeza M (2015) Physiological adaptations of yeasts living in cold environments and their potential applications. World J Microbiol Biotechnol 31:1467–1473. https://doi.org/10.1007/s11274-015-1900-8

Alcano M de J, Jahn RC, Scherer CD, Wigmann ÉF, Moraes VM, Garcia MV, Mallmann CA, Copetti MV (2016) Susceptibility of *Aspergillus* spp. to acetic and sorbic acids based on pH and effect of sub-inhibitory doses of sorbic acid on ochratoxin a production. Food Res Int 81:25–30. https://doi.org/10.1016/j.foodres.2015.12.020

Almeida B, Ohlmeier S, Almeida AJ, Madeo F, Leão C, Rodrigues F, Ludovico P (2009) Yeast protein expression profile during acetic acid-induced apoptosis indicates causal involvement of the TOR pathway. Proteomics 9:720–732. https://doi.org/10.1002/pmic.200700816

Amils R, González-Toril E, Aguilera A, Rodríguez N, Fernández-Remolar D, Gómez F, García-Moyano A, Malki M, Oggerin M, Sánchez-Andrea I, Sanz JL (2011) From Río Tinto to Mars: the terrestrial and extraterrestrial ecology of acidophiles. Adv Appl Microbiol 77:41–70. https://doi.org/10.1016/B978-0-12-387044-5.00002-9

Ariño J (2010) Integrative responses to high pH stress in *S. cerevisiae*. Omics J Integr Biol 14:517–523. https://doi.org/10.1089/omi.2010.0044

Babazadeh R, Adiels CB, Smedh M, Petelenz-Kurdziel E, Goksör M, Hohmann S (2013) Osmostress-induced cell volume loss delays yeast Hog1 signaling by limiting diffusion processes and by Hog1-specific effects. PLoS One 8:e80901. https://doi.org/10.1371/journal.pone.0080901

Babazadeh R, Furukawa T, Hohmann S, Furukawa K (2014) Rewiring yeast osmostress signalling through the MAPK network reveals essential and non-essential roles of Hog1 in osmoadaptation. Sci Rep 4:4697. https://doi.org/10.1038/srep04697

Ball P (2008) Water as an active constituent in cell biology. Chem Rev 108:74–108. https://doi.org/10.1021/cr068037a

Ballester-Tomás L, Prieto JA, Alepuz P, González A, Garre E, Randez-Gil F (2017) Inappropriate translation inhibition and P-body formation cause cold-sensitivity in tryptophan-auxotroph yeast mutants. Biochim Biophys Acta 1864:314–323. https://doi.org/10.1016/j.bbamcr.2016.11.012

Ballweg S, Ernst R (2017) Control of membrane fluidity: the OLE pathway in focus. Biol Chem 398:215–228. https://doi.org/10.1515/hsz-2016-0277

Batiza AF, Schulz T, Masson PH (1996) Yeast respond to hypotonic shock with a calcium pulse. J Biol Chem 271:23357–23362

Beese SE, Negishi T, Levin DE (2009) Identification of positive regulators of the yeast fps1 glycerol channel. PLoS Genet 5:e1000738. https://doi.org/10.1371/journal.pgen.1000738

Bergé J-P, Barnathan G (2005) Fatty acids from lipids of marine organisms: molecular biodiversity, roles as biomarkers, biologically active compounds, and economical aspects. Adv Biochem Eng Biotechnol 96:49–125

Bernardo RT, Cunha DV, Wang C, Pereira L, Silva S, Salazar SB, Schröder MS, Okamoto M, Takahashi-Nakaguchi A, Chibana H, Aoyama T, Sá-Correia I, Azeredo J, Butler G, Mira NP (2017) The CgHaa1-Regulon mediates response and tolerance to acetic acid stress in the human pathogen Candida glabrata. G3 Bethesda Md 7:1–18. https://doi.org/10.1534/g3.116.034660

Bilsland E, Molin C, Swaminathan S, Ramne A, Sunnerhagen P (2004) Rck 1 and Rck2 MAPKAP kinases and the HOG pathway are required for oxidative stress resistance. Mol Microbiol 53:1743–1756. https://doi.org/10.1111/j.1365-2958.2004.04238.x

Boothby TC, Pielak GJ (2017) Intrinsically disordered proteins and desiccation tolerance: elucidating functional and mechanistic underpinnings of anhydrobiosis. BioEssays News Rev Mol Cell Dev Biol 39. https://doi.org/10.1002/bies.201700119

Boothby TC, Tapia H, Brozena AH, Piszkiewicz S, Smith AE, Giovannini I, Rebecchi L, Pielak GJ, Koshland D, Goldstein B (2017) Tardigrades use intrinsically disordered proteins to survive desiccation. Mol Cell 65:975–984.e5. https://doi.org/10.1016/j.molcel.2017.02.018

Bravim F, da Silva LF, Souza DT, Lippman SI, Broach JR, Fernandes AAR, Fernandes PMB (2012) High hydrostatic pressure activates transcription factors involved in Saccharomyces cerevisiae stress tolerance. Curr Pharm Biotechnol 13:2712–2720

Bravim F, Mota MM, Fernandes AAR, Fernandes PMB (2016) High hydrostatic pressure leads to free radicals accumulation in yeast cells triggering oxidative stress. FEMS Yeast Res 16. https://doi.org/10.1093/femsyr/fow052

Brett CL, Kallay L, Hua Z, Green R, Chyou A, Zhang Y, Graham TR, Donowitz M, Rao R (2011) Genome-wide analysis reveals the vacuolar pH-stat of Saccharomyces cerevisiae. PLoS One 6:e17619. https://doi.org/10.1371/journal.pone.0017619

Brewster JL, Gustin MC (2014) Hog1: 20 years of discovery and impact. Sci Signal 7:re7. https://doi.org/10.1126/scisignal.2005458

Brodeur G, Yau E, Badal K, Collier J, Ramachandran KB, Ramakrishnan S (2011) Chemical and physicochemical pretreatment of lignocellulosic biomass: a review. Enzym Res 2011:787532. https://doi.org/10.4061/2011/787532

Brown AD, Simpson JR (1972) Water relations of sugar-tolerant yeasts: the role of intracellular polyols. J Gen Microbiol 72:589–591. https://doi.org/10.1099/00221287-72-3-589

Calahan D, Dunham M, DeSevo C, Koshland DE (2011) Genetic analysis of desiccation tolerance in Saccharomyces cerevisiae. Genetics 189:507–519. https://doi.org/10.1534/genetics.111.130369

Canadell D, García-Martínez J, Alepuz P, Pérez-Ortín JE, Ariño J (2015) Impact of high pH stress on yeast gene expression: a comprehensive analysis of mRNA turnover during stress responses. Biochim Biophys Acta 1849:653–664. https://doi.org/10.1016/j.bbagrm.2015.04.001

Capece A, Votta S, Guaragnella N, Zambuto M, Romaniello R, Romano P (2016) Comparative study of *Saccharomyces cerevisiae* wine strains to identify potential marker genes correlated to desiccation stress tolerance. FEMS Yeast Res 16. https://doi.org/10.1093/femsyr/fow015

Casado C, González A, Platara M, Ruiz A, Ariño J (2011) The role of the protein kinase a pathway in the response to alkaline pH stress in yeast. Biochem J 438:523–533. https://doi.org/10.1042/BJ20110607

Casamayor A, Serrano R, Platara M, Casado C, Ruiz A, Ariño J (2012) The role of the Snf1 kinase in the adaptive response of *Saccharomyces cerevisiae* to alkaline pH stress. Biochem J 444:39–49. https://doi.org/10.1042/BJ20112099

Castro CD, Meehan AJ, Koretsky AP, Domach MM (1995) In situ 31P nuclear magnetic resonance for observation of polyphosphate and catabolite responses of chemostat-cultivated *Saccharomyces cerevisiae* after alkalinization. Appl Environ Microbiol 61:4448–4453

Chen RE, Thorner J (2007) Function and regulation in MAPK signaling pathways: lessons learned from the yeast *Saccharomyces cerevisiae*. Biochim Biophys Acta 1773:1311–1340. https://doi.org/10.1016/j.bbamcr.2007.05.003

Chen XH, Xiao Z, Fitzgerald-Hayes M (1994) SCM2, a tryptophan permease in *Saccharomyces cerevisiae*, is important for cell growth. Mol Gen Genet MGG 244:260–268

Chen AK-L, Gelling C, Rogers PL, Dawes IW, Rosche B (2009) Response of *Saccharomyces cerevisiae* to stress-free acidification. J Microbiol Seoul Korea 47:1–8. https://doi.org/10.1007/s12275-008-0167-2

Chen Y, Stabryla L, Wei N (2016) Improved acetic acid resistance in *Saccharomyces cerevisiae* by overexpression of the WHI2 gene identified through inverse metabolic engineering. Appl Environ Microbiol 82:2156–2166. https://doi.org/10.1128/AEM.03718-15

Christ L, Raiborg C, Wenzel EM, Campsteijn C, Stenmark H (2017) Cellular functions and molecular mechanisms of the ESCRT membrane-scission machinery. Trends Biochem Sci 42:42–56. https://doi.org/10.1016/j.tibs.2016.08.016

Coelho M a Z, Belo I, Pinheiro R, Amaral AL, Mota M, Coutinho JP, Ferreira EC (2004) Effect of hyperbaric stress on yeast morphology: study by automated image analysis. Appl Microbiol Biotechnol 66:318–324. https://doi.org/10.1007/s00253-004-1648-9

Combina M, Pérez-Torrado R, Tronchoni J, Belloch C, Querol A (2012) Genome-wide gene expression of a natural hybrid between *Saccharomyces cerevisiae* and *S. kudriavzevii* under enological conditions. Int J Food Microbiol 157:340–345. https://doi.org/10.1016/j.ijfoodmicro.2012.06.001

Córcoles-Sáez I, Ballester-Tomas L, de la Torre-Ruiz MA, Prieto JA, Randez-Gil F (2012) Low temperature highlights the functional role of the cell wall integrity pathway in the regulation of growth in *Saccharomyces cerevisiae*. Biochem J 446:477–488. https://doi.org/10.1042/BJ20120634

Cornet M, Gaillardin C (2014) pH signaling in human fungal pathogens: a new target for antifungal strategies. Eukaryot Cell 13:342–352. https://doi.org/10.1128/EC.00313-13

Cottier F, Tan ASM, Chen J, Lum J, Zolezzi F, Poidinger M, Pavelka N (2015) The transcriptional stress response of Candida albicans to weak organic acids. G3 Bethesda Md 5:497–505. https://doi.org/10.1534/g3.114.015941

Covino R, Ballweg S, Stordeur C, Michaelis JB, Puth K, Wernig F, Bahrami A, Ernst AM, Hummer G, Ernst R (2016) A eukaryotic sensor for membrane lipid saturation. Mol Cell 63:49–59. https://doi.org/10.1016/j.molcel.2016.05.015

Crowe JH (2015) Anhydrobiosis: an unsolved problem with applications in human welfare. Subcell Biochem 71:263–280. https://doi.org/10.1007/978-3-319-19060-0_11

Crowe JH, Crowe LM, Chapman D (1984) Preservation of membranes in anhydrobiotic organisms: the role of trehalose. Science 223:701–703. https://doi.org/10.1126/science.223.4637.701

Crowe JH, Hoekstra FA, Crowe LM (1992) Anhydrobiosis. Annu Rev Physiol 54:579–599. https://doi.org/10.1146/annurev.ph.54.030192.003051

Crowe JH, Carpenter JF, Crowe LM (1998) The role of vitrification in anhydrobiosis. Annu Rev Physiol 60:73–103. https://doi.org/10.1146/annurev.physiol.60.1.73

Dang NX, Hincha DK (2011) Identification of two hydrophilins that contribute to the desiccation and freezing tolerance of yeast (*Saccharomyces cerevisiae*) cells. Cryobiology 62:188–193. https://doi.org/10.1016/j.cryobiol.2011.03.002

Davenport KR, Sohaskey M, Kamada Y, Levin DE, Gustin MC (1995) A second osmosensing signal transduction pathway in yeast. Hypotonic shock activates the PKC1 protein kinase-regulated cell integrity pathway. J Biol Chem 270:30157–30161

Davis DA (2009) How human pathogenic fungi sense and adapt to pH: the link to virulence. Curr Opin Microbiol 12:365–370. https://doi.org/10.1016/j.mib.2009.05.006

de Castro PA, Savoldi M, Bonatto D, Barros MH, Goldman MHS, Berretta AA, Goldman GH (2011) Molecular characterization of propolis-induced cell death in *Saccharomyces cerevisiae*. Eukaryot Cell 10:398–411. https://doi.org/10.1128/EC.00256-10

de Freitas JM, Bravim F, Buss DS, Lemos EM, Fernandes AAR, Fernandes PMB (2012) Influence of cellular fatty acid composition on the response of *Saccharomyces cerevisiae* to hydrostatic pressure stress. FEMS Yeast Res 12:871–878. https://doi.org/10.1111/j.1567-1364.2012. 00836.x

de Lima Alves F, Stevenson A, Baxter E, Gillion JLM, Hejazi F, Hayes S, Morrison IEG, Prior BA, McGenity TJ, Rangel DEN, Magan N, Timmis KN, Hallsworth JE (2015) Concomitant osmotic and chaotropicity-induced stresses in *Aspergillus wentii*: compatible solutes determine the biotic window. Curr Genet 61:457–477. https://doi.org/10.1007/s00294-015-0496-8

de Lucena RM, Elsztein C, Simões DA, de Morais MA (2012) Participation of CWI, HOG and Calcineurin pathways in the tolerance of *Saccharomyces cerevisiae* to low pH by inorganic acid. J Appl Microbiol 113:629–640. https://doi.org/10.1111/j.1365-2672.2012.05362.x

de Melo HF, Bonini BM, Thevelein J, Simões DA, Morais MA (2010) Physiological and molecular analysis of the stress response of *Saccharomyces cerevisiae* imposed by strong inorganic acid with implication to industrial fermentations. J Appl Microbiol 109:116–127. https://doi.org/10. 1111/j.1365-2672.2009.04633.x

de Nadal E, Posas F (2015) Osmostress-induced gene expression--a model to understand how stress-activated protein kinases (SAPKs) regulate transcription. FEBS J 282:3275–3285. https:// doi.org/10.1111/febs.13323

Deed RC, Fedrizzi B, Gardner RC (2017) Influence of fermentation temperature, yeast strain, and grape juice on the aroma chemistry and sensory profile of Sauvignon Blanc wines. J Agric Food Chem 65:8902–8912. https://doi.org/10.1021/acs.jafc.7b03229

Delarue M, Poterewicz G, Hoxha O, Choi J, Yoo W, Kayser J, Holt L, Hallatschek O (2017) SCWISh network is essential for survival under mechanical pressure. Proc Natl Acad Sci U S A 114:13465–13470. https://doi.org/10.1073/pnas.1711204114

Domitrovic T, Fernandes CM, Boy-Marcotte E, Kurtenbach E (2006) High hydrostatic pressure activates gene expression through Msn2/4 stress transcription factors which are involved in the acquired tolerance by mild pressure precondition in *Saccharomyces cerevisiae*. FEBS Lett 580:6033–6038. https://doi.org/10.1016/j.febslet.2006.10.007

Dong Y, Hu J, Fan L, Chen Q (2017) RNA-Seq-based transcriptomic and metabolomic analysis reveal stress responses and programmed cell death induced by acetic acid in *Saccharomyces cerevisiae*. Sci Rep 7:42659. https://doi.org/10.1038/srep42659

Dupont S, Beney L, Ritt J-F, Lherminier J, Gervais P (2010) Lateral reorganization of plasma membrane is involved in the yeast resistance to severe dehydration. Biochim Biophys Acta 1798:975–985. https://doi.org/10.1016/j.bbamem.2010.01.015

Dupont S, Lemetais G, Ferreira T, Cayot P, Gervais P, Beney L (2012) Ergosterol biosynthesis: a fungal pathway for life on land? Evol Int J Org Evol 66:2961–2968. https://doi.org/10.1111/j. 1558-5646.2012.01667.x

Dupont S, Rapoport A, Gervais P, Beney L (2014) Survival kit of *Saccharomyces cerevisiae* for anhydrobiosis. Appl Microbiol Biotechnol 98:8821–8834. https://doi.org/10.1007/s00253-014-6028-5

Duran R, Cary JW, Calvo AM (2010) Role of the osmotic stress regulatory pathway in morphogenesis and secondary metabolism in filamentous fungi. Toxins 2:367–381. https://doi.org/10.3390/toxins2040367

Dušková M, Ferreira C, Lucas C, Sychrová H (2015) Two glycerol uptake systems contribute to the high osmotolerance of *Zygosaccharomyces rouxii*. Mol Microbiol 97:541–559. https://doi.org/10.1111/mmi.13048

Enjalbert B, Smith DA, Cornell MJ, Alam I, Nicholls S, Brown AJP, Quinn J (2006) Role of the Hog1 stress-activated protein kinase in the global transcriptional response to stress in the fungal pathogen *Candida albicans*. Mol Biol Cell 17:1018–1032. https://doi.org/10.1091/mbc.E05-06-0501

Ernst R, Ejsing CS, Antonny B (2016) Homeoviscous adaptation and the regulation of membrane lipids. J Mol Biol 428:4776–4791. https://doi.org/10.1016/j.jmb.2016.08.013

Espindola A de S, Gomes DS, Panek AD, Eleutherio ECA (2003) The role of glutathione in yeast dehydration tolerance. Cryobiology 47:236–241

Farewell A, Neidhardt FC (1998) Effect of temperature on in vivo protein synthetic capacity in *Escherichia coli*. J Bacteriol 180:4704–4710

Fernandes AR, Mira NP, Vargas RC, Canelhas I, Sá-Correia I (2005) *Saccharomyces cerevisiae* adaptation to weak acids involves the transcription factor Haa1p and Haa1p-regulated genes. Biochem Biophys Res Commun 337:95–103. https://doi.org/10.1016/j.bbrc.2005.09.010

Fernández-Niño M, Marquina M, Swinnen S, Rodríguez-Porrata B, Nevoigt E, Ariño J (2015) The cytosolic pH of individual *Saccharomyces cerevisiae* cells is a key factor in acetic acid tolerance. Appl Environ Microbiol 81:7813–7821. https://doi.org/10.1128/AEM.02313-15

Ferreira C, van Voorst F, Martins A, Neves L, Oliveira R, Kielland-Brandt MC, Lucas C, Brandt A (2005) A member of the sugar transporter family, Stl1p is the glycerol/H+ symporter in *Saccharomyces cerevisiae*. Mol Biol Cell 16:2068–2076. https://doi.org/10.1091/mbc.E04-10-0884

Fletcher E, Feizi A, Kim S, Siewers V, Nielsen J (2015) RNA-seq analysis of *Pichia anomala* reveals important mechanisms required for survival at low pH. Microb Cell Factories 14:143. https://doi.org/10.1186/s12934-015-0331-4

Fletcher E, Feizi A, Bisschops MMM, Hallström BM, Khoomrung S, Siewers V, Nielsen J (2017) Evolutionary engineering reveals divergent paths when yeast is adapted to different acidic environments. Metab Eng 39:19–28. https://doi.org/10.1016/j.ymben.2016.10.010

França MB, Panek AD, Eleutherio ECA (2005) The role of cytoplasmic catalase in dehydration tolerance of *Saccharomyces cerevisiae*. Cell Stress Chaperones 10:167–170

Freitas P, Pereira SA, Santos MD, Alves SP, Bessa RJB, Delgadillo I, Saraiva JA (2016) Performance of raw bovine meat preservation by hyperbaric storage (quasi energetically costless) compared to refrigeration. Meat Sci 121:64–72. https://doi.org/10.1016/j.meatsci.2016.05.001

Furukawa K, Hoshi Y, Maeda T, Nakajima T, Abe K (2005) *Aspergillus nidulans* HOG pathway is activated only by two-component signalling pathway in response to osmotic stress. Mol Microbiol 56:1246–1261. https://doi.org/10.1111/j.1365-2958.2005.04605.x

Furukawa K, Sidoux-Walter F, Hohmann S (2009) Expression of the yeast aquaporin Aqy2 affects cell surface properties under the control of osmoregulatory and morphogenic signalling pathways. Mol Microbiol 74:1272–1286. https://doi.org/10.1111/j.1365-2958.2009.06933.x

Galafassi S, Toscano M, Vigentini I, Piškur J, Compagno C (2013) Osmotic stress response in the wine yeast *Dekkera bruxellensis*. Food Microbiol 36:316–319. https://doi.org/10.1016/j.fm.2013.06.011

Galafassi S, Toscano M, Vigentini I, Zambelli P, Simonetti P, Foschino R, Compagno C (2015) Cold exposure affects carbohydrates and lipid metabolism, and induces Hog1p phosphorylation in *Dekkera bruxellensis* strain CBS 2499. Antonie Van Leeuwenhoek 107:1145–1153. https://doi.org/10.1007/s10482-015-0406-6

Galeotti F, Maccari F, Fachini A, Volpi N (2018) Chemical composition and antioxidant activity of Propolis prepared in different forms and in different solvents useful for finished products. Foods Basel Switz 7. https://doi.org/10.3390/foods7030041

Galindo A, Calcagno-Pizarelli AM, Arst HN, Peñalva MÁ (2012) An ordered pathway for the assembly of fungal ESCRT-containing ambient pH signalling complexes at the plasma membrane. J Cell Sci 125:1784–1795. https://doi.org/10.1242/jcs.098897

Garre E, Raginel F, Palacios A, Julien A, Matallana E (2010) Oxidative stress responses and lipid peroxidation damage are induced during dehydration in the production of dry active wine yeasts. Int J Food Microbiol 136:295–303. https://doi.org/10.1016/j.ijfoodmicro.2009.10.018

Gasch AP, Spellman PT, Kao CM, Carmel-Harel O, Eisen MB, Storz G, Botstein D, Brown PO (2000) Genomic expression programs in the response of yeast cells to environmental changes. Mol Biol Cell 11:4241–4257

Gerday C, Aittaleb M, Arpigny JL, Baise E, Chessa JP, Garsoux G, Petrescu I, Feller G (1997) Psychrophilic enzymes: a thermodynamic challenge. Biochim Biophys Acta 1342:119–131

Gibney PA, Schieler A, Chen JC, Rabinowitz JD, Botstein D (2015) Characterizing the in vivo role of trehalose in *Saccharomyces cerevisiae* using the AGT1 transporter. Proc Natl Acad Sci U S A 112:6116–6121. https://doi.org/10.1073/pnas.1506289112

Giel-Pietraszuk M, Barciszewski J (2012) Hydrostatic and osmotic pressure study of the RNA hydration. Mol Biol Rep 39:6309–6318. https://doi.org/10.1007/s11033-012-1452-z

Godinho CP, Mira NP, Cabrito TR, Teixeira MC, Alasoo K, Guerreiro JF, Sá-Correia I (2017) Yeast response and tolerance to benzoic acid involves the Gcn4- and Stp1-regulated multidrug/ multixenobiotic resistance transporter Tpo1. Appl Microbiol Biotechnol 101:5005–5018. https://doi.org/10.1007/s00253-017-8277-6

Granot D, Snyder M (1991) Glucose induces cAMP-independent growth-related changes in stationary-phase cells of *Saccharomyces cerevisiae*. Proc Natl Acad Sci U S A 88:5724–5728

Grant WD (2004) Life at low water activity. Philos Trans R Soc Lond Ser B Biol Sci 359:1249–1266. Discussion 1266–1267. https://doi.org/10.1098/rstb.2004.1502

Gray MJ, Jakob U (2015) Oxidative stress protection by polyphosphate--new roles for an old player. Curr Opin Microbiol 24:1–6. https://doi.org/10.1016/j.mib.2014.12.004

Guerreiro JF, Mira NP, Sá-Correia I (2012) Adaptive response to acetic acid in the highly resistant yeast species *Zygosaccharomyces bailii* revealed by quantitative proteomics. Proteomics 12:2303–2318. https://doi.org/10.1002/pmic.201100457

Guerreiro JF, Muir A, Ramachandran S, Thorner J, Sá-Correia I (2016) Sphingolipid biosynthesis upregulation by TOR complex 2-Ypk1 signaling during yeast adaptive response to acetic acid stress. Biochem J 473:4311–4325. https://doi.org/10.1042/BCJ20160565

Guerreiro JF, Mira NP, Santos AXS, Riezman H, Sá-Correia I (2017) Membrane phosphoproteomics of yeast early response to acetic acid: role of Hrk1 kinase and lipid biosynthetic pathways, in particular sphingolipids. Front Microbiol 8:1302. https://doi.org/10.3389/fmicb.2017.01302

Gustin MC, Albertyn J, Alexander M, Davenport K (1998) MAP kinase pathways in the yeast *Saccharomyces cerevisiae*. Microbiol Mol Biol Rev MMBR 62:1264–1300

Häkkinen M, Sivasiddarthan D, Aro N, Saloheimo M, Pakula TM (2015) The effects of extracellular pH and of the transcriptional regulator PACI on the transcriptome of *Trichoderma reesei*. Microb Cell Factories 14:63. https://doi.org/10.1186/s12934-015-0247-z

Haro R, Garciadeblas B, Rodríguez-Navarro A (1991) A novel P-type ATPase from yeast involved in sodium transport. FEBS Lett 291:189–191

Hayashi M, Maeda T (2006) Activation of the HOG pathway upon cold stress in *Saccharomyces cerevisiae*. J Biochem (Tokyo) 139:797–803. https://doi.org/10.1093/jb/mvj089

Herrador A, Herranz S, Lara D, Vincent O (2010) Recruitment of the ESCRT machinery to a putative seven-transmembrane-domain receptor is mediated by an arrestin-related protein. Mol Cell Biol 30:897–907. https://doi.org/10.1128/MCB.00132-09

Higuchi Y, Mori H, Kubota T, Takegawa K (2018) Analysis of ambient pH stress response mediated by iron and copper intake in *Schizosaccharomyces pombe*. J Biosci Bioeng 125:92–96. https://doi.org/10.1016/j.jbiosc.2017.08.008

Hoekstra FA, Golovina EA, Buitink J (2001) Mechanisms of plant desiccation tolerance. Trends Plant Sci 6:431–438

Hohmann S (2002) Osmotic stress signaling and osmoadaptation in yeasts. Microbiol Mol Biol Rev MMBR 66:300–372

Hohmann S (2009) Control of high osmolarity signalling in the yeast *Saccharomyces cerevisiae*. FEBS Lett 583:4025–4029. https://doi.org/10.1016/j.febslet.2009.10.069

Hohmann S (2015) An integrated view on a eukaryotic osmoregulation system. Curr Genet 61:373–382. https://doi.org/10.1007/s00294-015-0475-0

Hong S-P, Carlson M (2007) Regulation of snf1 protein kinase in response to environmental stress. J Biol Chem 282:16838–16845. https://doi.org/10.1074/jbc.M700146200

Hottiger T, Boller T, Wiemken A (1987) Rapid changes of heat and desiccation tolerance correlated with changes of trehalose content in *Saccharomyces cerevisiae* cells subjected to temperature shifts. FEBS Lett 220:113–115

Huberman LB, Coradetti ST, Glass NL (2017) Network of nutrient-sensing pathways and a conserved kinase cascade integrate osmolarity and carbon sensing in *Neurospora crassa*. Proc Natl Acad Sci U S A 114:E8665–E8674. https://doi.org/10.1073/pnas.1707713114

Iturriaga G, Suárez R, Nova-Franco B (2009) Trehalose metabolism: from osmoprotection to signaling. Int J Mol Sci 10:3793–3810. https://doi.org/10.3390/ijms10093793

Iwahashi H (2015) Pressure-dependent gene activation in yeast cells. Subcell Biochem 72:407–422. https://doi.org/10.1007/978-94-017-9918-8_20

Izawa S, Sato M, Yokoigawa K, Inoue Y (2004) Intracellular glycerol influences resistance to freeze stress in *Saccharomyces cerevisiae*: analysis of a quadruple mutant in glycerol dehydrogenase genes and glycerol-enriched cells. Appl Microbiol Biotechnol 66:108–114. https://doi.org/10.1007/s00253-004-1624-4

Jacob JH, Hussein EI, Shakhatreh MAK, Cornelison CT (2017) Microbial community analysis of the hypersaline water of the Dead Sea using high-throughput amplicon sequencing. MicrobiologyOpen 6. https://doi.org/10.1002/mbo3.500

Jandric Z, Gregori C, Klopf E, Radolf M, Schüller C (2013) Sorbic acid stress activates the *Candida glabrata* high osmolarity glycerol MAP kinase pathway. Front Microbiol 4:350. https://doi.org/10.3389/fmicb.2013.00350

Ji Y, Yang F, Ma D, Zhang J, Wan Z, Liu W, Li R (2012) HOG-MAPK signaling regulates the adaptive responses of *Aspergillus fumigatus* to thermal stress and other related stress. Mycopathologia 174:273–282. https://doi.org/10.1007/s11046-012-9557-4

Jones PG, Inouye M (1996) RbfA, a 30S ribosomal binding factor, is a cold-shock protein whose absence triggers the cold-shock response. Mol Microbiol 21:1207–1218

Kandror O, Bretschneider N, Kreydin E, Cavalieri D, Goldberg AL (2004) Yeast adapt to near-freezing temperatures by STRE/Msn2,4-dependent induction of trehalose synthesis and certain molecular chaperones. Mol Cell 13:771–781

Karreman RJ, Dague E, Gaboriaud F, Quilès F, Duval JFL, Lindsey GG (2007) The stress response protein Hsp12p increases the flexibility of the yeast *Saccharomyces cerevisiae* cell wall. Biochim Biophys Acta 1774:131–137. https://doi.org/10.1016/j.bbapap.2006.10.009

Kawahata M, Masaki K, Fujii T, Iefuji H (2006) Yeast genes involved in response to lactic acid and acetic acid: acidic conditions caused by the organic acids in *Saccharomyces cerevisiae* cultures induce expression of intracellular metal metabolism genes regulated by Aft1p. FEMS Yeast Res 6:924–936. https://doi.org/10.1111/j.1567-1364.2006.00089.x

Kawarai T, Arai S, Furukawa S, Ogihara H, Yamasaki M (2006) High-hydrostatic-pressure treatment impairs actin cables and budding in *Saccharomyces cerevisiae*. J Biosci Bioeng 101:515–518. https://doi.org/10.1263/jbb.101.515

Kawazoe N, Kimata Y, Izawa S (2017) Acetic acid causes endoplasmic reticulum stress and induces the unfolded protein response in *Saccharomyces cerevisiae*. Front Microbiol 8:1192. https://doi.org/10.3389/fmicb.2017.01192

Kayikci Ö, Nielsen J (2015) Glucose repression in *Saccharomyces cerevisiae*. FEMS Yeast Res 15. https://doi.org/10.1093/femsyr/fov068

Kock C, Dufrêne YF, Heinisch JJ (2015) Up against the wall: is yeast cell wall integrity ensured by mechanosensing in plasma membrane microdomains? Appl Environ Microbiol 81:806–811. https://doi.org/10.1128/AEM.03273-14

Koppram R, Tomás-Pejó E, Xiros C, Olsson L (2014) Lignocellulosic ethanol production at high-gravity: challenges and perspectives. Trends Biotechnol 32:46–53. https://doi.org/10.1016/j.tibtech.2013.10.003

Kren A, Mamnun YM, Bauer BE, Schüller C, Wolfger H, Hatzixanthis K, Mollapour M, Gregori C, Piper P, Kuchler K (2003) War1p, a novel transcription factor controlling weak acid stress response in yeast. Mol Cell Biol 23:1775–1785

Kuanyshev N, Adamo GM, Porro D, Branduardi P (2017) The spoilage yeast *Zygosaccharomyces bailii*: foe or friend? Yeast Chichester Engl 34:359–370. https://doi.org/10.1002/yea.3238

Kumar S, Gummadi SN (2009) Osmotic adaptation in halotolerant yeast, *Debaryomyces nepalensis* NCYC 3413: role of osmolytes and cation transport. Extrem Life Extreme Cond 13:793–805. https://doi.org/10.1007/s00792-009-0267-x

Kuroki Y, Juvvadi PR, Arioka M, Nakajima H, Kitamoto K (2002) Cloning and characterization of vmaA, the gene encoding a 69-kDa catalytic subunit of the vacuolar H+-ATPase during alkaline pH mediated growth of *Aspergillus oryzae*. FEMS Microbiol Lett 209:277–282

Lages F, Silva-Graça M, Lucas C (1999) Active glycerol uptake is a mechanism underlying halotolerance in yeasts: a study of 42 species. Microbiol Read Engl 145. (Pt 9:2577–2585. https://doi.org/10.1099/00221287-145-9-2577

Lamb TM, Mitchell AP (2003) The transcription factor Rim101p governs ion tolerance and cell differentiation by direct repression of the regulatory genes NRG1 and SMP1 in *Saccharomyces cerevisiae*. Mol Cell Biol 23:677–686

Lamb TM, Xu W, Diamond A, Mitchell AP (2001) Alkaline response genes of *Saccharomyces cerevisiae* and their relationship to the RIM101 pathway. J Biol Chem 276:1850–1856. https://doi.org/10.1074/jbc.M008381200

Laroche C, Beney L, Marechal PA, Gervais P (2001) The effect of osmotic pressure on the membrane fluidity of *Saccharomyces cerevisiae* at different physiological temperatures. Appl Microbiol Biotechnol 56:249–254

Lemos ÁT, Ribeiro AC, Fidalgo LG, Delgadillo I, Saraiva JA (2017) Extension of raw watermelon juice shelf-life up to 58 days by hyperbaric storage. Food Chem 231:61–69. https://doi.org/10.1016/j.foodchem.2017.03.110

Lew RR, Levina NN, Walker SK, Garrill A (2004) Turgor regulation in hyphal organisms. Fungal Genet Biol FG B 41:1007–1015. https://doi.org/10.1016/j.fgb.2004.07.007

Lin X, Qi Y, Yan D, Liu H, Chen X, Liu L (2017) CgMED3 changes membrane sterol composition to help Candida glabrata tolerate low-pH stress. Appl Environ Microbiol 83. https://doi.org/10.1128/AEM.00972-17

Lindberg L, Santos AX, Riezman H, Olsson L, Bettiga M (2013) Lipidomic profiling of *Saccharomyces cerevisiae* and Zygosaccharomyces bailii reveals critical changes in lipid composition in response to acetic acid stress. PLoS One 8:e73936. https://doi.org/10.1371/journal.pone.0073936

Longo V, Ždralević M, Guaragnella N, Giannattasio S, Zolla L, Timperio AM (2015) Proteome and metabolome profiling of wild-type and YCA1-knock-out yeast cells during acetic acid-induced programmed cell death. J Proteome 128:173–188. https://doi.org/10.1016/j.jprot.2015.08.003

López-Martínez G, Rodríguez-Porrata B, Margalef-Català M, Cordero-Otero R (2012) The STF2p hydrophilin from *Saccharomyces cerevisiae* is required for dehydration stress tolerance. PLoS One 7:e33324. https://doi.org/10.1371/journal.pone.0033324

López-Martínez G, Margalef-Català M, Salinas F, Liti G, Cordero-Otero R (2015) ATG18 and FAB1 are involved in dehydration stress tolerance in *Saccharomyces cerevisiae*. PLoS One 10: e0119606. https://doi.org/10.1371/journal.pone.0119606

Los DA, Murata N (2004) Membrane fluidity and its roles in the perception of environmental signals. Biochim Biophys Acta 1666:142–157. https://doi.org/10.1016/j.bbamem.2004.08.002

Lucena-Agell D, Galindo A, Arst HN, Peñalva MA (2015) *Aspergillus nidulans* ambient pH signaling does not require endocytosis. Eukaryot Cell 14:545–553. https://doi.org/10.1128/EC.00031-15

Ludovico P, Sousa MJ, Silva MT, Leão C, Côrte-Real M (2001) *Saccharomyces cerevisiae* commits to a programmed cell death process in response to acetic acid. Microbiol Read Engl 147:2409–2415. https://doi.org/10.1099/00221287-147-9-2409

Ma D, Li R (2013) Current understanding of HOG-MAPK pathway in *Aspergillus fumigatus*. Mycopathologia 175:13–23. https://doi.org/10.1007/s11046-012-9600-5

Maeda T, Takekawa M, Saito H (1995) Activation of yeast PBS2 MAPKK by MAPKKKs or by binding of an SH3-containing osmosensor. Science 269:554–558

Mager WH, Siderius M (2002) Novel insights into the osmotic stress response of yeast. FEMS Yeast Res 2:251–257

Mahmud SA, Hirasawa T, Shimizu H (2010) Differential importance of trehalose accumulation in *Saccharomyces cerevisiae* in response to various environmental stresses. J Biosci Bioeng 109:262–266. https://doi.org/10.1016/j.jbiosc.2009.08.500

Malachowski AN, Yosri M, Park G, Bahn Y-S, He Y, Olszewski MA (2016) Systemic approach to virulence gene network analysis for gaining new insight into Cryptococcal virulence. Front Microbiol 7:1652. https://doi.org/10.3389/fmicb.2016.01652

Malakar D, Dey A, Ghosh AK (2006) Protective role of S-adenosyl-L-methionine against hydrochloric acid stress in *Saccharomyces cerevisiae*. Biochim Biophys Acta 1760:1298–1303. https://doi.org/10.1016/j.bbagen.2006.07.004

Marotta DH, Nantel A, Sukala L, Teubl JR, Rauceo JM (2013) Genome-wide transcriptional profiling and enrichment mapping reveal divergent and conserved roles of Sko1 in the *Candida albicans* osmotic stress response. Genomics 102:363–371. https://doi.org/10.1016/j.ygeno.2013.06.002

Martin CE, Oh C-S, Jiang Y (2007) Regulation of long chain unsaturated fatty acid synthesis in yeast. Biochim Biophys Acta 1771:271–285. https://doi.org/10.1016/j.bbalip.2006.06.010

Matsuki H, Goto M, Tada K, Tamai N (2013) Thermotropic and barotropic phase behavior of phosphatidylcholine bilayers. Int J Mol Sci 14:2282–2302. https://doi.org/10.3390/ijms14022282

Matsushika A, Suzuki T, Goshima T, Hoshino T (2017) Evaluation of *Saccharomyces cerevisiae* GAS1 with respect to its involvement in tolerance to low pH and salt stress. J Biosci Bioeng 124:164–170. https://doi.org/10.1016/j.jbiosc.2017.03.004

Matsuyama S, Llopis J, Deveraux QL, Tsien RY, Reed JC (2000) Changes in intramitochondrial and cytosolic pH: early events that modulate caspase activation during apoptosis. Nat Cell Biol 2:318–325. https://doi.org/10.1038/35014006

Miermont A, Waharte F, Hu S, McClean MN, Bottani S, Léon S, Hersen P (2013) Severe osmotic compression triggers a slowdown of intracellular signaling, which can be explained by molecular crowding. Proc Natl Acad Sci U S A 110:5725–5730. https://doi.org/10.1073/pnas.1215367110

Mika JT, Poolman B (2011) Macromolecule diffusion and confinement in prokaryotic cells. Curr Opin Biotechnol 22:117–126. https://doi.org/10.1016/j.copbio.2010.09.009

Mille Y, Girard J-P, Beney L, Gervais P (2005) Air drying optimization of *Saccharomyces cerevisiae* through its water-glycerol dehydration properties. J Appl Microbiol 99:376–382. https://doi.org/10.1111/j.1365-2672.2005.02615.x

Mira NP, Lourenço AB, Fernandes AR, Becker JD, Sá-Correia I (2009) The RIM101 pathway has a role in *Saccharomyces cerevisiae* adaptive response and resistance to propionic acid and other weak acids. FEMS Yeast Res 9:202–216. https://doi.org/10.1111/j.1567-1364.2008.00473.x

Mira NP, Becker JD, Sá-Correia I (2010a) Genomic expression program involving the Haa1p-regulon in *Saccharomyces cerevisiae* response to acetic acid. Omics J Integr Biol 14:587–601. https://doi.org/10.1089/omi.2010.0048

Mira NP, Palma M, Guerreiro JF, Sá-Correia I (2010b) Genome-wide identification of *Saccharomyces cerevisiae* genes required for tolerance to acetic acid. Microb Cell Factories 9:79. https://doi.org/10.1186/1475-2859-9-79

Mira NP, Teixeira MC, Sá-Correia I (2010c) Adaptive response and tolerance to weak acids in *Saccharomyces cerevisiae*: a genome-wide view. Omics J Integr Biol 14:525–540. https://doi.org/10.1089/omi.2010.0072

Miura T, Minegishi H, Usami R, Abe F (2006) Systematic analysis of HSP gene expression and effects on cell growth and survival at high hydrostatic pressure in *Saccharomyces cerevisiae*. Extrem Life Extreme Cond 10:279–284. https://doi.org/10.1007/s00792-005-0496-6

Molina AM, Swiegers JH, Varela C, Pretorius IS, Agosin E (2007) Influence of wine fermentation temperature on the synthesis of yeast-derived volatile aroma compounds. Appl Microbiol Biotechnol 77:675–687. https://doi.org/10.1007/s00253-007-1194-3

Mollapour M, Piper PW (2001) The ZbYME2 gene from the food spoilage yeast *Zygosaccharomyces bailii* confers not only YME2 functions in *Saccharomyces cerevisiae*, but also the capacity for catabolism of sorbate and benzoate, two major weak organic acid preservatives. Mol Microbiol 42:919–930

Mollapour M, Piper PW (2006) Hog1p mitogen-activated protein kinase determines acetic acid resistance in *Saccharomyces cerevisiae*. FEMS Yeast Res 6:1274–1280. https://doi.org/10.1111/j.1567-1364.2006.00118.x

Mollapour M, Piper PW (2007) Hog1 mitogen-activated protein kinase phosphorylation targets the yeast Fps1 aquaglyceroporin for endocytosis, thereby rendering cells resistant to acetic acid. Mol Cell Biol 27:6446–6456. https://doi.org/10.1128/MCB.02205-06

Moreira SA, Fernandes PAR, Duarte R, Santos DI, Fidalgo LG, Santos MD, Queirós RP, Delgadillo I, Saraiva JA (2015) A first study comparing preservation of a ready-to-eat soup under pressure (hyperbaric storage) at 25°C and 30°C with refrigeration. Food Sci Nutr 3:467–474. https://doi.org/10.1002/fsn3.212

Murata Y, Homma T, Kitagawa E, Momose Y, Sato MS, Odani M, Shimizu H, Hasegawa-Mizusawa M, Matsumoto R, Mizukami S, Fujita K, Parveen M, Komatsu Y, Iwahashi H (2006) Genome-wide expression analysis of yeast response during exposure to 4 degrees C. Extrem Life Extreme Cond 10:117–128. https://doi.org/10.1007/s00792-005-0480-1

Naicker MC, Seul Jo I, Im H (2012) Identification of chaperones in freeze tolerance in *Saccharomyces cerevisiae*. J Microbiol Seoul Korea 50:882–887. https://doi.org/10.1007/s12275-012-2411-z

Nakagawa Y, Sakumoto N, Kaneko Y, Harashima S (2002) Mga2p is a putative sensor for low temperature and oxygen to induce OLE1 transcription in *Saccharomyces cerevisiae*. Biochem Biophys Res Commun 291:707–713. https://doi.org/10.1006/bbrc.2002.6507

Nakayama Y, Yoshimura K, Iida H (2012) Organellar mechanosensitive channels in fission yeast regulate the hypo-osmotic shock response. Nat Commun 3:1020. https://doi.org/10.1038/ncomms2014

Nishizawa M, Tanigawa M, Hayashi M, Maeda T, Yazaki Y, Saeki Y, Toh-e A (2010) Pho85 kinase, a cyclin-dependent kinase, regulates nuclear accumulation of the Rim101 transcription factor in the stress response of *Saccharomyces cerevisiae*. Eukaryot Cell 9:943–951. https://doi.org/10.1128/EC.00247-09

Nombela C, Gil C, Chaffin WL (2006) Non-conventional protein secretion in yeast. Trends Microbiol 14:15–21. https://doi.org/10.1016/j.tim.2005.11.009

Oliveira BM, Barrio E, Querol A, Pérez-Torrado R (2014) Enhanced enzymatic activity of glycerol-3-phosphate dehydrogenase from the cryophilic *Saccharomyces kudriavzevii*. PLoS One 9: e87290. https://doi.org/10.1371/journal.pone.0087290

Orij R, Postmus J, Ter Beek A, Brul S, Smits GJ (2009) In vivo measurement of cytosolic and mitochondrial pH using a pH-sensitive GFP derivative in *Saccharomyces cerevisiae* reveals a relation between intracellular pH and growth. Microbiol Read Engl 155:268–278. https://doi.org/10.1099/mic.0.022038-0

Ost KS, O'Meara TR, Huda N, Esher SK, Alspaugh JA (2015) The *Cryptococcus neoformans* alkaline response pathway: identification of a novel rim pathway activator. PLoS Genet 11: e1005159. https://doi.org/10.1371/journal.pgen.1005159

Palma M, Dias PJ, Roque F de C, Luzia L, Guerreiro JF, Sá-Correia I (2017) The *Zygosaccharomyces bailii* transcription factor Haa1 is required for acetic acid and copper stress responses suggesting subfunctionalization of the ancestral bifunctional protein Haa1/Cup2. BMC Genomics 18:75. https://doi.org/10.1186/s12864-016-3443-2

Palma M, Guerreiro JF, Sá-Correia I (2018) Adaptive response and tolerance to acetic acid in *Saccharomyces cerevisiae* and *Zygosaccharomyces bailii*: a physiological genomics perspective. Front Microbiol 9:274. https://doi.org/10.3389/fmicb.2018.00274

Panadero J, Pallotti C, Rodríguez-Vargas S, Randez-Gil F, Prieto JA (2006) A downshift in temperature activates the high osmolarity glycerol (HOG) pathway, which determines freeze tolerance in *Saccharomyces cerevisiae*. J Biol Chem 281:4638–4645. https://doi.org/10.1074/jbc.M512736200

Park JI, Grant CM, Attfield PV, Dawes IW (1997) The freeze-thaw stress response of the yeast *Saccharomyces cerevisiae* is growth phase specific and is controlled by nutritional state via the RAS-cyclic AMP signal transduction pathway. Appl Environ Microbiol 63:3818–3824

Peñalva MA, Arst HN (2004) Recent advances in the characterization of ambient pH regulation of gene expression in filamentous fungi and yeasts. Annu Rev Microbiol 58:425–451. https://doi.org/10.1146/annurev.micro.58.030603.123715

Peñalva MA, Lucena-Agell D, Arst HN (2014) Liaison alcaline: Pals entice non-endosomal ESCRTs to the plasma membrane for pH signaling. Curr Opin Microbiol 22:49–59. https://doi.org/10.1016/j.mib.2014.09.005

Pérez-Sampietro M, Herrero E (2014) The PacC-family protein Rim101 prevents selenite toxicity in *Saccharomyces cerevisiae* by controlling vacuolar acidification. Fungal Genet Biol FG B 71:76–85. https://doi.org/10.1016/j.fgb.2014.09.001

Persson E, Halle B (2008) Cell water dynamics on multiple time scales. Proc Natl Acad Sci U S A 105:6266–6271. https://doi.org/10.1073/pnas.0709585105

Petitjean M, Teste M-A, François JM, Parrou J-L (2015) Yeast tolerance to various stresses relies on the Trehalose-6P synthase (Tps1) protein, not on trehalose. J Biol Chem 290:16177–16190. https://doi.org/10.1074/jbc.M115.653899

Petrezsélyová S, López-Malo M, Canadell D, Roque A, Serra-Cardona A, Marqués MC, Vilaprinyó E, Alves R, Yenush L, Ariño J (2016) Regulation of the Na+/K+-ATPase Ena1 expression by Calcineurin/Crz1 under high pH stress: a quantitative study. PLoS One 11: e0158424. https://doi.org/10.1371/journal.pone.0158424

Piccirillo S, White MG, Murphy JC, Law DJ, Honigberg SM (2010) The Rim101p/PacC pathway and alkaline pH regulate pattern formation in yeast colonies. Genetics 184:707–716. https://doi.org/10.1534/genetics.109.113480

Pick U, Bental M, Chitlaru E, Weiss M (1990) Polyphosphate-hydrolysis--a protective mechanism against alkaline stress? FEBS Lett 274:15–18

Piper PW (1999) Yeast superoxide dismutase mutants reveal a pro-oxidant action of weak organic acid food preservatives. Free Radic Biol Med 27:1219–1227

Piper PW (2011) Resistance of yeasts to weak organic acid food preservatives. Adv Appl Microbiol 77:97–113. https://doi.org/10.1016/B978-0-12-387044-5.00004-2

Piper P, Mahé Y, Thompson S, Pandjaitan R, Holyoak C, Egner R, Mühlbauer M, Coote P, Kuchler K (1998) The pdr12 ABC transporter is required for the development of weak organic acid resistance in yeast. EMBO J 17:4257–4265. https://doi.org/10.1093/emboj/17.15.4257

Piper P, Calderon CO, Hatzixanthis K, Mollapour M (2001) Weak acid adaptation: the stress response that confers yeasts with resistance to organic acid food preservatives. Microbiol Read Engl 147:2635–2642. https://doi.org/10.1099/00221287-147-10-2635

Platara M, Ruiz A, Serrano R, Palomino A, Moreno F, Ariño J (2006) The transcriptional response of the yeast Na(+)-ATPase ENA1 gene to alkaline stress involves three main signaling pathways. J Biol Chem 281:36632–36642. https://doi.org/10.1074/jbc.M606483200

Portillo F (2000) Regulation of plasma membrane H(+)-ATPase in fungi and plants. Biochim Biophys Acta 1469:31–42

Potts M (2001) Desiccation tolerance: a simple process? Trends Microbiol 9:553–559

Qi Y, Liu H, Yu J, Chen X, Liu L (2017) Med15B regulates acid stress response and tolerance in *Candida glabrata* by altering membrane lipid composition. Appl Environ Microbiol 83. https://doi.org/10.1128/AEM.01128-17

Rane HS, Bernardo SM, Hayek SR, Binder JL, Parra KJ, Lee SA (2014) The contribution of *Candida albicans* vacuolar ATPase subunit V_1B, encoded by VMA2, to stress response, autophagy, and virulence is independent of environmental pH. Eukaryot Cell 13:1207–1221. https://doi.org/10.1128/EC.00135-14

Rapoport A, Turchetti B, Buzzini P (2016) Application of anhydrobiosis and dehydration of yeasts for non-conventional biotechnological goals. World J Microbiol Biotechnol 32:104. https://doi.org/10.1007/s11274-016-2058-8

Ratnakumar S, Tunnacliffe A (2006) Intracellular trehalose is neither necessary nor sufficient for desiccation tolerance in yeast. FEMS Yeast Res 6:902–913. https://doi.org/10.1111/j.1567-1364.2006.00066.x

Ratnakumar S, Hesketh A, Gkargkas K, Wilson M, Rash BM, Hayes A, Tunnacliffe A, Oliver SG (2011) Phenomic and transcriptomic analyses reveal that autophagy plays a major role in desiccation tolerance in *Saccharomyces cerevisiae*. Mol BioSyst 7:139–149. https://doi.org/10.1039/c0mb00114g

Rep M, Krantz M, Thevelein JM, Hohmann S (2000) The transcriptional response of *Saccharomyces cerevisiae* to osmotic shock. Hot1p and Msn2p/Msn4p are required for the induction of subsets of high osmolarity glycerol pathway-dependent genes. J Biol Chem 275:8290–8300

Roche J, Caro JA, Norberto DR, Barthe P, Roumestand C, Schlessman JL, Garcia AE, García-Moreno BE, Royer CA (2012) Cavities determine the pressure unfolding of proteins. Proc Natl Acad Sci U S A 109:6945–6950. https://doi.org/10.1073/pnas.1200915109

Rodicio R, Heinisch JJ (2010) Together we are strong--cell wall integrity sensors in yeasts. Yeast Chichester Engl 27:531–540. https://doi.org/10.1002/yea.1785

Rodrigues F, Sousa MJ, Ludovico P, Santos H, Côrte-Real M, Leão C (2012) The fate of acetic acid during glucose co-metabolism by the spoilage yeast *Zygosaccharomyces bailii*. PLoS One 7: e52402. https://doi.org/10.1371/journal.pone.0052402

Rodríguez-Porrata B, Carmona-Gutierrez D, Reisenbichler A, Bauer M, Lopez G, Escoté X, Mas A, Madeo F, Cordero-Otero R (2012) Sip 18 hydrophilin prevents yeast cell death during desiccation stress. J Appl Microbiol 112:512–525. https://doi.org/10.1111/j.1365-2672.2011.05219.x

Rodríguez-Vargas S, Sánchez-García A, Martínez-Rivas JM, Prieto JA, Randez-Gil F (2007) Fluidization of membrane lipids enhances the tolerance of *Saccharomyces cerevisiae* to freezing and salt stress. Appl Environ Microbiol 73:110–116. https://doi.org/10.1128/AEM.01360-06

Roque A, Petrezsélyová S, Serra-Cardona A, Ariño J (2016) Genome-wide recruitment profiling of transcription factor Crz1 in response to high pH stress. BMC Genomics 17:662. https://doi.org/10.1186/s12864-016-3006-6

Russell NJ (2008) Membrane components and cold sensing. In: Margesin R, Schinner F, Marx J-C, Gerday C (eds) Psychrophiles: from biodiversity to biotechnology. Springer, Berlin, pp 177–190

Sahara T, Goda T, Ohgiya S (2002) Comprehensive expression analysis of time-dependent genetic responses in yeast cells to low temperature. J Biol Chem 277:50015–50021. https://doi.org/10.1074/jbc.M209258200

Saito H (2010) Regulation of cross-talk in yeast MAPK signaling pathways. Curr Opin Microbiol 13:677–683. https://doi.org/10.1016/j.mib.2010.09.001

Sangwan V, Orvar BL, Beyerly J, Hirt H, Dhindsa RS (2002) Opposite changes in membrane fluidity mimic cold and heat stress activation of distinct plant MAP kinase pathways. Plant J Cell Mol Biol 31:629–638

Saxena A, Sitaraman R (2016) Osmoregulation in *Saccharomyces cerevisiae* via mechanisms other than the high-osmolarity glycerol pathway. Microbiol Read Engl 162:1511–1526. https://doi.org/10.1099/mic.0.000360

Schaber J, Adrover MA, Eriksson E, Pelet S, Petelenz-Kurdziel E, Klein D, Posas F, Goksör M, Peter M, Hohmann S, Klipp E (2010) Biophysical properties of *Saccharomyces cerevisiae* and their relationship with HOG pathway activation. Eur Biophys J EBJ 39:1547–1556. https://doi.org/10.1007/s00249-010-0612-0

Schade B, Jansen G, Whiteway M, Entian KD, Thomas DY (2004) Cold adaptation in budding yeast. Mol Biol Cell 15:5492–5502. https://doi.org/10.1091/mbc.E04-03-0167

Schüller C, Mamnun YM, Mollapour M, Krapf G, Schuster M, Bauer BE, Piper PW, Kuchler K (2004) Global phenotypic analysis and transcriptional profiling defines the weak acid stress response regulon in *Saccharomyces cerevisiae*. Mol Biol Cell 15:706–720. https://doi.org/10.1091/mbc.e03-05-0322

Schutt KL, Moseley JB (2017) Transient activation of fission yeast AMPK is required for cell proliferation during osmotic stress. Mol Biol Cell 28:1804–1814. https://doi.org/10.1091/mbc.E17-04-0235

Serra-Cardona A, Petrezsélyová S, Canadell D, Ramos J, Ariño J (2014) Coregulated expression of the Na+/phosphate Pho89 transporter and Ena1 Na+-ATPase allows their functional coupling under high-pH stress. Mol Cell Biol 34:4420–4435. https://doi.org/10.1128/MCB.01089-14

Serra-Cardona A, Canadell D, Ariño J (2015) Coordinate responses to alkaline pH stress in budding yeast. Microb Cell Graz Austria 2:182–196. https://doi.org/10.15698/mic2015.06.205

Serrano R, Bernal D, Simón E, Ariño J (2004) Copper and iron are the limiting factors for growth of the yeast *Saccharomyces cerevisiae* in an alkaline environment. J Biol Chem 279:19698–19704. https://doi.org/10.1074/jbc.M313746200

Serrano R, Martín H, Casamayor A, Ariño J (2006) Signaling alkaline pH stress in the yeast *Saccharomyces cerevisiae* through the Wsc1 cell surface sensor and the Slt2 MAPK pathway. J Biol Chem 281:39785–39795. https://doi.org/10.1074/jbc.M604497200

Shamrock VJ, Lindsey GG (2008) A compensatory increase in trehalose synthesis in response to desiccation stress in *Saccharomyces cerevisiae* cells lacking the heat shock protein Hsp12p. Can J Microbiol 54:559–568. https://doi.org/10.1139/w08-044

Sharma P, Meena N, Aggarwal M, Mondal AK (2005) *Debaryomyces hansenii*, a highly osmo-tolerant and halo-tolerant yeast, maintains activated Dhog1p in the cytoplasm during its growth under severe osmotic stress. Curr Genet 48:162–170. https://doi.org/10.1007/s00294-005-0010-9

Simões T, Mira NP, Fernandes AR, Sá-Correia I (2006) The SPI1 gene, encoding a glycosylphosphatidylinositol-anchored cell wall protein, plays a prominent role in the development of yeast resistance to lipophilic weak-acid food preservatives. Appl Environ Microbiol 72:7168–7175. https://doi.org/10.1128/AEM.01476-06

Singh J, Kumar D, Ramakrishnan N, Singhal V, Jervis J, Garst JF, Slaughter SM, DeSantis AM, Potts M, Helm RF (2005) Transcriptional response of *Saccharomyces cerevisiae* to desiccation and rehydration. Appl Environ Microbiol 71:8752–8763. https://doi.org/10.1128/AEM.71.12.8752-8763.2005

Solé C, Nadal-Ribelles M, de Nadal E, Posas F (2015) A novel role for lncRNAs in cell cycle control during stress adaptation. Curr Genet 61:299–308. https://doi.org/10.1007/s00294-014-0453-y

Sousa M, Duarte AM, Fernandes TR, Chaves SR, Pacheco A, Leão C, Côrte-Real M, Sousa MJ (2013) Genome-wide identification of genes involved in the positive and negative regulation of acetic acid-induced programmed cell death in *Saccharomyces cerevisiae*. BMC Genomics 14:838. https://doi.org/10.1186/1471-2164-14-838

Stratford M, Nebe-von-Caron G, Steels H, Novodvorska M, Ueckert J, Archer DB (2013a) Weak-acid preservatives: pH and proton movements in the yeast *Saccharomyces cerevisiae*. Int J Food Microbiol 161:164–171. https://doi.org/10.1016/j.ijfoodmicro.2012.12.013

Stratford M, Steels H, Nebe-von-Caron G, Novodvorska M, Hayer K, Archer DB (2013b) Extreme resistance to weak-acid preservatives in the spoilage yeast *Zygosaccharomyces bailii*. Int J Food Microbiol 166:126–134. https://doi.org/10.1016/j.ijfoodmicro.2013.06.025

Strydom M, Kirschbaum AF, Tromp A (1982) Ergosterol concentration of several different *Saccharomyces cerevisiae* yeast strains. South Afr J Enol Vitic 3:23–28. https://doi.org/10.21548/3-1-2391

Suescún-Bolívar LP, Thomé PE (2015) Osmosensing and osmoregulation in unicellular eukaryotes. World J Microbiol Biotechnol 31:435–443. https://doi.org/10.1007/s11274-015-1811-8

Suzuki T, Sugiyama M, Wakazono K, Kaneko Y, Harashima S (2012) Lactic-acid stress causes vacuolar fragmentation and impairs intracellular amino-acid homeostasis in *Saccharomyces cerevisiae*. J Biosci Bioeng 113:421–430. https://doi.org/10.1016/j.jbiosc.2011.11.010

Takabatake A, Kawazoe N, Izawa S (2015) Plasma membrane proteins Yro2 and Mrh1 are required for acetic acid tolerance in *Saccharomyces cerevisiae*. Appl Microbiol Biotechnol 99:2805–2814. https://doi.org/10.1007/s00253-014-6278-2

Takagi H, Sakai K, Morida K, Nakamori S (2000) Proline accumulation by mutation or disruption of the proline oxidase gene improves resistance to freezing and desiccation stresses in *Saccharomyces cerevisiae*. FEMS Microbiol Lett 184:103–108

Tamás MJ, Luyten K, Sutherland FC, Hernandez A, Albertyn J, Valadi H, Li H, Prior BA, Kilian SG, Ramos J, Gustafsson L, Thevelein JM, Hohmann S (1999) Fps1p controls the accumulation and release of the compatible solute glycerol in yeast osmoregulation. Mol Microbiol 31:1087–1104

Tan H, Dong J, Wang G, Xu H, Zhang C, Xiao D (2014) Enhanced freeze tolerance of baker's yeast by overexpressed trehalose-6-phosphate synthase gene (TPS1) and deleted trehalase genes in frozen dough. J Ind Microbiol Biotechnol 41:1275–1285. https://doi.org/10.1007/s10295-014-1467-7

Tapia H, Koshland DE (2014) Trehalose is a versatile and long-lived chaperone for desiccation tolerance. Curr Biol CB 24:2758–2766. https://doi.org/10.1016/j.cub.2014.10.005

Tapia H, Young L, Fox D, Bertozzi CR, Koshland D (2015) Increasing intracellular trehalose is sufficient to confer desiccation tolerance to *Saccharomyces cerevisiae*. Proc Natl Acad Sci U S A 112:6122–6127. https://doi.org/10.1073/pnas.1506415112

Tatebayashi K, Tanaka K, Yang H-Y, Yamamoto K, Matsushita Y, Tomida T, Imai M, Saito H (2007) Transmembrane mucins Hkr1 and Msb2 are putative osmosensors in the SHO1 branch of yeast HOG pathway. EMBO J 26:3521–3533. https://doi.org/10.1038/sj.emboj.7601796

Tatebayashi K, Yamamoto K, Nagoya M, Takayama T, Nishimura A, Sakurai M, Momma T, Saito H (2015) Osmosensing and scaffolding functions of the oligomeric four-transmembrane domain osmosensor Sho1. Nat Commun 6:6975. https://doi.org/10.1038/ncomms7975

Thammavongs B, Denou E, Missous G, Guéguen M, Panoff J-M (2008) Response to environmental stress as a global phenomenon in biology: the example of microorganisms. Microbes Environ 23:20–23

Thewes S (2014) Calcineurin-Crz1 signaling in lower eukaryotes. Eukaryot Cell 13:694–705. https://doi.org/10.1128/EC.00038-14

Tilburn J, Sarkar S, Widdick DA, Espeso EA, Orejas M, Mungroo J, Peñalva MA, Arst HN (1995) The Aspergillus PacC zinc finger transcription factor mediates regulation of both acid- and alkaline-expressed genes by ambient pH. EMBO J 14:779–790

Tronchoni J, Medina V, Guillamón JM, Querol A, Pérez-Torrado R (2014) Transcriptomics of cryophilic *Saccharomyces kudriavzevii* reveals the key role of gene translation efficiency in cold stress adaptations. BMC Genomics 15:432. https://doi.org/10.1186/1471-2164-15-432

Tulha J, Lima A, Lucas C, Ferreira C (2010) *Saccharomyces cerevisiae* glycerol/H+ symporter Stl1p is essential for cold/near-freeze and freeze stress adaptation. A simple recipe with high biotechnological potential is given. Microb Cell Factor 9:82. https://doi.org/10.1186/1475-2859-9-82

Turk M, Gostinčar C (2018) Glycerol metabolism genes in *Aureobasidium pullulans* and *Aureobasidium subglaciale*. Fungal Biol 122:63–73. https://doi.org/10.1016/j.funbio.2017.10. 005

Ullah A, Orij R, Brul S, Smits GJ (2012) Quantitative analysis of the modes of growth inhibition by weak organic acids in *Saccharomyces cerevisiae*. Appl Environ Microbiol 78:8377–8387. https://doi.org/10.1128/AEM.02126-12

Ullah A, Chandrasekaran G, Brul S, Smits GJ (2013) Yeast adaptation to weak acids prevents futile energy expenditure. Front Microbiol 4:142. https://doi.org/10.3389/fmicb.2013.00142

Umebayashi K, Nakano A (2003) Ergosterol is required for targeting of tryptophan permease to the yeast plasma membrane. J Cell Biol 161:1117–1131. https://doi.org/10.1083/jcb.200303088

van de Vossenberg JL, Driessen AJ, Konings WN (1998) The essence of being extremophilic: the role of the unique archaeal membrane lipids. Extrem Life Extreme Cond 2:163–170

van der Rest ME, Kamminga AH, Nakano A, Anraku Y, Poolman B, Konings WN (1995) The plasma membrane of *Saccharomyces cerevisiae*: structure, function, and biogenesis. Microbiol Rev 59:304–322

Verghese J, Abrams J, Wang Y, Morano KA (2012) Biology of the heat shock response and protein chaperones: budding yeast (*Saccharomyces cerevisiae*) as a model system. Microbiol Mol Biol Rev MMBR 76:115–158. https://doi.org/10.1128/MMBR.05018-11

Vicent I, Navarro A, Mulet JM, Sharma S, Serrano R (2015) Uptake of inorganic phosphate is a limiting factor for *Saccharomyces cerevisiae* during growth at low temperatures. FEMS Yeast Res 15. https://doi.org/10.1093/femsyr/fov008

Viladevall L, Serrano R, Ruiz A, Domenech G, Giraldo J, Barceló A, Ariño J (2004) Characterization of the calcium-mediated response to alkaline stress in *Saccharomyces cerevisiae*. J Biol Chem 279:43614–43624. https://doi.org/10.1074/jbc.M403606200

Vilela-Moura A, Schuller D, Mendes-Faia A, Silva RD, Chaves SR, Sousa MJ, Côrte-Real M (2011) The impact of acetate metabolism on yeast fermentative performance and wine quality: reduction of volatile acidity of grape musts and wines. Appl Microbiol Biotechnol 89:271–280. https://doi.org/10.1007/s00253-010-2898-3

Virgilio S, Cupertino FB, Ambrosio DL, Bertolini MC (2017) Regulation of the reserve carbohydrate metabolism by alkaline pH and calcium in *Neurospora crassa* reveals a possible cross-regulation of both signaling pathways. BMC Genomics 18:457. https://doi.org/10.1186/s12864-017-3832-1

Wang H, Liang Y, Zhang B, Zheng W, Xing L, Li M (2011) Alkaline stress triggers an immediate calcium fluctuation in *Candida albicans* mediated by Rim101p and Crz1p transcription factors. FEMS Yeast Res 11:430–439. https://doi.org/10.1111/j.1567-1364.2011.00730.x

Welch AZ, Gibney PA, Botstein D, Koshland DE (2013) TOR and RAS pathways regulate desiccation tolerance in *Saccharomyces cerevisiae*. Mol Biol Cell 24:115–128. https://doi.org/10.1091/mbc.E12-07-0524

Westfall PJ, Patterson JC, Chen RE, Thorner J (2008) Stress resistance and signal fidelity independent of nuclear MAPK function. Proc Natl Acad Sci U S A 105:12212–12217. https://doi.org/10.1073/pnas.0805797105

Wiegel J (2011) Anaerobic alkaliphiles and alkaliphilic poly-extremophiles. In: Horikoshi K (ed) Extremophiles handbook. Springer Japan, Tokyo, pp 81–97

Xu W, Smith FJ, Subaran R, Mitchell AP (2004) Multivesicular body-ESCRT components function in pH response regulation in *Saccharomyces cerevisiae* and *Candida albicans*. Mol Biol Cell 15:5528–5537. https://doi.org/10.1091/mbc.E04-08-0666

Yazawa H, Iwahashi H, Kamisaka Y, Kimura K, Uemura H (2009) Production of polyunsaturated fatty acids in yeast *Saccharomyces cerevisiae* and its relation to alkaline pH tolerance. Yeast Chichester Engl 26:167–184. https://doi.org/10.1002/yea.1659

Yenush L (2016) Potassium and sodium transport in yeast. Adv Exp Med Biol 892:187–228. https://doi.org/10.1007/978-3-319-25304-6_8

Yun J, Lee DG (2016) A novel fungal killing mechanism of propionic acid. FEMS Yeast Res 16. https://doi.org/10.1093/femsyr/fow089

Zambuto M, Romaniello R, Guaragnella N, Romano P, Votta S, Capece A (2017) Identification by phenotypic and genetic approaches of an indigenous *Saccharomyces cerevisiae* wine strain with high desiccation tolerance. Yeast Chichester Engl 34:417–426. https://doi.org/10.1002/yea.3245

Zvyagilskaya R, Parchomenko O, Abramova N, Allard P, Panaretakis T, Pattison-Granberg J, Persson BL (2001) Proton- and sodium-coupled phosphate transport systems and energy status of *Yarrowia lipolytica* cells grown in acidic and alkaline conditions. J Membr Biol 183:39–50

How Do Yeast and Other Fungi Recognize and Respond to Genome Perturbations?

3

Adrianna Skoneczna, Kamil Krol, and Marek Skoneczny

Abstract

The integrity of a cellular genome may be endangered in many ways. However, cells are not powerless and indifferent to their future fate; they are prepared to solve various problems concerning their genome to enable their survival. Maintaining a stable genome is a complicated task, even without any external problems such as genotoxic agents, irradiation, or any other harmful treatment to which cells are exposed in their environment. Inaccuracies in replication alone may lead to changes in DNA sequence, i.e., to mutations. When cells are exposed to stressful conditions that cause DNA lesions, the frequency of mutations increases. Moreover, unrepaired DNA damage may lead to the replication fork block that delays the replication round until the damage is repaired. However, if for some reason the DNA damage in the replication fork remains unrepaired, it may lead to DNA breakage causing DNA rearrangements or to fork collapse followed by cell death. Consequently, DNA damage is dangerous for replicating cells. What happens when DNA damage appears after a replication round has been successfully finished? Various DNA lesions may disturb the transcription efficiency of the affected genes. DNA strand breaks may generate problems during chromosome segregation because such lesions may lead to an unequal DNA distribution to daughter cells.

The DNA lesions are generally unwanted; however, during certain phases of the cell cycle, some classes of damage can be particularly dangerous. Thus, cellular resources have to recognize this type of damage and adjust the DNA damage response not only to this particular damage but also to the actual cell cycle phase. Sometimes similar DNA lesions may require different remedies. As a result, cells have developed a complicated protein network to maintain the

A. Skoneczna (✉) · K. Krol · M. Skoneczny
Institute of Biochemistry and Biophysics, Polish Academy of Sciences, Warsaw, Poland
e-mail: ada@ibb.waw.pl

© Springer Nature Switzerland AG 2018
M. Skoneczny (ed.), *Stress Response Mechanisms in Fungi*,
https://doi.org/10.1007/978-3-030-00683-9_3

genome integrity. In this chapter, some of the solutions applied by yeast and other fungi in response to genotoxic challenges will be presented.

3.1   Introduction

DNA contains all the information necessary for cells to live. Not surprisingly, evolution has provided cells with the ability to faithfully replicate DNA, to distribute chromosomes equally to daughter cells, and to repair damaged DNA. If everything goes well, errors in DNA rarely appear due to a number of safeguards. One of them is a rule that only well-prepared cells, i.e., cells that are prepared to produce the deoxynucleotides (dNTPs) needed for replication and the enzymes capable of carrying out this process and that have error-free template DNA are allowed to enter the S phase of the cell cycle. The checkpoint G1/S is responsible for going through this checklist and giving permission for replication START (Woollard and Nurse 1995; Bøe et al. 2012), as reviewed by de Bruin and Wittenberg (2009), Latif et al. (2001), Boye et al. (2009), and Reed (1996). With the usual availability of dNTPs for DNA synthesis (I), the presence of functional molecular mechanisms that ensure proper selection of the inserted nucleotide (II), the specific exonuclease $3' \rightarrow 5'$ activity of replicative DNA polymerase (i.e., proofreading activity of replicases) (III), and an additional control level that corrects improperly inserted nucleotides (IV), replication errors appear approximately once per 10^9 inserted nucleotides (reviewed in Skoneczna et al. 2015).

There are two replication error correction systems known thus far. The first one, the mismatch repair (MMR) system, can remove an incorrectly inserted deoxynucleotide and replace it with the proper one, and it can also remove a needlessly inserted deoxynucleotide or supplement sequences with an overlooked deoxynucleotide, as reviewed by Li (2008), Spampinato et al. (2009), and Kolodner and Marsischky (1999). The second one, the ribonucleotide excision repair system (RER), is able to remove a ribonucleotide inserted into DNA and replace it with a correct deoxynucleotide, as reviewed by Vaisman and Woodgate (2015). Thus, under conditions of undisturbed DNA replication, DNA is synthesized faithfully, and mistakes appear rarely. Moreover, if replication errors do not result in crucial amino acid changes in a protein that is essential for cell viability, they can be tolerated. The frequency of spontaneous replication errors is not a serious problem in yeast and fungi, with genome lengths ranging from less than 9 Mbp (as in the methylotrophic yeast *Hansenula polymorpha*) through 13–14 Mbp (as in the common budding and fission yeast model cells *Saccharomyces cerevisiae* and *Schizosaccharomyces pombe*, respectively, or the human pathogen *Candida albicans*), to 40 Mbp (as in the bread mold *Neurospora crassa*), and over 152 Mbp (as in the plant pathogen *Zopfia rhizophila*). A replication error frequency in the range of 10^{-9} means that in yeast and fungi grown under normal conditions, replication errors appear once per many replication rounds (Mohanta and Bae 2015; Odds et al. 2004; Goffeau et al. 1996; Wood et al. 2002).

The situation changes when the DNA template is damaged. The replication of such DNA frequently results in an erroneous deoxynucleotide insertion opposite the damage site or in DNA synthesis blockage. To avoid such a scenario, upon DNA damage detection, repair pathways are activated in S phase of the cell cycle, and DNA synthesis is slowed down (Iyer and Rhind 2017), which causes the elongation of single-stranded DNA (ssDNA) fragments in replication forks. Such ssDNA fragments present in a replication fork and coated with replication protein A complex (RPA) can activate the DNA damage response as effectively as the ssDNA gaps that are created by nucleotide excision during various types of DNA repair (Sogo et al. 2002; Mojas et al. 2007; Reha-Krantz et al. 2011; Deshpande et al. 2017). Moreover, the DNA damage response is so perfectly organized that distinct DNA repair pathways are mobilized to match the type of damage found in DNA. When DNA damage is extensive, or when appropriate corrective mechanisms are not yet available, cell cycle arrest occurs, persisting until the problem is solved. Extending the time spent on repair often yields successful results, and DNA synthesis can be resumed and completed immediately after recovery from the checkpoint (Diao et al. 2017; Yeung and Durocher 2011; Hishida et al. 2010; Chaudhury and Koepp 2017). However, if for some reason the repair does not occur, then the following solutions are possible: a delay in the repair of DNA damage by omitting the synthesis of the damaged fragment of DNA template enabled by firing of late origins of replication (I) (O'Neill et al. 2007), extension of cell cycle arrest until the DNA can be repaired, e.g., during the next phase of the cell cycle when the other repair pathways are activated (II) (Guo and Jinks-Robertson 2013; Brandão et al. 2014), or permanent cell cycle arrest, which abolishes the reproductive potential of the cell and leads to senescence (III) (Zyrina et al. 2015; Deshpande et al. 2011) or even a cell death throughout a mitotic catastrophe event (IV) (Vitale et al. 2011; Zyrina et al. 2015).

If unrepaired DNA persists until cell division, it often causes further trouble. DNA breaks might be the cause of unequal segregation of chromosomes, leading to the loss of part of the essential genetic information in the daughter cell and subsequently to its death. The control mechanisms that function at this point of the cell cycle attempt to prevent the division of cells with damaged or incompletely replicated DNA. It is controlled by the G2/M phase checkpoint, reviewed by Latif et al. (2001), Zámborszky et al. (2014), Yasutis and Kozminski (2013), Pomerening (2009), and Callegari and Kelly (2007). If previously performed DNA repair leads to gross chromosomal rearrangements (Myung and Kolodner 2003; Paek et al. 2010; Motegi et al. 2006; Hwang et al. 2005), or if previous imperfections in chromosome segregation (Hartwell and Smith 1985; Klein 2001; Bodi et al. 1991) or strong environmental pressure leads to aneuploidy (Howlett and Schiestl 2000; Pavelka et al. 2010; Sheltzer et al. 2011), the imbalance will occur during the next DNA division. This imbalance, resulting from different chromosome numbers, rearranged chromatids or altered chromosomes sizes, can be detected as a tension force acting across kinetochore, leading to activation of the spindle assembly checkpoint (SAC), which stops further division steps, i.e., anaphase progression and mitotic exit (Rancati et al. 2005).

Thus, the cellular response to DNA damage is tailored to the type of problem and to the phase of the cell cycle in which they appear. Yeast and fungi do not differ in this respect from other eukaryotic cells. In contrast, they often serve as model organisms in which these topics can be studied.

3.2 The DNA Damage Response in Yeast and Other Fungi

We have already mentioned that the cellular response to DNA damage depends on the phase of the cell cycle. Why is this happening? This can be explained using genotoxic stress as an example. In an asynchronous culture exposed to a mutagen, e.g., methyl methanesulfonate (MMS), cells are in different phases of the cell cycle. All cells are exposed to the same concentration of this toxic compound, but will it be equally harmful for all of them? Will all cells contain the same amount of it? This will be determined by the following: mutagen stability (I); mutagen uptake rate (II); intracellular deposition of the mutagen, which determines whether the mutagen reaches its intracellular target (III); mutagen export rate (IV); and the presence of possible detoxification pathways (V). The DNA damage level caused by the mutagen will be determined not only by its concentration within the cell (I) but also the efficiency of its movement inside the cell, in particular its ability to enter the nucleus (II), the state of chromatin condensation, and hence the accessibility of DNA to the mutagen (III), the availability and efficiency of DNA repair systems (IV), and cellular memory of earlier contact with this type of mutagen that results in the adaptive response (V). Moreover, cells are often exposed to another stress in addition to the genotoxic one, and in such conditions, an intensified reaction to stress is observed.

Returning to our example, let us consider which cells of the asynchronous culture exposed to MMS will be more susceptible to the alkylation damage instigated by this mutagen. It seems obvious that the most affected cells would be those that are currently in S phase, in which intense transcription and replication require DNA unwinding. The unwound DNA can easily be modified/damaged, especially in fast-growing cells that are rapidly taking up nutrients from the environment and, in parallel, also the mutagen. Therefore, the mutagen concentration in such cells may be higher than in less metabolically active ones, unless intense growth is accompanied by an increase in the activity of detoxification systems. Nevertheless, the DNA of S phase cells is more vulnerable than that of cells in other stages of the cell cycle (Daigaku et al. 2006; Davidson et al. 2012; Yang et al. 2010; Chen and Kolodner 1999), resulting in a greater sensitivity to genotoxic stress of exponentially growing cells than stationary ones (Krol et al. 2015; Sousa-Lopes et al. 2004). In contrast, in mitotic cells, in which DNA is tightly packed and partially shielded with proteins, the mutagen will cause less damage because it has no access to the DNA. Additionally, unlike exponentially growing cells, senescent cells do not take up biomolecules as vigorously from the environment, and therefore they should be less affected by the mutagen.

Sometimes a mutagen affects all cells in an asynchronous population to the same extent. This happens, for example, during osmotic stress caused by a high concentration of NaCl or during the exposure of cells either to ultraviolet (UV) radiation or

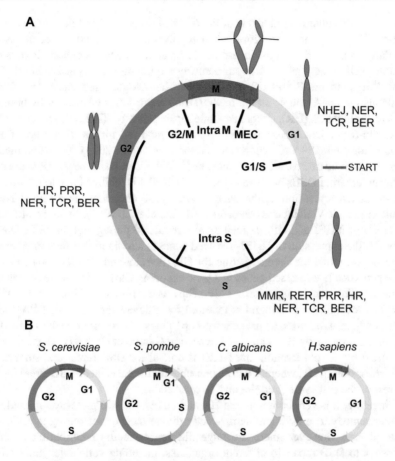

Fig. 3.1 The cell cycle phase length influences the repair ability of the cell. (**a**) The cell cycle of the yeast *Saccharomyces cerevisiae* with the DNA repair pathways assigned to a certain cell cycle phase. *MEC* mitotic exit checkpoint. (**b**) Adapted from Côte et al. (2009). The differences in length of cell cycle phases between various yeast/fungi species

to ionizing radiation. Each of these stresses leads to DNA breaks. In addition, UV irradiation causes a generation of photoproducts, while ionizing radiation causes mostly oxidative damage (Nikitaki et al. 2015). Nevertheless, despite the same dose of stressor, the response of individual cells in the population may be different. Even the exact same amount of DNA damage may be variously perceived and processed in cells in different phases of the cell cycle due to different availabilities of DNA repair pathways during individual phases of the cell cycle (Fig. 3.1). This effect may be explained in part by the periodic expression in the cell cycle of genes encoding some DNA repair proteins or proteins that are responsible for DNA damage response control (Andaluz et al. 1999; Limbo et al. 2007; de Bruin and Wittenberg 2009; Cho et al. 1998). The level of transcription of the *CaLIG4* (*CaCDC9*) gene encoding DNA ligase in *C. albicans* varies during the cell cycle. While newly budded daughter cells contain basal levels of the *CaLIG4* transcript, it peaks when cells are in late G1 phase and then drops when replication starts (Andaluz et al. 1999). Similarly to *C. albicans* cells, in haploid *S. cerevisiae* cells, a nonhomologous end joining (NHEJ), i.e., the DNA repair pathway relying on Lig4, is active in G1 (Gao et al. 2016). In the S and G2 phases, when homologous sequences are already available, the DNA double-strand break (DSB) repair job is taken over by homologous recombination (HR), a more accurate repair pathway. This is tightly controlled by cyclin-dependent kinase (CDK), which is required for the first step of HR, namely, for the ssDNA resection step (Aylon and Kupiec 2005). The same mechanism seems to be true for all eukaryotes. In diploid *S. cerevisiae* cells, HR is the only pathway repairing DSBs because in these cells NHEJ is inhibited due to the mating-type-mediated repression of the *NEJ1* (*LIF2*) gene encoding the essential NHEJ regulator (Frank-Vaillant and Marcand 2001; Kegel et al. 2001). In *Sc. pombe* cells, the levels of NHEJ and HR are reciprocally regulated throughout the cell cycle. In early G1, the expression of NHEJ-involved genes is tenfold higher than in other cell cycle stages, while the reverse is true for HR (Ferreira and Cooper 2004). One of these periodically expressed genes is a gene encoding Ctp1, a fission yeast ortholog of mammalian tumor suppressor CtIP. Ctp1 is recruited to DSBs in an MRN complex-dependent manner and is required for efficient formation of RPA-coated ssDNA adjacent to DSBs. Consequently, Ctp1 plays an essential role during the first step of DSB repair by HR, i.e., $5' \rightarrow 3'$ resection (Limbo et al. 2007). These data are supported by the observation that NHEJ is crucial for DSB repair and survival of DSB stress during G1, whereas it is dispensable for the survival of DSB stress during the rest of the cell cycle (Ferreira and Cooper 2004).

Therefore, it is very common that HR and NHEJ, the two pathways utilized as a cellular remedy to a DSB problem, strictly divide their duties during specific cell cycle phases. However, there are huge differences in the contribution of these pathways to DSB repair in different organisms, including vertebrates and various fungal species. Where do these differences come from? Could they simply reflect the differences in length of individual cell cycle phases in various yeast species (Fig. 3.1) (Côte et al. 2009)? We must consider herein that the yeast cell cycle model is different from that of pseudohyphal and hyphal cells (Berman 2006), which is the likely explanation for why mechanisms of homologous integration are efficient only in *Saccharomyces* sp. and *Ashbya gossypii* (Steiner et al. 1995). However, the participation of HR pathways in DSB repair and subsequent homologous integration efficiency is markedly increased in cells with a NHEJ pathway deficiency. This useful practical aspect of fungal research, which helps in strain construction, was observed, e.g., in *Neurospora crassa* and *Aspergillus nidulans*, among others (Ninomiya et al. 2004; Nayak et al. 2006; Kück and Hoff 2010). Moreover, the utilization of one DSB repair system when the other is not functional is typical not only for filamentous fungi but also for *S. cerevisiae* and higher eukaryote cells.

It is worth mentioning here that the length of the particular cell cycle phase is influenced by a range of factors. It is well-known that a change in nitrogen source in the growth medium not only influences the whole cell cycle length but also determines the length of individual phases of the cell cycle (Rivin and Fangman

1980). Budding yeast cells growing on ammonia have a much shorter G1 phase than cells growing on threonine. S phase is extended the most in cells growing on methionine as a nitrogen source. The G2 phase length in the cell cycle is the shortest in cells growing on methionine and longest in those growing on threonine (Rivin and Fangman 1980). Additionally, other environmental conditions can influence the duration of the cell cycle phases, which is not limited to the type and abundance of nutrients but encompasses the cellular response to various environmental stresses that do not necessarily, or directly, cause DNA damage. However, the strong link between the length of the cell cycle phase and nutritional limitation can also be treated as a stress-related response. The common knowledge is that starved cells are delayed in G1 phase. It is interesting that the mechanism responsible for this effect is conserved from yeast to human cells to such an extent that human GCN2 can functionally substitute for budding yeast Gcn2 (Dever et al. 1993). In the fission yeast, upon nutrient deprivation, uncharged tRNAs accumulate and bind a histidyl-tRNA synthetase-like (HisRS) domain in GCN2, which in turn leads to a conformational change that activates this kinase. The activation involves GCN1 protein that ensures delivery of the uncharged tRNA onto GCN2 (Anda et al. 2017). In the budding yeast, Gcn1 has been shown not only to transfer uncharged tRNAs to the HisRS domain of Gcn2 but also to recruit it to the translating ribosome (Sattlegger and Hinnebusch 2000, 2005). Activated GCN2 kinase regulates translation initiation in response to amino acid starvation by phosphorylation of Ser 52 (Ser 51 in *S. cerevisiae*) of the eukaryotic translation initiation factor 2α (eIF2α). This eIF2α phosphorylation leads to a general downregulation of translation but enhances the translation of specific stress response mRNAs (Hinnebusch 1993; Dang Do et al. 2009). Significantly, GCN2 is activated by some other stresses, such as UV irradiation and oxidative stress, without the accumulation of uncharged tRNAs. However, the mechanism underlying GCN2 activation remains the same (Anda et al. 2017).

In *S. cerevisiae* cells, G1 phase is also extended upon osmotic stress caused by 0.4 M NaCl treatment. The osmotic stress activates Hog1 kinase, which delays cell cycle progression by direct phosphorylation of the CDK inhibitor Sic1 and by downregulation of the transcription of the G1 cyclin genes *CLN2* and *CLB5* through the activity of Whi5 and Msa1 mediated by Hog1 phosphorylation (Adrover et al. 2011; González-Novo et al. 2015).

Extension of the G1 phase length, even if not related to DNA damage, as in the examples presented above, determines the DNA repair options, limiting them to those specific for a particular phase of the cell cycle. Therefore, one can expect that the frequency of usage of the NHEJ pathway during DSB repair in starved cells, which is delayed in G1, will be elevated in comparison to non-starved cells. In contrast, the factors that result in S phase extension are expected to cause preferred usage of HR for DSB repair.

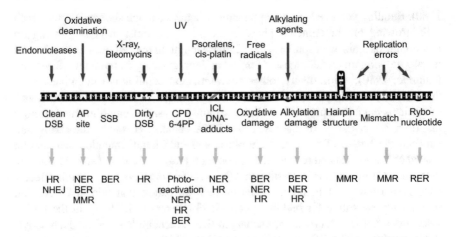

Fig. 3.2 Sources of DNA damage and molecular pathways responsible for its repair. *UV* ultraviolet irradiation, *AP site* abasic site, *DSB* double-strand break, *SSB* single-strand break, *CPD* cyclobutane bi-pyrimidine dimers, *6-4PP* (6-4) pyrimidone photoproducts, *ICL* inter-strand cross-links, *HR* homologous recombination, *NHEJ* nonhomologous end joining, *NER* nucleotide excision repair, *BER* base excision repair, *MMR* mismatch repair, *RER* ribonucleotide excision repair

3.2.1   DNA Damage and DNA Damage Signal

There are many circumstances in which DNA damage occurs. Among the stresses that are well-known for being destructive to DNA are, for example, alkylation stress, oxidative stress, γ-irradiation, UV irradiation, and radiomimetic or other genotoxic compound treatment. All of these stresses generate a range of lesions in DNA, which are specific for a particular stress (Fig. 3.2). Among these lesions, some are predominant and the most toxic or the most mutagenic. For example, 8-oxo-7,8-dihydroguanine (8-oxo-G) is the hallmark of oxidative stress and the major cause of mutations (C→A) that arise in DNA in response to this stress because its presence in DNA frequently leads to misincorporation during DNA synthesis (Haracska et al. 2003). In cells exposed to UV light, two major classes of damage can be produced at bi-pyrimidine sites: cyclobutane bi-pyrimidine dimers (CPD) and (6-4) pyrimidone photoproducts (6-4PP) (Cadet et al. 2005). If unrepaired, this kind of damage frequently leads to a replication block. The toxicity of chemicals from the bleomycin family is mostly attributed to their ability to induce DSBs, likely arising from the formation of two adjacent DNA strand breaks; however, they can also induce the formation of a clustered DNA lesion via reaction of the broken end with a proximate cytosine base (Regulus et al. 2007). Such "dirty" DNA breaks are unrepairable without resection, so they strongly depend on HR. If this repair pathway cannot be used, such damage will cause cell death, and therefore bleomycins are widely used in anticancer therapy.

A variety of environmental stresses, such as chemicals, UV, or ionizing radiation, cause DNA damage directly, although some environmental stresses lead to genotoxic stress in a less direct manner. The disturbance of iron homeostasis may cause an inefficiency in DNA repair since many enzymes that are crucial for DNA metabolism use iron-sulfur groups as essential cofactors (Puig et al. 2017; Kaniak-Golik and Skoneczna 2015). This group of enzymes includes DNA polymerases, various DNA repair enzymes (such as helicases, nucleases, demethylases, or glycosylases), and ribonucleotide reductase. Arsenic may cause DNA breaks directly or act indirectly by generating oxidative DNA damage that can be converted to replication-dependent DSBs or by interfering with DNA repair pathways in both budding and fission yeast (Litwin et al. 2013; Guo et al. 2016). The source of DSBs may also be hyperosmolarity caused by increased concentrations of NaCl in the medium (Kültz and Chakravarty 2001). Additionally, various cellular processes, such as energy metabolism, may cause DNA damage. This can lead to the generation of reactive oxygen and nitrogen species, which can attack cellular DNA (Nikitaki et al. 2015). Another endogenous source of damage is replication. The frequent source of DSB during replication is constituted by SSB in a template or another DNA lesion that leads to a replication block (reviewed in Aguilera and Gómez-González 2008).

However, regardless of the cause of the DNA damage and the kind of damage that has occurred, a simple evolutionary solution is the use of a common element that accompanies different DNA lesions as a damage signal. The signal is the appearance of ssDNA in cells. Fragments of ssDNA appear during the slowdown of replication or transcription when the template DNA is damaged (I); ssDNA gaps remain in DNA after an unsuccessful replication round, during which the replication block has been omitted by a restart of replication from an adjacent undamaged replication origin (II); they are generated during the repair of DNA damage, e.g., during nucleotide excision repair (NER), MMR, or RER (III). ssDNA is coated with RPA (a complex consisting of three subunits: Rfa1, 2, and 3) to protect it from nucleases, but RPA can also coordinate the recruitment and exchange of genome maintenance factors to regulate DNA replication, recombination, and repair (Maréchal and Zou 2015; Seeber et al. 2016; Sabatinos and Forsburg 2015).

3.2.2 Cellular Response to DSBs

One of the best known cellular DNA damage responses is the DSB response during S phase. As already mentioned in Sect. 3.2.1 this kind of damage is preferably repaired via HR during DNA synthesis. DSB repair requires the generation of ssDNA around the break, which is promoted by DNA resection by $5' \rightarrow 3'$ exonucleases. The first step of resection depends on an MRX complex [Mre11-Rad50-Xrs2 in *S. cerevisiae*, Rad32-Rad50-Nbs1 in *Sc. pombe* (for yeast orthologues, see Table 3.1)] and Sae2 endonuclease (Ctp1 in *Sc. pombe*), which together remove oligonucleotides from the DNA ends to form an early intermediate. The second step involves Exo1 nuclease and Sgs1 DNA helicase, which process this intermediate to generate longer tracts of ssDNA that may serve as a substrate for Rad51 (Mimitou and Symington 2008; Zhu et al. 2008). When bound to DNA, Rad51 protein protects DNA and forms a filament structure (Mimitou and Symington 2008; Zhu et al. 2008). The Rad51 filament forms the invasive strand that is responsible for a homology sequence search, i.e., initiates DSB repair via

Table 3.1 DNA damage checkpoint proteins

Saccharomyces cerevisiae	Schizosaccharomyces pombe	Candida albicans	Homo sapiens	Molecular function	Checkpoint function
Rfa1, Rfa2, Rfa3	Rad11 (Ssb1)	Rfa1, Rfa2	RPA	Replication protein A complex	Sensor/signaling/target
Mre11-Rad50-Xrs2	Rad32-Rad50-Nbs1	Mre11	MRE11-RAD50-NBS1	MRX/MRN complex DSB repair Tel1/ATM activator	Sensor
Ddc1-Rad17-Mec3	Rad9-Rad1-Hus1	Ddc1-Rad17-?	RAD9-RAD1-HUS1	9-1-1 complex, Mec1/ATR activator	Sensor
RFC-Rad24	RFC-Rad17	RFC-Rad24	RFC-RAD17	Checkpoint clamp loader	Sensor
Dpb11	Rad4/Cut5	Dpb11	TopBP1	BRCT domain adaptor, replication initiation Mec1/ATR activator	Sensor
Pol2-Dpb2-Dpb3-Dpb4	Cdc20-Dpb2-Dpb3-Dpb4	Pol2-Dpb2-Hfl1-Dpb4	POLE-POLE2-POLE3-POLE4	DNA polymerase ε	Sensor/mediator
RFC-Ctf18-Dcc1-Ctf8	RFC-Ctf18-Dcc1-Ctf8	RFC-Ctf18-Dcc1-Ctf8	RFC-CTF18-DCC1-CTF8	RFC-Ctf18 complex, PCNA clamp loader	Mediator
Rad9	Crb2/Rhp9	Rad9	53BP1, MDC1, BRCA1	Rad53/Dun1 activator	Mediator
Mrc1	Mrc1	Mrc1	CLASPIN	Rad53 activator	Mediator
Mec1	Rad3	Mec1	ATR	Kinase PIKK	Sensor/signaling
Ddc2	Rad26	?	ATRIP	Mec1/ATR binding partner	Signaling
Tel1	Tel1	Tel1	ATM	Kinase	Sensor/signaling
Rad53	Cds1	Rad53	CHK2	Kinase	Signaling/effector
Chk1	Chk1	?	CHK1	Kinase	Signaling/effector

Dun1	?	Dun1	?	Kinase	Effector/target
Nrm1	Nrm1	Nrm1	Rb	Co-repressor	Effector
Mbf1-Swi6	Mbf1-Cdc10	Mbf1-Swi6	EDF1-?	MBF complex, transcription factor of G1/S transition genes	Effector
Swi4-Swi6	Res2-Cdc10	Swi4-Swi6	?	SBF complex; transcription factor of DNA replication and repair genes	Effector
Dna2	Dna2	Dna2	DNA2	Okazaki fragment processing DSB repair Mec1/ATR activator	Target
Sgs1	Rqh1	Sgs1	BLM, WRN	Rad53 activator	Mediator/target
Cdc5	Plo1	Cdc5	PLK1	Polo kinase	Target
Pds1	Cut2	?	Securin	Securin	Target
Esp1	Cut1	Esp1	Separase	Separase	Target
Cdc20	Slp1	Cdc20	p55^{CDC}/CDC20	APC-targeting subunit	Target
Sae2	Ctp1		CtIP/RBBP8		Signaling/target

homologous recombination (reviewed in Mehta and Haber 2014; Jasin and Rothstein 2013).

Recently obtained data have shown that neither is the resection step easy, nor is the splitting of labor between NHEJ and HR proteins during DSB repair as restrictive as originally thought. In G1 phase of the cell cycle, Yku70-Yku80 protein inhibits DNA end resection and is stimulated by the MRX complex to channel DSB repair to NHEJ (Chen et al. 2001). In G2 phase, the Yku70-Yku80 protein inhibits resection only when overexpressed, but not when present at physiological levels (Clerici et al. 2008). What will happen if NHEJ, which is dependent on Yku70-Yku80 proteins, does not occur in a timely fashion? The DNA ends are still shielded from exonucleases by the Yku70-Yku80 complex. However, at S/G2, the S267 of Sae2 becomes phosphorylated in a CDK-dependent fashion (Huertas et al. 2008). Phosphorylated Sae2 stimulates the endonucleolytic activity of MRX, and DNA is cleaved nearby DSB stabilized by the Yku70-Yku80 complex. The resultant substrate is suitable for subsequent processing by Exo1 and initiates repair via HR (Shibata et al. 2014; Reginato et al. 2017; Wang et al. 2017b). Thus, in S/G2 cells, Yku70-Yku80 actually stimulates the resection needed for HR.

To acquire more time for DNA repair, regardless of the pathway chosen, cells may activate a DNA damage checkpoint. The damage detection is so sensitive that even a single unrepaired DSB causes yeast cells to arrest through the action of yeast Mec1 and Ddc2 proteins (Rad3 and Rad26 in *Sc. pombe*, respectively) (Paciotti et al. 2000; Melo et al. 2001). Mec1-Ddc2, a key DNA-damage-sensing kinase, is recruited to ssDNA in a RPA-dependent manner (Deshpande et al. 2017). However, this recruitment is not sufficient to activate Mec1. Mec1 activation depends on an additional signal that involves a 9-1-1 complex (composed of Ddc1, Mec3, and Rad17 in budding yeast and of Rad9-Rad1-Hus1 in fission yeast, hence the name of this complex), the replication factor Dpb11, and the DNA helicase/nuclease Dna2 (Melo et al. 2001; Wanrooij and Burgers 2015). The 9-1-1 complex is a replication sliding clamp that functions to recruit the substrates of Mec1. Phosphorylation of Rad9 and the checkpoint effector kinase Rad53 that guide the DDR (Allen et al. 1994; Nielsen et al. 2013), reviewed in Bartek et al. (2001) and Toh and Lowndes (2003), is reduced in 9-1-1 and *rad24Δ* mutants (Emili 1998) and leads to a strong checkpoint defect. The 9-1-1 complex is loaded on DNA by the checkpoint clamp loader known as the Rad24-RFC complex. It can be loaded onto DNA using two different configurations: at the 5′ end of RPA-coated ssDNA or at both the 5′ and 3′ junctions of naked DNA at gaps left on the lagging strand after replication (Majka et al. 2006). This complex subsequently stimulates the Mec1 kinase and/or the checkpoint sensor Dpb11 (Rad4 in *Sc. pombe*), depending on the cell cycle phase (Furuya et al. 2004; Majka et al. 2006a; Navadgi-Patil and Burgers 2009; Puddu et al. 2011). Cell cycle arrest requires, in addition to the 9-1-1 clamp, Rad9 protein (human 53BP1 orthologue, Crb2 in *Sc. pombe*) (Harrison and Haber 2006). Rad9 is hyperphosphorylated by both Mec1 and Tel1 in response to DNA damage (Emili 1998; Vialard et al. 1998). Activated Rad9 stimulates Mec1-dependent phosphorylation of the effector kinases such as Chk1 and Rad53 (Cds1 in *Sc. pombe*) and stimulates trans-autophosphorylation of Rad53 needed for full activation of this protein (Pellicioli et al. 1999; Lee et al. 2003b). Thus, the DNA damage-initiated checkpoint signal is amplified (I) to trigger a G2/M checkpoint arrest prior to mitosis by Pds1 phosphorylation, which inhibits its APC-mediated degradation (Sanchez et al. 1999); (II) to induce Dun1-dependent expression of DNA damage-inducible genes (Bastos de Oliveira et al. 2012); and (III) to regulate ribonucleotide reductase (RNR), which helps to rescue the DNA damage stress and to preserve genome integrity (Sanvisens et al. 2013; Skoneczna et al. 2015).

A second branch of the DNA damage checkpoint involves another phosphoinositide 3-kinase-related kinase (PIKK kinase), Tel1, which associates with DSB ends even in the absence of resection through its association with the MRX complex (Nakada et al. 2003; Falck et al. 2005). The MRX complex and Tel1 play a role in tethering DSB ends together; however, even in their absence, the broken ends mostly remain together (Lee et al. 2008). This pathway depends on the kinase activity of Tel1 and Dun1, and it involves the damage-induced phosphorylation of Sae2 and histone H2AX proteins. Tel1- and Sae2-dependent tethering of DSB ends and their 5′ to 3′ degradation prevents NHEJ and favors HR, thus preserving chromosome integrity despite DNA damage (Lee et al. 2008).

Mec1 is normally responsible for imposing cell cycle arrest after DNA damage during S phase, whereas Tel1 operates more often after replication in G2 phase. However, hyperactive alleles of Tel1 do cause a damage-dependent G2/M arrest when Mec1 is absent (Baldo et al. 2008). In fact, both Mec1 and Tel1 redundantly control DNA repair through the phosphorylation of a number of proteins, which in itself is not surprising taking into account that they phosphorylate the same consensus sequence. Among their targets are proteins that contribute to replisome function and are involved in DNA repair, e.g., Rfa1 and Rfa2, the RPA complex subunits, Mrc1 and Tof1 subunits of replication-pausing checkpoint complex, replicative DNA polymerases, CMG (Cdc42/Mcm2-7/GINS) helicase complex subunits, Ctf4 cohesin, and HR proteins (Hu et al. 2001; Alcasabas et al. 2001; Bartrand et al. 2004; Smolka et al. 2007; Chen et al. 2010; Randell et al. 2010; De Piccoli et al. 2012; Zhou et al. 2016; Willis et al. 2016).

3.2.3   Cellular Response to Replication Block

While analyzing the DSB response in terms of the kinetics of this process, the Mre11 nuclease and the Tel1 kinase are the first proteins detected at DSBs. Next, the Rfa1 ssDNA-binding protein relocalizes to the DNA break and recruits other key checkpoint proteins. Later, the HR machinery assembles at the site, but only in the S and G2 phases (Lisby et al. 2004). However, if the DNA damage is not repaired until the next replication round, it will most likely cause a stall of the replication fork. This situation can be artificially induced by exposing cells to hydroxyurea, a compound that causes replication block. In hydroxyurea-treated cells, the order of events is different. Mre11 and recombination proteins are not recruited to hydroxyurea-stalled replication forks unless the forks collapse (Lisby et al. 2004). Both the DNA damage response (DDR) and DNA replication stress response (DRR) are activated by the sensor kinase Mec1 and converge on the effector kinase Rad53. However, DDR operates throughout the whole cell cycle and depends on the checkpoint mediator Rad9 to activate Rad53, whereas DRR is limited to S phase, and its activation is mediated by Mrc1 and other replisome components transmitting the signal of replication barriers (reviewed recently in Pardo et al. 2017). Both signaling pathways, DDR and DRR, are directed by the MRX complex, which in conjunction with Sae2 detects and processes DNA ends exposed by RPA bound to ssDNA (Lisby et al. 2004). However, the RPA-coated ssDNA, which is necessary to recruit Mec1 via its partner Ddc2 (Rouse and Jackson 2002; Zou et al. 2003) that stimulates 9-1-1 complex loading near the $5'$ junction (Majka et al. 2006), can be naturally found at Okazaki fragments on the lagging strand of the replication fork. Similarly, DDR triggers activation of the 9-1-1 complex via Mec1-dependent phosphorylation (Majka et al. 2006a; Navadgi-Patil and Burgers 2009), which allows recruitment of the replication initiation factor Dpb11 (Puddu et al. 2008; Araki et al. 1995). Dpb11 is in turn responsible for recruitment of the adaptor protein Rad9 during DDR, but during DRR, Dpb4, the nonessential subunit of DNA polymerase ε (Pol ε), promotes activation of Mec1 (Puddu et al. 2011; Navas et al. 1996), yet another non-catalytic Pol ε subunit, Dpb2, has been shown to participate in the DNA replication checkpoint but in a Nrm1-dependent manner (Dmowski et al. 2017). Dpb11 interacts with one additional fork protein that is needed for the S phase checkpoint after the hydroxyurea-induced block, an *ARS* ssDNA-binding protein denoted Sld2 (Kamimura et al. 1998). It also controls the association between Pol α and Pol ε and the *ARS*, which contribute to late origin firing during DRR (Masumoto et al. 2000). A few other replication fork proteins have also been shown to be checkpoint sensors during DRR, including the replisome protein Ctf4, Mcm6, and Sld3 subunits of the CMG helicase complex (Tanaka et al. 2009; Komata et al. 2009; Deegan et al. 2016). Mec1 activation also depends on Dna2, a factor that is involved in the maturation of lagging strand DNA synthesis (Kumar and Burgers 2013). Thus, during DRR, the signal activating S phase arrest may come from various replication fork elements reflecting actual fork problems, but the signal generally derives from the 9-1-1 complex on the lagging strand, whereas it is generated by Pol ε on the leading strand. Subsequently, Ddc1, Dpb11, Dna2, and perhaps other proteins initiate S phase checkpoint signaling by redundantly activating Mec1 kinase (Kumar and Burgers 2013; Wanrooij et al. 2016); reviewed in (Pardo et al. 2017).

The checkpoint signal initiated by the Mec1 activation is then amplified to reach all cellular targets (Table 3.1, Fig. 3.3). The signal is transduced to the effector kinases Rad53 and Chk1, which are transiently recruited to the damage sites in DNA and/or to the stalled replication fork and then released to amplify DDR/DRR throughout the nucleus (Sanchez et al. 1996, 1999; Zhou et al. 2016). When activated, these kinases influence a range of biological processes (Fig. 3.3), including progression of the cell cycle, repression of late-firing replication origins, dNTP pools, transcription of DDR genes, and DNA damage repair and replication resumption, which together ensure the maintenance of genome integrity.

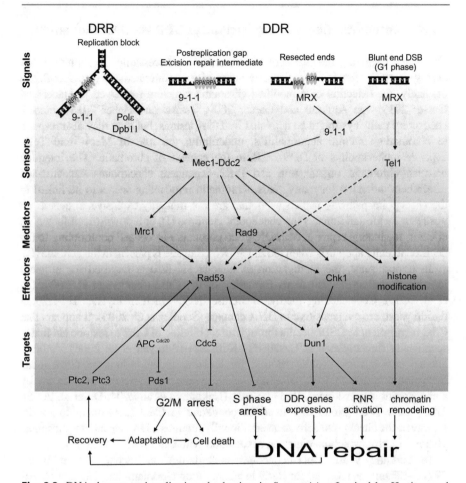

Fig. 3.3 DNA damage and replication checkpoints in *S. cerevisiae*. Inspired by Harrison and Haber (2006) and Reha-Krantz et al. (2011). RPA-coated ssDNA can be generated during slowed replication, replication re-firing, excision repair, or DNA end resection of DSBs. The DNA damage signal triggers the Mec1-Ddc2-dependent DNA damage checkpoint kinase cascade. The blunt-ended DNA activates a DNA damage response through the Tel1 protein kinase and its associated MRX complex. When Mec1 is absent, Tel1 can substitute for it by activating the S phase checkpoint through Rad53 and other kinases. Among important outputs of DNA damage signaling are the histone modifications and chromatin remodeling facilitating DNA repair events, G2/M arrest, and the induction of damage-inducible genes and RNR activation. When damage has been repaired, Rad53 kinase is dephosphorylated, and cells recover from the stress. Even when DNA damage persists, cells may resume the cell cycle and adapt. When damage is too high, apoptosis is executed. The 9-1-1, the complex composed of Ddc1, Mec3, and Rad17 that recognizes the free DNA end adjacent to the ssDNA gap; MRX, the Mre11-Rad50-Xrs2 complex involved in DSB recognition and DNA end resection; Pol ε, DNA polymerase ε; *RNR*, ribonucleotide reductase

3.2.4   Chromatin Remodeling Facilitates DDR via DNA Movement

The signal triggering the checkpoint response is not necessarily the primary DNA damage itself. It often involves the recognition of repair intermediates (see Sect. 3.2.2) or, likely, the detection of a modified chromatin structure (reviewed in Hauer and Gasser 2017; van Attikum and Gasser 2009; Tsabar and Haber 2013). Indeed, checkpoint pathways might be activated by DNA lesions, but they may also respond to chromatin structure abnormalities, underlining the role of Mec1- and Tel1-dependent checkpoints in DDR/DRR in the context of chromatin. The interplay between chromatin organization and PIKK-dependent checkpoints has attracted researchers' attention for many years. Chromatin remodeling seems to be linked to the checkpoint pathway in two ways that are not mutually exclusive. It acts as an upstream factor that initiates the checkpoint response (I), or a downstream factor, an effector, facilitating repair at both DNA and chromatin levels and contributing to the decision to enter apoptosis when too much DNA damage is present in the genome (II).

In fission yeast, the methylation of histone H4-K20 is required for efficient ionizing radiation-induced focus formation of the Crb2 (Sanders et al. 2004; Du et al. 2006). Cells lacking methyltransferase Set9, which methylates H4-K20 in fission yeast, are oversensitive to DNA damage (Sanders et al. 2004). It appears that Crb2 recruitment to DSBs occurs through direct binding of Crb2 to histone H4 that is specifically dimethylated at Lys20 (H4-K20me2) (Botuyan et al. 2006). Another histone modification that is required for ionizing radiation-induced Crb2 focus formation is histone H2A C-terminal phosphorylation at conserved SQE motifs by a checkpoint sensor kinase, Rad3 or Tel1 (Nakamura et al. 2004; Du et al. 2006). Thus, these two histone modifications cooperate in Crb2 recruitment to DSBs. However, the checkpoint activation additionally requires CDK-dependent phosphorylation of Crb2, which mediates an association with Cut5 (Du et al. 2006).

In budding yeast, Dot1-dependent methylation of histone H3 at lysine 79 (H3-K79me) is required for Rad9 localization on chromatin and efficient checkpoint signaling (Wysocki et al. 2005; Grenon et al. 2007). Rad9 interacts with H3-K79me via its Tudor domain, and this interaction is crucial for two independent steps of the DNA damage response: (I) checkpoint activation and (II) DNA repair. However, these functions are separated in time; their role in checkpoint activation is limited to G1 phase, but their role in DNA repair is restricted to S/G2 (Wysocki et al. 2005; Grenon et al. 2007). Moreover, H3-K79me seems to be crucial for proficient nucleotide excision repair (NER) upon UV irradiation. In the H3K79R and H3K4,79R mutants, chromatin is silenced due to increased binding of Sir complexes, which reduces the DNA lesion accessibility to repair enzymes and subsequently impairs NER (Chaudhuri et al. 2009).

Methylation is not the only posttranslational histone modification that contributes to DDR (Rossetto et al. 2010). The others are, e.g., Rad6-dependent histone H2B ubiquitination (Wang et al. 2017a; Robzyk et al. 2000), Mec1-/Tel1-dependent phosphorylation of histone H2A (Downs et al. 2000), or histone H2A, H2B, H3, and H4 acetylation, which occurs in response to DSBs. The yeast NuA4 histone acetyltransferase and INO80 and SWR1, Swi2-family ATP-dependent remodelers, are recruited to DSBs, where they directly interact with histone H2A through their shared subunit Arp4 (Downs et al. 2004; van Attikum et al. 2007). NuA4 is the first of the modifiers to appear at DSBs. It acetylates histones H4 and H2A and promotes relaxation of the chromatin at the DSB (Downs et al. 2004; Bird et al. 2002). In *S. cerevisiae*, NuA4 also acetylates the histone variant H2A.Z (Htz1), which promotes error-prone bypass of MMS-induced DNA lesions in a Rev3-dependent fashion, i.e., it stimulates the trans-lesion synthesis (TLS)-mediated DNA damage tolerance pathway (Renaud-Young et al. 2015). In turn, the acetylation at lysines 9,14 of histone H3 by histone acetyltransferase Gcn5 in response to UV-damaged DNA contributes to NER (Waters et al. 2015). Among the factors influencing the Gcn5-mediated histone modification, the complex of NER proteins, Rad16-Rad7, with Abf1, the NER-specific SWI/SNF factor, and the yeast histone variant of H2AZ (Htz1) have been indicated (Waters et al. 2015; Reed et al. 1999).

Various chromatin-remodeling complexes are promptly recruited to DSBs in budding yeast. One of them, the RSC complex, stimulates the phosphorylation of histone H2A, which spreads far away from DSBs and enables the recruitment of two other complexes, namely, INO80 and SWR1, to the damage site in DNA. All three chromatin-remodeling complexes contribute to the processing of DSBs. The SWR1 complex catalyzes the replacement of H2A with its variant H2A.Z (Htz1) (Mizuguchi et al. 2004), which is required for DSB relocation to the nuclear periphery (Horigome et al. 2014). In both budding and fission yeast, this process is needed for future repair and is determined by the nuclear periphery region, to which DSB was relocated. During the S/G2 phase of the cell cycle, DSBs are often anchored to the nuclear envelop protein Mps3, and their repair also requires resection, Rad51-, INO80-, Siz1-dependent SUMO modification of H2A.Z, and DNA-damage checkpoint activation (Kalocsay et al. 2009; Oza et al. 2009; Swartz et al. 2014). Persistent DSBs, including those generated at collapsed replication forks or in subtelomeric regions, are anchored to a nuclear pore complex (Nagai et al. 2008; Horigome et al. 2014; Therizols et al. 2006). Their repair at this location is independent of the cell-cycle stage, requiring neither the recombinase Rad51 nor INO80 and is independent of extensive resection (Horigome et al. 2014). Notably, the SWR1 chromatin remodeler and its transient deposition of H2A.Z at breaks contribute to the peripheral relocation to either site of anchorage; however, the outcomes of relocation are distinct. Mps3 appears to help suppress illegitimate recombination, while the nuclear pore complex promotes alternative repair pathways, such as template switching at a broken replication fork or break-induced replication (Horigome et al. 2016).

In response to DNA damage, the RSC and INO80 chromatin remodelers are responsible for nucleosome eviction. The cellular levels of histones drop 20–40% in response to DNA damage (Hauer et al. 2017). The nucleosome eviction stimulates the repair of different DNA lesions, such as DNA breaks, UV-induced lesions, and oxidative or alkylation damage. For example, CPDs, 6-4PPs, and N-methylpurine are repaired at faster rates in nucleosome-free regions and in the linker DNA than in the nucleosome core (Smerdon and Thoma 1990; Wellinger and Thoma 1997; Tijsterman et al. 1999; Powell et al. 2003; Ferreiro et al. 2004). In accordance with these data, the RSC complex promotes efficient base excision repair (BER) in chromatin after methyl methanesulfonate treatment (Czaja et al. 2014), whereas INO80 is recruited to the UV-induced DNA lesions, where it interacts with NER proteins and contributes to DNA repair (Sarkar et al. 2010). The INO80 complex also facilitates efficient homologous recombination upon DNA damages (Kawashima et al. 2007). Histone loss from chromatin occurs in a proteasome-mediated fashion, favoring the initiation of DNA end resection and affecting the efficiency of loading the Yku70-Yku80 complex, resulting in enhanced recombination rates. It has been proposed that a generalized reduction in nucleosome occupancy is an integral part of the DNA damage response in yeast. It increases chromatin mobility and facilitates identification of the homologous DNA sequence needed for DSB repair via HR (Hauer et al. 2017). While all three chromatin remodelers, RSC, INO80, and SWR1, affect the efficiency of yeast Yku70-Yku80 complex loading, only SWR1 seems to affect NHEJ (van Attikum et al. 2007). Interestingly, nucleosome eviction at a DSB is delayed in a SWI/SNF mutant (Wiest et al. 2017). Moreover, SWI/SNF is required for recruitment of the MRX complex at DSB ends, thereby contributing to the timely initiation of DNA end resection during HR (Wiest et al. 2017).

Another chromatin remodeling factor that is recruited to DSBs is Fun30. Similar to INO80 and RSC, Fun30 promotes the repair of these lesions by facilitating end resection; however, in contrast to them, Fun30 does not cause histone eviction. In budding yeast, Fun30 is cell cycle regulated. Its abundance peaks in the G2/M phase of the cell cycle, so its role in the DNA damage response is discernible mainly in this phase (Siler et al. 2017). When phosphorylated by CDK1, it can be recruited to chromatin via interacting with the already DSB-localized Dpb11 and form a ternary complex with the 9-1-1 complex, which is required for efficient long-range resection of DSBs (Bantele et al. 2017; Eapen et al. 2012). Fun30 is not required for later steps in HR, but like its homolog Rdh54, it is required to allow the adaptation of DNA damage checkpoint-arrested cells with persistent DSBs to resume cell cycle progression (Eapen et al. 2012). In contrast to Fun30, Rdh54, the chromatin remodeler with translocase activity on dsDNA, plays a role in HR. It has been suggested that Rdh54 can act in the avoidance of replication fork pausing by removing certain proteins from dsDNA, as has been shown for Rad51 (Chi et al. 2006). Thereby, it regulates restart at replication block sites by employing one of the two post-replicative repair sub-pathways that are dependent on the t helicase Rad5, fork regression, or template switching or by using the Rad5-independent but Rdh54-dependent HR-mediated pathway promoting DNA strand invasion (Syed et al. 2016; Nimonkar et al. 2012).

3.2.5   DDR/DRR Proteins Act in Repair Foci

One of the hallmarks of DDR and DRR is protein relocalization (Lisby et al. 2004; Tkach et al. 2012). Many proteins that respond to DNA lesions concentrate at the damaged site into microscopically detectable foci (Lisby et al. 2001; Gasior et al. 1998; Agarwal and Roeder 2000; Saugar et al. 2017; Matsumoto et al. 2005; Yimit et al. 2015; Hombauer et al. 2011). The local concentration of these proteins increases the chances for interactions between them and with the DNA lesion, thus increasing the effectiveness of responses to DNA damage. Analyses of cells suffering multiple DSBs demonstrated that a single DNA repair focus is able to simultaneously recruit more than one DSB (Lisby et al. 2003). Moreover, data suggest that the co-localization of DDR/DRR proteins rather than DNA damage per se is critical for DNA damage signaling (Bonilla et al. 2008).

3.2.5.1 DSB Repair

The proteins that are initially recruited to the DNA lesion determine the later steps of DNA repair. For example, DSBs are repaired by either the NHEJ or HR pathway, as comprehensively reviewed in Emerson and Bertuch (2016) and Mehta and Haber (2014). Which of these two repair pathways will be used is determined in the very first repair step—DNA end resection. The generation of 3′ ssDNA overhangs at the break that are initiated by the action of the MRX complex and Sae2 directs repair toward HR. Extension of the ssDNA region in the next resection step relies on the exonucleases Exo1 and/or Dna2 or Sgs1 (Lydeard et al. 2010; Ngo et al. 2014). The RPA complex binds to these ssDNA overhangs and, in concert with Rad52, helps to load Rad51 to form a ssDNA-Rad51 nucleoprotein filament, i.e., an invasive strand. The structure of the DNA surrounding the DSB is modified not only due to histone modifications or even its eviction but also by enrichment of cohesion in this region. Interestingly, even a single DSB can reactivate cohesion in a genome-wide manner (Ström et al. 2007). Cohesin is recruited to DSBs and stalls replication forks in a checkpoint-related fashion, where it contributes to DSB repair (Unal et al. 2007; Ström et al. 2007; Heidinger-Pauli et al. 2009). The Rad51 nucleofilament searches for a homologous DNA sequence, e.g., on the sister chromatid or a homologous chromosome. The finding of a homologous sequence will result in strand invasion, followed by DNA synthesis, exchange of the copied genetic information and resolution of the final DNA product. The last step is a ligation step, which completes this error-free repair event. The key players in HR are members of the Rad52 epistasis group [e.g., Rad51, Rad52, Rad55, Rad57, etc.; reviewed in Symington (2002)], structure-selective endonucleases [e.g., a Holliday junction resolvase Sgs1-Top3-Rmi1, Yen1, Mms4-Mus81, and Slx1-Slx4 or other structure-selective endonucleases as a ssDNA endonuclease Rad1-Rad10; reviewed in Wyatt and West (2014)], or structure-selective enzymes (DNA helicases Mph1, Srs2, and Pif1 or ds/ssDNA junction recognition complex Msh2-Msh3), DNA replicases (DNA polymerase δ or ε), and DNA ligase, as reviewed in Mehta and Haber (2014) and Skoneczna et al. (2015). Interestingly, some of the enzymes mentioned herein that might be expected to work interchangeably have recently been shown to co-localize in the same repair foci, e.g., Mus81-Mms4, Rad1-Rad10, and Slx1-Slx4 (Saugar et al. 2017). These results shed new light on how the repair pathways work in vivo, emphasizing that HR is not a uniform pathway. Depending on the proteins that are involved in the repair, various scenarios may come into play with different outcomes, e.g., products with a crossover or without it (see Fig. 3.4 for details). Moreover, in contrast to common belief, HR may even lead to an error-prone product, as in the case of break-induced replication or single-strand annealing.

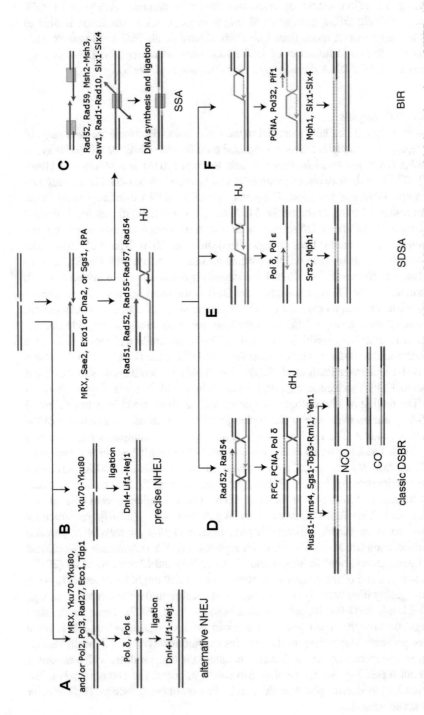

Fig. 3.4 The DSB repair pathways in budding yeast. Inspired by Mehta and Haber (2014). (**a**) Nonhomologous end joining sub-pathways (NHEJ). (**a**) Precise NHEJ—ligation of unperturbed DNA ends. (**b**) Aberrant NHEJ—MRX-dependent repair after initial end resection. (**c**–**f**) Homologous recombination (HR) sub-pathways. (**c**) SSA single-strand annealing (gray squares indicate repeated sequences in DNA). (**d**) DSBR classical double-strand break repair, the main

NHEJ is initiated when the ends of a DSB are bound by the Yku70-Yku80 heterodimer. This complex holds the DNA ends together to enable their religation by Dnl4-Lif1, a DNA ligase IV complex (Wilson et al. 1997). Dnl4 interacts with Lif1 (human XRCC4 ortholog) via BRCT domains in its C-terminus, and this interaction is required for the stabilization and activation of Dnl4 (Doré et al. 2006; Zhang et al. 2007). Dnl4 and Lif1 both interact with the Nej1 protein, which also contributes to NHEJ (Deshpande and Wilson 2007). Since most DSBs occurring in vivo have terminal structures that cannot be directly ligated, binding of the Yku70-Yku80 complex to DNA ends is sometimes accompanied by short processing of the DNA ends by the MRX complex. In such cases, Dnl4-Lif1 seals the break when ssDNA overhangs anneal at DNA regions containing the microhomology sequence, resulting in error-prone DSB repair (Chen et al. 2001). This process requires the $5'$ flap endonuclease Rad27 (orthologue of human FEN1) and DNA polymerase δ or ε, which fill the gap that can form by imprecise pairing of overhanging $3'$ DNA ends (Tseng and Tomkinson 2004; Galli et al. 2015). Other DNA polymerases are also involved in the imprecise NHEJ, each with a distinct role in the process. For instance, the proofreading activity of Pol2, a catalytic subunit of DNA polymerase ε, is important for the recession of $3'$ flaps that can form during imprecise pairing, whereas Pol3, a catalytic subunit of DNA polymerase δ, is required for NHEJ-mediated chromosomal rearrangements, and Pol1 modulates both imprecise end joining and more complex chromosomal rearrangements (Tseng et al. 2008) (see Fig. 3.4). In addition to Pol2, Pol3, and Rad27, several other factors assist in the end processing at DNA breaks in the budding yeast during NHEJ, including the $5'$ exonuclease Exo1 and the tyrosyl DNA phosphodiesterase Tdp1. Tpd1 regulates the accuracy of NHEJ repair junction formation (Bahmed et al. 2010). Exo1 likely supports strand annealing in NHEJ by exposing short microhomologous sequences and reversing nucleotide additions due to DNA synthesis (Bahmed et al. 2011), despite its well-known function in long-range resection in HR (Zhu et al. 2008).

3.2.5.2 Excision Repair Pathways

All excision repair pathways involve the recognition of DNA damage, dual incisions of the DNA phosphodiester backbone in the damaged DNA strand, excision of the DNA lesion, filling of the gap by DNA polymerase using the complementary DNA strand as a template, and finally a ligation step and restoration of the chromatin structure. The repair pathway used in a given condition is determined by the type of DNA lesion, stage of chromatin in the neighborhood of the lesion, and accessibility of DNA repair protein.

The nucleotide excision repair pathway is specialized to remove DNA lesions that distort the DNA helix (Figs. 3.2 and 3.5), e.g., 6-4PPs and CPDs that arise in DNA

Fig. 3.4 (continued) path of DSB repair by HR, associated with the formation of double-stranded structures, the solution of which may result in crossover (CO) or noncrossover (NCO) products. (**e**) *SDSA* synthesis-dependent strand annealing. (**f**) *BIR* break-induced replication, *HJ* Holliday junction, *dHJ* double Holliday junction, *MRX* Mre11-Rad50-Xrs2-complex

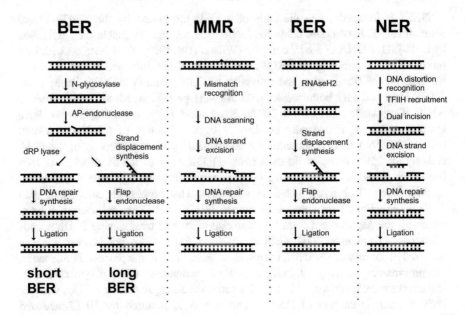

Fig. 3.5 The excision repair pathways in budding yeast. Excision repair pathways are involved in the removal of DNA damage or DNA synthesis errors and in the restoration of the original DNA sequence. The base excision repair (BER) removes a chemical modification of bases that do not distort the DNA double helix. The nucleotide excision repair removes DNA lesions that distort the DNA helix. The mismatch repair removes mis-paired deoxyribonucleotides. The ribonucleotide excision repair removes ribonucleotides incorporated into DNA. *BER* base excision repair, *NER* nucleotide excision repair, *MMR* mismatch repair, *RER* ribonucleotide excision repair, *short BER* short-patch BER, *long BER* long-patch BER

upon UV irradiation (Thoma 1999), DNA inter-strand cross-links (ICL), or DNA adducts formed upon treatment with cross-linking agents, such as psolarens or cis-platinum (Lehoczký et al. 2007). However, NER also contributes to the removal of abasic sites (AP site) in yeast (Torres-Ramos et al. 2000). At this point, it should be mentioned that the photoproducts might be alternatively repaired by a photolyase in a photoreactivation process, i.e., light-dependent direct reversal of damage (Sancar et al. 2000; Guintini et al. 2015).

When a bulky lesion appears in DNA, e.g., following UV irradiation, early-acting NER factors, i.e., damage-binding proteins, such as the Rad4-Rad23 complex and RPA, are involved in initiation of the repair process (reviewed in Prakash and Prakash 2000). Then, Rad14 binds the damage site as a damage recognition factor, the prerequisite to recruit the Rad1-Rad10 endonuclease complex. The latter complex functions in a downstream step incising DNA 5′ to the site of DNA damage (Mardiros et al. 2011; Guzder et al. 2006). In addition to the damage-binding factors, the reaction requires Rad3 and Rad25, subunits of both a transcription factor TFIIH and a nucleotide excision factor 3 (NEF3), which possess DNA helicase activity. DNA unwinding is essential to create a bubble structure during NER. Following bubble formation, two endonucleases, the Rad1–Rad10 complex and Rad2, incise the damaged DNA strand on the 5′- and 3′-side of the lesion, respectively. The Rad7–Rad16-Elc1 complex stimulates the incision reaction. Thus, during eukaryotic NER, DNA is incised on both sides of the lesion, resulting in the removal of approximately 25–30 nucleotides from the damaged strand. This process is followed by filling of the gap via DNA synthesis and the final step of repair, ligation (reviewed in Prakash and Prakash 2000). It should be mentioned here that NER is believed to work slightly differently when DNA damage occurs in an actively transcribing chromatin region. Therefore, two sub-pathways of NER are recognized: a global genome repair (GGR) and a transcription coupled repair (TCR) where lesions in the transcribed strands are processed (Tornaletti et al. 1999; Woudstra et al. 2002; Li and Smerdon 2004; reviewed in Svejstrup 2002; Spivak 2016). There are some differences between the GGR and TCR pathways, e.g., during the DNA damage recognition step. While TCR involves the DNA-dependent ATPase Rad26 and RNA polymerase subunits Rpb4 or Rpb9, GGR requires Rad4, Rad16, and Rad7 DNA-binding proteins for this step (reviewed in Svejstrup 2002; Ataian and Krebs 2006).

The base excision repair pathway is specialized to remove DNA lesions resulting from oxidation, alkylation, and deamination of bases in DNA (Figs. 3.2 and 3.5). The chemical modification of bases results in a minimal, if any, distortion of the DNA double helix. However, such damage frequently leads to spontaneous mutations (Thomas et al. 1997; Kunz et al. 1994; Baute and Depicker 2008; Eyler et al. 2017; Bauer et al. 2015). The bases have many potential modification sites, so a variety of DNA damage types may appear upon exogenous or endogenous stress (Evans et al. 2004; Fu et al. 2012; Jena 2012). BER is initiated by the recognition and hydrolysis of the damaged base by a DNA glycosylase leaving an AP site, cleavage of the resulting AP site, cleaning-up of the DNA end to enable filling the DNA gap by DNA polymerase to replace the excised nucleotide, and sealing of the remaining DNA nick by DNA ligase 1 (reviewed in Svilar et al. 2011; Boiteux and Jinks-Robertson 2013; Gros et al. 2002). In *S. cerevisiae*, the sequential repair steps are executed by a set of enzymes that include some specialized DNA N-glycosylases (e.g., 3-methyl-adenine DNA glycosylase Mag1, uracil-DNA glycosylase Ung1, N-glycosylase/AP-lyase Ogg1 excising 8-oxo-G located opposite cytosine or thymine in DNA, or DNA N-glycosylase/AP-lyase Ntg1 or Ntg2), AP-endonucleases and 3′-phosphodiesterases (Apn1, Apn2), and the 3′-flap endonuclease Rad1-Rad10, which are involved in the removal of 3′-blocked ends, a structure that is specific to endonuclease Rad27 involved in the repair of 5′-blocked ends, DNA 3′- phosphatase Tpp1, DNA polymerases ε or δ, and ligase Cdc9 (Wu et al. 1999; Unk et al. 2000; Boiteux et al. 2002; Boiteux and Guillet 2004). Depending on the type of initial lesion and recruitment of lesion enzymes, BER can be completed by either short-patch or long-patch pathways. The difference between these two alternative pathways is the size of the patch being repaired, which is a single nucleotide during a short-patch BER (known also as single-nucleotide BER) and two or more nucleotides during a long-patch BER (reviewed in Boiteux and Jinks-Robertson 2013; Dogliotti et al. 2001; Fortini and Dogliotti 2007; Bauer et al. 2015).

The ribonucleotide excision repair pathway is used to remove ribonucleotides from DNA (Figs. 3.2 and 3.5). This type of replication error is very common, occurring on average once per 2 kb of newly synthesized DNA (Sparks et al. 2012). Ribonucleotides are incorporated into DNA due to a high cellular excess of ribonucleotide triphosphates in comparison to deoxyribonucleotide triphosphates leading to misinsertions, even though their incorporation is partially limited by the selectivity properties of DNA polymerases. The first step of RER requires RNase H2, the same enzyme that is used to remove RNA primers from Okazaki fragments. RNase H2 incises the ribonucleotide, followed by strand displacement synthesis of DNA by DNA polymerase δ or ε. Then, the ssDNA fragment is excised by the flap endonuclease Fen1, or Exo1, and finally repair is completed by DNA ligase I (reviewed by Sparks et al. 2012; Vaisman and Woodgate 2015). The PCNA clamp and its loader RFC also participate in RER. PCNA interacts with all proteins engaged in RER, giving them access to the replication fork (Sparks et al. 2012).

The mismatch repair pathway can correct replication errors such as DNA mis-pairs resulting from insertions of improper nucleotides during DNA synthesis, or insertions or deletions, leading to the formation of hairpin-like structures that result from either primer-template misalignment during replication or DNA polymerase slippage on repetitive sequences in DNA (reviewed in Bebenek and Kunkel 2000; Lehner et al. 2012; Skoneczna and Skoneczny 2017) (Figs. 3.2 and 3.5). In *S. cerevisiae*, for recognition of structural abnormalities in the DNA helix caused either by mis-paired bases or hairpin structures, specialized protein complexes, Msh2-Msh3 or Msh2-Msh6, are responsible, respectively (reviewed in Spampinato et al. 2009). The asymmetric binding of Mlh1-Pms1, Mlh1-Mlh2, or Mlh1-Mlh3 complexes marks the nascent versus template DNA strands, as the nascent strand is recognized by its discontinuity associated with DNA replication. Then, the distortion-containing segment of the nascent strand can be removed by a flap endonuclease, e.g., Exo1. The DNA gap is filled up by a DNA polymerase using the original template, and the DNA is sealed. MMR also requires the participation of RPA coating ssDNA, the PCNA sliding clamp providing access to DNA for all proteins involved in MMR, and a PCNA loader, the RFC complex.

It should be mentioned here that Msh2-Msh6 complexes and all three Mlh1-heterocomplexes are involved in the resolution of HR intermediates since they bind to Holliday junctions with high affinities, thereby contributing to the reciprocal meiotic recombination (Marsischky et al. 1999; Argueso et al. 2003). Moreover, because the Msh2-Msh6 complex efficiently recognizes and binds not only to mis-paired bases but also, e.g., the 8-oxo-G mis-paired with A or C, MMR is also involved in the repair of oxidative DNA damage. Moreover, data suggest that MMR is the major mechanism by which the misincorporation of A opposite 8-oxo-G is corrected (Ni et al. 1999). In fact, cells seem to contain more overlapping and redundant oxidative damage recognition systems. In addition to BER, NER, recombination, and lesion bypass pathways have also been implied in the removal of, or tolerance to, oxidative DNA damage in *S. cerevisiae* (Doetsch et al. 2001). Thus, different repair pathways, with overlapping specificities, may contribute to the repair of the same DNA lesion (Fig. 3.2).

3.2.6   Stress Adaptation and Recovery

After successful repair, the checkpoint response is deactivated. This process is known as "recovery." However, the checkpoint response may also be deactivated despite some persistent DNA damage. This is referred to as "adaptation." The ability of yeast cells to recover from, or adapt to, the checkpoint arrest suggests that the arrest signal may be deactivated in different manners and, potentially, with different outcomes.

Several mechanisms contribute to checkpoint recovery:

(I) Completion of the Repair
The removal of DNA lesions turns off the signal that activates checkpoint arrest. However, because the ssDNA regions possess huge checkpoint activation potential, it is crucial to limit their size and number in the genome to recover from the checkpoint arrest. For this reason, Mec1-dependent phosphorylation of Sae2, which confines its nuclease activity in DNA end resection, thereby limiting ssDNA formation (Baroni et al. 2004; Clerici et al. 2006), and the stimulation of Srs2-dependent disassembly of Rad51 filaments, leading to a decomposition of structures that stabilize ssDNA (Vaze et al. 2002; Yeung and Durocher 2011), both contribute to the checkpoint recovery.

(II) Disassembly of the Repair Foci
Since different DNA repair pathways employ distinct DNA-binding structures to recognize the damage site and/or to gain access to lesions in DNA, the mechanisms of repair focus decomposition are distinct. The Rad51 filament that recruits other HR components is removed from DNA by the Rad51 translocase Srs2 (Krejci et al. 2003; Veaute et al. 2003). Serving as a platform for many DNA repair proteins, the DNA sliding clamps, PCNA, and the 9-1-1 complex are unloaded from DNA by their respective clamp loaders, one of the RFC-like complexes (reviewed by Skoneczna et al. 2015).

(III) Inactivation Of Checkpoint Activators
Since many components of the DNA damage checkpoint are phosphorylated upon DNA damage, the recruitment of specific protein phosphatases to dephosphorylate and subsequently deactivate checkpoint effectors, transducers, and targets is a rational evolutionary solution (Leroy et al. 2003; O'Neill et al. 2007; Keogh et al. 2006), reviewed in Heideker et al. (2007). Thus, phosphatases play an important role in checkpoint inactivation. In fission yeast, Dis2 phosphatase controls the timing of checkpoint recovery by abrogating Chk1 phosphorylation, which inactivates this kinase (den Elzen and O'Connell 2004). The budding yeast protein phosphatases Ptc2, Ptc3, Pph3, and Glc7 have been implicated in dephosphorylating Rad53, the protein kinase crucial for the DNA damage response, upon various genotoxic stresses (Leroy et al. 2003; Smolka et al. 2006; Guillemain et al. 2007; Bazzi et al. 2010), reviewed in Travesa et al. (2008). Interestingly, a nucleosome assembly factor, Asf1, was recently shown to facilitate complete Rad53 dephosphorylation following DSB repair (Tsabar et al. 2016). Phosphatases also contribute to the histone modification

stage. Pph3 and Glc7 phosphatases counteract histone H2AX phosphorylation on serine 129 in *S. cerevisiae* cells, enabling recovery from DSB and hydroxyurea-induced stress, respectively (Keogh et al. 2006; Vázquez-Martin et al. 2008; Bazzi et al. 2010). Inactivation of checkpoint activators does not necessarily have to be carried out by the simple reversion of activation events. Phosphorylation of checkpoint activation proteins can also be inactivated by proteolysis (Mailand et al. 2006; Mamely et al. 2006; Peschiaroli et al. 2006).

(IV) Links to Other Cellular Pathways

Many checkpoint targets, such those involved in cell cycle regulation, the stress response, or transcriptional control, are subject to multiple regulators that may indirectly favor or oppose the action of the DNA damage checkpoint. Transient upregulation of DNA damage response genes, the expression of which depends, among others, on the Swi6 transcription cofactor, can be turned off by the presence of transcription repressors such as Nrm1 or Whi5 (Travesa et al. 2013) or by the sequestration of Swi6 from the nucleus (Atencio et al. 2014). The dNTP abundance is also increased after activation of the DNA damage checkpoint because excess dNTPs stimulate DNA repair synthesis. This increase is due to increased ribonucle-otide reductase complex (RNR) activity upon checkpoint activation. Since RNR is a very tightly regulated enzyme and its activity depends on a range of factors, it is under feedback control. Several mechanisms contribute to this control, such as transcriptional changes, subunit localization control, cofactors and prosthetic group availability, and allosteric regulation, among others (reviewed in Skoneczna and Skoneczny 2017; Woolstencroft et al. 2006).

As one can expect, some mutations prevent checkpoint recovery, e.g., budding yeast *srs2Δ*, *sae2Δ* or *ptc3Δ*, and *ptc2Δ* (Vaze et al. 2002; Clerici et al. 2006; Leroy et al. 2003). Other mutations slow down recovery, e.g., *rad51Δ* (Vaze et al. 2002) or *ckb1Δ* and *ckb2Δ*, mutants lacking regulatory subunits β of casein kinase (Guillemain et al. 2007). In contrast, various adaptation-defective mutants, such as *yku70Δ*, *rdh54Δ*, or the conditional mutant in Polo-like kinase *cdc5-ad*, are recovery proficient (Lee et al. 2001; Vaze et al. 2002; Toczyski et al. 1997). Interestingly, *S. cerevisiae* deleted for genes encoding the Golgi-associated retrograde protein transport (GARP) complex are both adaptation- and recovery-defective (Eapen and Haber 2013) due to mislocalization of the mitotic regulator securin Pds1 to the vacuole in GARP mutants such as *vps51Δ*. Pds1 then undergoes degradation by the vacuolar protease Prb1. In addition, in GARP mutants, the separase Esp1 is excluded from the nucleus, which leads to pre-anaphase cell cycle arrest (Eapen and Haber 2013).

As we have already mentioned, cells that suffer from persistent DNA damage, after long-lasting checkpoint arrest (more than 12 h), may finally adapt and reenter the cell cycle. Studies on adaptation have revealed numerous proteins that allow checkpoint inactivation, including proteins that function in DSB repair or chromatin remodeling, for instance, Yku70 and Yku80, Sae2, Srs2, Rad51, Rpd3, Rdh54, and Fun30, and proteins involved in checkpoint control, such as the phosphatases Ptc2 or Ptc3, Polo-like kinase Cdc5, or Ckb1and Ckb2 subunits of casein kinase (Toczyski et al. 1997; Rawal et al. 2016; Mersaoui et al. 2015; Tao et al. 2013; Eapen et al.

2012; Clerici et al. 2006; Lee et al. 1998, 2001, 2003a; Leroy et al. 2003; Vaze et al. 2002). Recent data have also shown that posttranslational modification of the RPA subunit Rfa2 N-terminus determines whether cells proceed into mitosis when unrepaired DNA lesions are sustained in their genome (Ghospurkar et al. 2015). Authors refer to adaptation as to a "last-resort" mechanism for cell survival. Indeed, it is a last chance to survive; however, the costs of choosing to escape from checkpoint arrest route are high. Unfortunately, adaptation promotes the mis-segregation of chromosomes, leading to increases in chromosome loss and translocations events and thus causing genome instability (Galgoczy and Toczyski 2001; Kaye et al. 2004). Why is that? When cells with unrepaired DNA damage escape cell cycle arrest, they may take advantage of the opportunity offered by the next phase of the cell cycle, i.e., to use other pathways to repair the DNA damage. However, there is also the risk that unrepaired DNA damage may be converted to DSBs during the next replication round. If DSBs remain unrepaired until the next mitosis, mis-segregation may occur (Aguilera and Gómez-González 2008; Kaye et al. 2004; Sandell and Zakian 1993; Rawal et al. 2016).

3.3   Conclusion and Perspectives

Eukaryotic cells, including yeast and other fungi, are, like any other cells, constantly at risk of DNA damage. Unfortunately for them as our "guinea pig," but beneficial to us, evolution has conserved most of the mechanisms sustaining the functional genome. Thus, the yeast has become a favorite model organism employed in genome maintenance research. The range of uses of *S. cerevisiae* and *Sc. pombe* cells is extraordinary, as exemplified, for instance, by the long list of toxic compounds that have been tested on yeast (Kemmer et al. 2009; Lum et al. 2004; Parsons et al. 2004, 2006). Moreover, yeast cells are amenable to genetic manipulation, their genome sequence has been known for many years, and the available collections of yeast deletion mutants have been utilized in a range of genome-wide approaches, including those linked to the genome stability topic (Yuen et al. 2007; Kanellis et al. 2007; Alabrudzinska et al. 2011; Stirling et al. 2011; McKinney et al. 2013; Krol et al. 2015; Zheng et al. 2016). Thus, a wealth of knowledge has been gathered with the help of this organism. Fission yeasts have been invaluable in studies on the functioning of the cell cycle and the control of cell division. Research in budding yeast, in turn, has laid the foundation for the entire field of research regarding replication control, cellular responses to genotoxic stress, and the replication block; elucidated the rules that guide the choice of DNA repair pathway and revealed the molecular mechanisms of the repair pathways; and helped to uncover the mechanisms that lead to genome instability.

Most importantly, due to the conserved nature of genome maintenance mechanisms, all knowledge collected using yeast or fungi may be applied to mammalian cells, which is particularly convenient because some human diseases, e.g., cancer, are linked to genome instability. Understanding the mechanisms that lead to genetic instability in yeast may thus shed some light on disease development and may also help to find appropriate genetic markers that allow disease identification at an early stage, hopefully also helping to identify remedies. Currently, yeasts are more often used as a physiological model of the "disease-affected cell." In such studies, the mutation in the human gene linked to a certain disease is recreated in its yeast homologue. The changes in cell phenotypes and possible alterations in various biochemical pathways are then analyzed, allowing resolution of the molecular mechanism and physiological outcomes of the mutated gene, as well as screening for chemicals that may potentially cure the effects of this mutation. Understanding the molecular basis of the disease and limiting the number of chemicals that may be used during further experiments in mammals shortens the path to the development of an effective medicine. Yeasts also frequently serve as a cellular model to test the response mechanisms to various therapeutics and in cytotoxicity tests.

Thus, the contribution of yeast to human knowledge and well-being is great. From a research perspective, we can only appreciate one additional bright site of yeast research, it does not burden the conscience as much as when working with higher Eukaryota. We also appreciate that yeast model cells smell nice in the laboratory, recalling pleasant aspects of life such as bread. Thus, let us learn from yeast as much as possible!

Acknowledgments This work was supported by Polish National Science Center grant 2016/21/B/NZ3/03641.

References

Adrover MÀ, Zi Z, Duch A, Schaber J, González-Novo A, Jimenez J, Nadal-Ribelles M, Clotet J, Klipp E, Posas F (2011) Time-dependent quantitative multicomponent control of the G_1-S network by the stress-activated protein kinase Hog1 upon osmostress. Sci Signal 4:ra63. https://doi.org/10.1126/scisignal.2002204

Agarwal S, Roeder GS (2000) Zip3 provides a link between recombination enzymes and synaptonemal complex proteins. Cell 102:245–255

Aguilera A, Gómez-González B (2008) Genome instability: a mechanistic view of its causes and consequences. Nat Rev Genet 9:204–217. https://doi.org/10.1038/nrg2268

Alabrudzinska M, Skoneczny M, Skoneczna A (2011) Diploid-specific genome stability genes of *S. cerevisiae*: genomic screen reveals haploidization as an escape from persisting DNA rearrangement stress. PLoS One 6:e21124. https://doi.org/10.1371/journal.pone.0021124

Alcasabas AA, Osborn AJ, Bachant J, Hu F, Werler PJ, Bousset K, Furuya K, Diffley JF, Carr AM, Elledge SJ (2001) Mrc1 transduces signals of DNA replication stress to activate Rad53. Nat Cell Biol 3:958–965. https://doi.org/10.1038/ncb1101-958

Allen JB, Zhou Z, Siede W, Friedberg EC, Elledge SJ (1994) The SAD1/RAD53 protein kinase controls multiple checkpoints and DNA damage-induced transcription in yeast. Genes Dev 8:2401–2415

Anda S, Zach R, Grallert B (2017) Activation of Gcn2 in response to different stresses. PLoS One 12:e0182143. https://doi.org/10.1371/journal.pone.0182143

Andaluz E, Ciudad A, Rubio Coque J, Calderone R, Larriba G (1999) Cell cycle regulation of a DNA ligase-encoding gene (CaLIG4) from *Candida albicans*. Yeast Chichester Engl 15:1199–1210. https://doi.org/10.1002/(SICI)1097-0061(19990915)15:12<1199::AID-YEA447>3.0.CO;2-S

Araki H, Leem SH, Phongdara A, Sugino A (1995) Dpb11, which interacts with DNA polymerase II(epsilon) in *Saccharomyces cerevisiae*, has a dual role in S-phase progression and at a cell cycle checkpoint. Proc Natl Acad Sci USA 92:11791–11795

Argueso JL, Kijas AW, Sarin S, Heck J, Waase M, Alani E (2003) Systematic mutagenesis of the *Saccharomyces cerevisiae* MLH1 gene reveals distinct roles for Mlh1p in meiotic crossing over and in vegetative and meiotic mismatch repair. Mol Cell Biol 23:873–886

Ataian Y, Krebs JE (2006) Five repair pathways in one context: chromatin modification during DNA repair. Biochem Cell Biol Biochim Biol Cell 84:490–504. https://doi.org/10.1139/o06-075

Atencio D, Barnes C, Duncan TM, Willis IM, Hanes SD (2014) The yeast Ess1 prolyl isomerase controls Swi6 and Whi5 nuclear localization. G3 Bethesda Md 4:523–537. https://doi.org/10.1534/g3.113.008763

Aylon Y, Kupiec M (2005) Cell cycle-dependent regulation of double-strand break repair: a role for the CDK. Cell Cycle Georget Tex 4:259–261

Bahmed K, Nitiss KC, Nitiss JL (2010) Yeast Tdp1 regulates the fidelity of nonhomologous end joining. Proc Natl Acad Sci USA 107:4057–4062. https://doi.org/10.1073/pnas.0909917107

Bahmed K, Seth A, Nitiss KC, Nitiss JL (2011) End-processing during non-homologous end-joining: a role for exonuclease 1. Nucleic Acids Res 39:970–978. https://doi.org/10.1093/nar/gkq886

Baldo V, Testoni V, Lucchini G, Longhese MP (2008) Dominant TEL1-hy mutations compensate for Mec1 lack of functions in the DNA damage response. Mol Cell Biol 28:358–375. https://doi.org/10.1128/MCB.01214-07

Bantele SC, Ferreira P, Gritenaite D, Boos D, Pfander B (2017) Targeting of the Fun30 nucleosome remodeller by the Dpb11 scaffold facilitates cell cycle-regulated DNA end resection. eLife 6. https://doi.org/10.7554/eLife.21687

Baroni E, Viscardi V, Cartagena-Lirola H, Lucchini G, Longhese MP (2004) The functions of budding yeast Sae2 in the DNA damage response require Mec1- and Tel1-dependent phosphorylation. Mol Cell Biol 24:4151–4165

Bartek J, Falck J, Lukas J (2001) CHK2 kinase--a busy messenger. Nat Rev Mol Cell Biol 2:877–886. https://doi.org/10.1038/35103059

Bartrand AJ, Iyasu D, Brush GS (2004) DNA stimulates Mec1-mediated phosphorylation of replication protein A. J Biol Chem 279:26762–26767. https://doi.org/10.1074/jbc.M312353200

Bastos de Oliveira FM, Harris MR, Brazauskas P, de Bruin RAM, Smolka MB (2012) Linking DNA replication checkpoint to MBF cell-cycle transcription reveals a distinct class of G1/S genes. EMBO J 31:1798–1810. https://doi.org/10.1038/emboj.2012.27

Bauer NC, Corbett AH, Doetsch PW (2015) The current state of eukaryotic DNA base damage and repair. Nucleic Acids Res 43:10083–10101. https://doi.org/10.1093/nar/gkv1136

Baute J, Depicker A (2008) Base excision repair and its role in maintaining genome stability. Crit Rev Biochem Mol Biol 43:239–276. https://doi.org/10.1080/10409230802309905

Bazzi M, Mantiero D, Trovesi C, Lucchini G, Longhese MP (2010) Dephosphorylation of gamma H2A by Glc7/protein phosphatase 1 promotes recovery from inhibition of DNA replication. Mol Cell Biol 30:131–145. https://doi.org/10.1128/MCB.01000-09

Bebenek K, Kunkel TA (2000) Streisinger revisited: DNA synthesis errors mediated by substrate misalignments. Cold Spring Harb Symp Quant Biol 65:81–91

Berman J (2006) Morphogenesis and cell cycle progression in *Candida albicans*. Curr Opin Microbiol 9:595–601. https://doi.org/10.1016/j.mib.2006.10.007

Bird AW, Yu DY, Pray-Grant MG, Qiu Q, Harmon KE, Megee PC, Grant PA, Smith MM, Christman MF (2002) Acetylation of histone H4 by Esa1 is required for DNA double-strand break repair. Nature 419:411–415. https://doi.org/10.1038/nature01035

Bodi Z, Gysler-Junker A, Kohli J (1991) A quantitative assay to measure chromosome stability in *Schizosaccharomyces pombe*. Mol Gen Genet MGG 229:77–80

Bøe CA, Krohn M, Rødland GE, Capiaghi C, Maillard O, Thoma F, Boye E, Grallert B (2012) Induction of a G1-S checkpoint in fission yeast. Proc Natl Acad Sci USA 109:9911–9916. https://doi.org/10.1073/pnas.1204901109

Boiteux S, Guillet M (2004) Abasic sites in DNA: repair and biological consequences in *Saccharomyces cerevisiae*. DNA Repair 3:1–12

Boiteux S, Jinks-Robertson S (2013) DNA repair mechanisms and the bypass of DNA damage in *Saccharomyces cerevisiae*. Genetics 193:1025–1064. https://doi.org/10.1534/genetics.112. 145219

Boiteux S, Gellon L, Guibourt N (2002) Repair of 8-oxoguanine in *Saccharomyces cerevisiae*: interplay of DNA repair and replication mechanisms. Free Radic Biol Med 32:1244–1253

Bonilla CY, Melo JA, Toczyski DP (2008) Colocalization of sensors is sufficient to activate the DNA damage checkpoint in the absence of damage. Mol Cell 30:267–276. https://doi.org/10. 1016/j.molcel.2008.03.023

Botuyan MV, Lee J, Ward IM, Kim J-E, Thompson JR, Chen J, Mer G (2006) Structural basis for the methylation state-specific recognition of histone H4-K20 by 53BP1 and Crb2 in DNA repair. Cell 127:1361–1373. https://doi.org/10.1016/j.cell.2006.10.043

Boye E, Skjølberg HC, Grallert B (2009) Checkpoint regulation of DNA replication. Methods Mol Biol Clifton NJ 521:55–70. https://doi.org/10.1007/978-1-60327-815-7_4

Brandão LN, Ferguson R, Santoro I, Jinks-Robertson S, Sclafani RA (2014) The role of Dbf4-dependent protein kinase in DNA polymerase ζ-dependent mutagenesis in *Saccharomyces cerevisiae*. Genetics 197:1111–1122. https://doi.org/10.1534/genetics.114.165308

Cadet J, Sage E, Douki T (2005) Ultraviolet radiation-mediated damage to cellular DNA. Mutat Res 571:3–17. https://doi.org/10.1016/j.mrfmmm.2004.09.012

Callegari AJ, Kelly TJ (2007) Shedding light on the DNA damage checkpoint. Cell Cycle Georget Tex 6:660–666. https://doi.org/10.4161/cc.6.6.3984

Chaudhuri S, Wyrick JJ, Smerdon MJ (2009) Histone H3 Lys79 methylation is required for efficient nucleotide excision repair in a silenced locus of *Saccharomyces cerevisiae*. Nucleic Acids Res 37:1690–1700. https://doi.org/10.1093/nar/gkp003

Chaudhury I, Koepp DM (2017) Degradation of Mrc1 promotes recombination-mediated restart of stalled replication forks. Nucleic Acids Res 45:2558–2570. https://doi.org/10.1093/nar/ gkw1249

Chen C, Kolodner RD (1999) Gross chromosomal rearrangements in *Saccharomyces cerevisiae* replication and recombination defective mutants. Nat Genet 23:81–85. https://doi.org/10.1038/ 12687

Chen L, Trujillo K, Ramos W, Sung P, Tomkinson AE (2001) Promotion of Dnl4-catalyzed DNA end-joining by the Rad50/Mre11/Xrs2 and Hdf1/Hdf2 complexes. Mol Cell 8:1105–1115

Chen S, Albuquerque CP, Liang J, Suhandynata RT, Zhou H (2010) A proteome-wide analysis of kinase-substrate network in the DNA damage response. J Biol Chem 285:12803–12812. https:// doi.org/10.1074/jbc.M110.106989

Chi P, Kwon Y, Seong C, Epshtein A, Lam I, Sung P, Klein HL (2006) Yeast recombination factor Rdh54 functionally interacts with the Rad51 recombinase and catalyzes Rad51 removal from DNA. J Biol Chem 281:26268–26279. https://doi.org/10.1074/jbc.M602983200

Cho RJ, Campbell MJ, Winzeler EA, Steinmetz L, Conway A, Wodicka L, Wolfsberg TG, Gabrielian AE, Landsman D, Lockhart DJ, Davis RW (1998) A genome-wide transcriptional analysis of the mitotic cell cycle. Mol Cell 2:65–73. https://doi.org/10.1016/S1097-2765(00)80114-8

Clerici M, Mantiero D, Lucchini G, Longhese MP (2006) The *Saccharomyces cerevisiae* Sae2 protein negatively regulates DNA damage checkpoint signalling. EMBO Rep 7:212–218. https://doi.org/10.1038/sj.embor.7400593

Clerici M, Mantiero D, Guerini I, Lucchini G, Longhese MP (2008) The Yku70-Yku80 complex contributes to regulate double-strand break processing and checkpoint activation during the cell cycle. EMBO Rep 9:810–818. https://doi.org/10.1038/embor.2008.121

Côte P, Hogues H, Whiteway M (2009) Transcriptional analysis of the *Candida albicans* cell cycle. Mol Biol Cell 20:3363–3373. https://doi.org/10.1091/mbc.E09-03-0210

Czaja W, Mao P, Smerdon MJ (2014) Chromatin remodelling complex RSC promotes base excision repair in chromatin of *Saccharomyces cerevisiae*. DNA Repair 16:35–43. https://doi. org/10.1016/j.dnarep.2014.01.002

Daigaku Y, Mashiko S, Mishiba K, Yamamura S, Ui A, Enomoto T, Yamamoto K (2006) Loss of heterozygosity in yeast can occur by ultraviolet irradiation during the S phase of the cell cycle. Mutat Res 600:177–183. https://doi.org/10.1016/j.mrfmmm.2006.04.001

Dang Do AN, Kimball SR, Cavener DR, Jefferson LS (2009) eIF2α kinases GCN2 and PERK modulate transcription and translation of distinct sets of mRNAs in mouse liver. Physiol Genomics 38:328–341. https://doi.org/10.1152/physiolgenomics.90396.2008

Davidson MB, Katou Y, Keszthelyi A, Sing TL, Xia T, Ou J, Vaisica JA, Thevakumaran N, Marjavaara L, Myers CL, Chabes A, Shirahige K, Brown GW (2012) Endogenous DNA replication stress results in expansion of dNTP pools and a mutator phenotype. EMBO J 31:895–907. https://doi.org/10.1038/emboj.2011.485

de Bruin RAM, Wittenberg C (2009) All eukaryotes: before turning off G1-S transcription, please check your DNA. Cell Cycle Georget Tex 8:214–217. https://doi.org/10.4161/cc.8.2.7412

De Piccoli G, Katou Y, Itoh T, Nakato R, Shirahige K, Labib K (2012) Replisome stability at defective DNA replication forks is independent of S phase checkpoint kinases. Mol Cell 45:696–704. https://doi.org/10.1016/j.molcel.2012.01.007

Deegan TD, Yeeles JT, Diffley JF (2016) Phosphopeptide binding by Sld3 links Dbf4-dependent kinase to MCM replicative helicase activation. EMBO J 35:961–973. https://doi.org/10.15252/embj.201593552

den Elzen NR, O'Connell MJ (2004) Recovery from DNA damage checkpoint arrest by PP1-mediated inhibition of Chk1. EMBO J 23:908–918. https://doi.org/10.1038/sj.emboj.7600105

Deshpande RA, Wilson TE (2007) Modes of interaction among yeast Nej1, Lif1 and Dnl4 proteins and comparison to human XLF, XRCC4 and Lig4. DNA Repair 6:1507–1516. https://doi.org/10.1016/j.dnarep.2007.04.014

Deshpande AM, Ivanova IG, Raykov V, Xue Y, Maringele L (2011) Polymerase epsilon is required to maintain replicative senescence. Mol Cell Biol 31:1637–1645. https://doi.org/10.1128/MCB.00144-10

Deshpande I, Seeber A, Shimada K, Keusch JJ, Gut H, Gasser SM (2017) Structural basis of Mec1-Ddc2-RPA assembly and activation on single-stranded DNA at sites of damage. Mol Cell 68:431–445.e5. https://doi.org/10.1016/j.molcel.2017.09.019

Dever TE, Chen JJ, Barber GN, Cigan AM, Feng L, Donahue TF, London IM, Katze MG, Hinnebusch AG (1993) Mammalian eukaryotic initiation factor 2 alpha kinases functionally substitute for GCN2 protein kinase in the GCN4 translational control mechanism of yeast. Proc Natl Acad Sci USA 90:4616–4620

Diao L-T, Chen C-C, Dennehey B, Pal S, Wang P, Shen Z-J, Deem A, Tyler JK (2017) Delineation of the role of chromatin assembly and the Rtt101Mms1 E3 ubiquitin ligase in DNA damage checkpoint recovery in budding yeast. PLoS One 12:e0180556. https://doi.org/10.1371/journal.pone.0180556

Dmowski M, Rudzka J, Campbell JL, Jonczyk P, Fijałkowska IJ (2017) Mutations in the non-catalytic subunit Dpb2 of DNA polymerase epsilon affect the Nrm1 branch of the DNA replication checkpoint. PLoS Genet 13:e1006572. https://doi.org/10.1371/journal.pgen.1006572

Doetsch PW, Morey NJ, Swanson RL, Jinks-Robertson S (2001) Yeast base excision repair: interconnections and networks. Prog Nucleic Acid Res Mol Biol 68:29–39

Dogliotti E, Fortini P, Pascucci B, Parlanti E (2001) The mechanism of switching among multiple BER pathways. Prog Nucleic Acid Res Mol Biol 68:3–27

Doré AS, Furnham N, Davies OR, Sibanda BL, Chirgadze DY, Jackson SP, Pellegrini L, Blundell TL (2006) Structure of an Xrcc4-DNA ligase IV yeast ortholog complex reveals a novel BRCT interaction mode. DNA Repair 5:362–368. https://doi.org/10.1016/j.dnarep.2005.11.004

Downs JA, Lowndes NF, Jackson SP (2000) A role for Saccharomyces cerevisiae histone H2A in DNA repair. Nature 408:1001–1004. https://doi.org/10.1038/35050000

Downs JA, Allard S, Jobin-Robitaille O, Javaheri A, Auger A, Bouchard N, Kron SJ, Jackson SP, Côté J (2004) Binding of chromatin-modifying activities to phosphorylated histone H2A at DNA damage sites. Mol Cell 16:979–990. https://doi.org/10.1016/j.molcel.2004.12.003

Du L-L, Nakamura TM, Russell P (2006) Histone modification-dependent and -independent pathways for recruitment of checkpoint protein Crb2 to double-strand breaks. Genes Dev 20:1583–1596. https://doi.org/10.1101/gad.1422606

Eapen VV, Haber JE (2013) DNA damage signaling triggers the cytoplasm-to-vacuole pathway of autophagy to regulate cell cycle progression. Autophagy 9:440–441. https://doi.org/10.4161/auto.23280

Eapen VV, Sugawara N, Tsabar M, Wu W-H, Haber JE (2012) The *Saccharomyces cerevisiae* chromatin remodeler Fun30 regulates DNA end resection and checkpoint deactivation. Mol Cell Biol 32:4727–4740. https://doi.org/10.1128/MCB.00566-12

Emerson CH, Bertuch AA (2016) Consider the workhorse: nonhomologous end-joining in budding yeast. Biochem Cell Biol Biochim Biol Cell:1–11. https://doi.org/10.1139/bcb-2016-0001

Emili A (1998) MEC1-dependent phosphorylation of Rad9p in response to DNA damage. Mol Cell 2:183–189

Evans MD, Dizdaroglu M, Cooke MS (2004) Oxidative DNA damage and disease: induction, repair and significance. Mutat Res 567:1–61. https://doi.org/10.1016/j.mrrev.2003.11.001

Eyler DE, Burnham KA, Wilson TE, O'Brien PJ (2017) Mechanisms of glycosylase induced genomic instability. PLoS One 12:e0174041. https://doi.org/10.1371/journal.pone.0174041

Falck J, Coates J, Jackson SP (2005) Conserved modes of recruitment of ATM, ATR and DNA-PKcs to sites of DNA damage. Nature 434:605–611. https://doi.org/10.1038/nature03442

Ferreira MG, Cooper JP (2004) Two modes of DNA double-strand break repair are reciprocally regulated through the fission yeast cell cycle. Genes Dev 18:2249–2254. https://doi.org/10.1101/gad.315804

Ferreiro JA, Powell NG, Karabetsou N, Kent NA, Mellor J, Waters R (2004) Cbf1p modulates chromatin structure, transcription and repair at the *Saccharomyces cerevisiae* MET16 locus. Nucleic Acids Res 32:1617–1626. https://doi.org/10.1093/nar/gkh324

Fortini P, Dogliotti E (2007) Base damage and single-strand break repair: mechanisms and functional significance of short- and long-patch repair subpathways. DNA Repair 6:398–409. https://doi.org/10.1016/j.dnarep.2006.10.008

Frank-Vaillant M, Marcand S (2001) NHEJ regulation by mating type is exercised through a novel protein, Lif2p, essential to the ligase IV pathway. Genes Dev 15:3005–3012. https://doi.org/10.1101/gad.206801

Fu D, Calvo JA, Samson LD (2012) Balancing repair and tolerance of DNA damage caused by alkylating agents. Nat Rev Cancer 12:104–120. https://doi.org/10.1038/nrc3185

Furuya K, Poitelea M, Guo L, Caspari T, Carr AM (2004) Chk1 activation requires Rad9 S/TQ-site phosphorylation to promote association with C-terminal BRCT domains of Rad4TOPBP1. Genes Dev 18:1154–1164. https://doi.org/10.1101/gad.291104

Galgoczy DJ, Toczyski DP (2001) Checkpoint adaptation precedes spontaneous and damage-induced genomic instability in yeast. Mol Cell Biol 21:1710–1718. https://doi.org/10.1128/MCB.21.5.1710-1718.2001

Galli A, Chan CY, Parfenova L, Cervelli T, Schiestl RH (2015) Requirement of POL3 and POL4 on non-homologous and microhomology-mediated end joining in rad50/xrs2 mutants of *Saccharomyces cerevisiae*. Mutagenesis 30:841–849. https://doi.org/10.1093/mutage/gev046

Gao S, Honey S, Futcher B, Grollman AP (2016) The non-homologous end-joining pathway of *S. cerevisiae* works effectively in G1-phase cells, and religates cognate ends correctly and non-randomly. DNA Repair 42:1–10. https://doi.org/10.1016/j.dnarep.2016.03.013

Gasior SL, Wong AK, Kora Y, Shinohara A, Bishop DK (1998) Rad52 associates with RPA and functions with rad55 and rad57 to assemble meiotic recombination complexes. Genes Dev 12:2208–2221

Ghospurkar PL, Wilson TM, Severson AL, Klein SJ, Khaku SK, Walther AP, Haring SJ (2015) The DNA damage response and checkpoint adaptation in *Saccharomyces cerevisiae*: distinct roles for the replication protein A2 (Rfa2) N-terminus. Genetics 199:711–727. https://doi.org/10.1534/genetics.114.173211

Goffeau A, Barrell BG, Bussey H, Davis RW, Dujon B, Feldmann H, Galibert F, Hoheisel JD, Jacq C, Johnston M, Louis EJ, Mewes HW, Murakami Y, Philippsen P, Tettelin H, Oliver SG (1996) Life with 6000 genes. Science 274:546, 563–567

González-Novo A, Jiménez J, Clotet J, Nadal-Ribelles M, Cavero S, de Nadal E, Posas F (2015) Hog1 targets Whi5 and Msa1 transcription factors to downregulate cyclin expression upon stress. Mol Cell Biol 35:1606–1618. https://doi.org/10.1128/MCB.01279-14

Grenon M, Costelloe T, Jimeno S, O'Shaughnessy A, Fitzgerald J, Zgheib O, Degerth L, Lowndes NF (2007) Docking onto chromatin via the *Saccharomyces cerevisiae* Rad9 Tudor domain. Yeast Chichester Engl 24:105–119. https://doi.org/10.1002/yea.1441

Gros L, Saparbaev MK, Laval J (2002) Enzymology of the repair of free radicals-induced DNA damage. Oncogene 21:8905–8925. https://doi.org/10.1038/sj.onc.1206005

Guillemain G, Ma E, Mauger S, Miron S, Thai R, Guérois R, Ochsenbein F, Marsolier-Kergoat M-C (2007) Mechanisms of checkpoint kinase Rad53 inactivation after a double-strand break in *Saccharomyces cerevisiae*. Mol Cell Biol 27:3378–3389. https://doi.org/10.1128/MCB.00863-06

Guintini L, Charton R, Peyresaubes F, Thoma F, Conconi A (2015) Nucleosome positioning, nucleotide excision repair and photoreactivation in *Saccharomyces cerevisiae*. DNA Repair 36:98–104. https://doi.org/10.1016/j.dnarep.2015.09.012

Guo X, Jinks-Robertson S (2013) Roles of exonucleases and translesion synthesis DNA polymerases during mitotic gap repair in yeast. DNA Repair 12:1024–1030. https://doi.org/10.1016/j.dnarep.2013.10.001

Guo L, Ganguly A, Sun L, Suo F, Du L-L, Russell P (2016) Global fitness profiling identifies arsenic and cadmium tolerance mechanisms in fission yeast. G3 Bethesda Md 6:3317–3333. https://doi.org/10.1534/g3.116.033829

Guzder SN, Sommers CH, Prakash L, Prakash S (2006) Complex formation with damage recognition protein Rad14 is essential for *Saccharomyces cerevisiae* Rad1-Rad10 nuclease to perform its function in nucleotide excision repair in vivo. Mol Cell Biol 26:1135–1141. https://doi.org/10.1128/MCB.26.3.1135-1141.2006

Haracska L, Prakash S, Prakash L (2003) Yeast DNA polymerase ζ is an efficient extender of primer ends opposite from 7,8-dihydro-8-Oxoguanine and O6-methylguanine. Mol Cell Biol 23:1453–1459. https://doi.org/10.1128/MCB.23.4.1453-1459.2003

Harrison JC, Haber JE (2006) Surviving the breakup: the DNA damage checkpoint. Annu Rev Genet 40:209–235. https://doi.org/10.1146/annurev.genet.40.051206.105231

Hartwell LH, Smith D (1985) Altered fidelity of mitotic chromosome transmission in cell cycle mutants of *S. cerevisiae*. Genetics 110:381–395

Hauer MH, Gasser SM (2017) Chromatin and nucleosome dynamics in DNA damage and repair. Genes Dev 31:2204–2221. https://doi.org/10.1101/gad.307702.117

Hauer MH, Seeber A, Singh V, Thierry R, Sack R, Amitai A, Kryzhanovska M, Eglinger J, Holcman D, Owen-Hughes T, Gasser SM (2017) Histone degradation in response to DNA damage enhances chromatin dynamics and recombination rates. Nat Struct Mol Biol 24:99–107. https://doi.org/10.1038/nsmb.3347

Heideker J, Lis ET, Romesberg FE (2007) Phosphatases, DNA damage checkpoints and checkpoint deactivation. Cell Cycle Georget Tex 6:3058–3064. https://doi.org/10.4161/cc.6.24.5100

Heidinger-Pauli JM, Unal E, Koshland D (2009) Distinct targets of the Eco1 acetyltransferase modulate cohesion in S phase and in response to DNA damage. Mol Cell 34:311–321. https://doi.org/10.1016/j.molcel.2009.04.008

Hinnebusch AG (1993) Gene-specific translational control of the yeast GCN4 gene by phosphorylation of eukaryotic initiation factor 2. Mol Microbiol 10:215–223. https://doi.org/10.1111/j.1365-2958.1993.tb01947.x

Hishida T, Hirade Y, Haruta N, Kubota Y, Iwasaki H (2010) Srs2 plays a critical role in reversible G2 arrest upon chronic and low doses of UV irradiation via two distinct homologous recombination-dependent mechanisms in postreplication repair-deficient cells. Mol Cell Biol 30:4840–4850. https://doi.org/10.1128/MCB.00453-10

Hombauer H, Campbell CS, Smith CE, Desai A, Kolodner RD (2011) Visualization of eukaryotic DNA mismatch repair reveals distinct recognition and repair intermediates. Cell 147:1040–1053. https://doi.org/10.1016/j.cell.2011.10.025

Horigome C, Oma Y, Konishi T, Schmid R, Marcomini I, Hauer MH, Dion V, Harata M, Gasser SM (2014) SWR1 and INO80 chromatin remodelers contribute to DNA double-strand break perinuclear anchorage site choice. Mol Cell 55:626–639. https://doi.org/10.1016/j.molcel.2014.06.027

Horigome C, Bustard DE, Marcomini I, Delgoshaie N, Tsai-Pflugfelder M, Cobb JA, Gasser SM (2016) PolySUMOylation by Siz2 and Mms21 triggers relocation of DNA breaks to nuclear pores through the Slx5/Slx8 STUbL. Genes Dev 30:931–945. https://doi.org/10.1101/gad.277665.116

Howlett NG, Schiestl RH (2000) Simultaneous measurement of the frequencies of intrachromosomal recombination and chromosome gain using the yeast DEL assay. Mutat Res 454:53–62

Hu F, Alcasabas AA, Elledge SJ (2001) Asf1 links Rad53 to control of chromatin assembly. Genes Dev 15:1061–1066. https://doi.org/10.1101/gad.873201

Huertas P, Cortés-Ledesma F, Sartori AA, Aguilera A, Jackson SP (2008) CDK targets Sae2 to control DNA-end resection and homologous recombination. Nature 455:689–692. https://doi.org/10.1038/nature07215

Hwang J-Y, Smith S, Myung K (2005) The Rad1-Rad10 complex promotes the production of gross chromosomal rearrangements from spontaneous DNA damage in *Saccharomyces cerevisiae*. Genetics 169:1927–1937. https://doi.org/10.1534/genetics.104.039768

Iyer DR, Rhind N (2017) Replication fork slowing and stalling are distinct, checkpoint-independent consequences of replicating damaged DNA. PLoS Genet 13:e1006958. https://doi.org/10.1371/journal.pgen.1006958

Jasin M, Rothstein R (2013) Repair of strand breaks by homologous recombination. Cold Spring Harb Perspect Biol 5:a012740. https://doi.org/10.1101/cshperspect.a012740

Jena NR (2012) DNA damage by reactive species: mechanisms, mutation and repair. J Biosci 37:503–517

Kalocsay M, Hiller NJ, Jentsch S (2009) Chromosome-wide Rad51 spreading and SUMO-H2A.Z-dependent chromosome fixation in response to a persistent DNA double-strand break. Mol Cell 33:335–343. https://doi.org/10.1016/j.molcel.2009.01.016

Kamimura Y, Masumoto H, Sugino A, Araki H (1998) Sld2, which interacts with Dpb11 in *Saccharomyces cerevisiae*, is required for chromosomal DNA replication. Mol Cell Biol 18:6102–6109

Kanellis P, Gagliardi M, Banath JP, Szilard RK, Nakada S, Galicia S, Sweeney FD, Cabelof DC, Olive PL, Durocher D (2007) A screen for suppressors of gross chromosomal rearrangements identifies a conserved role for PLP in preventing DNA lesions. PLoS Genet 3:e134. https://doi.org/10.1371/journal.pgen.0030134

Kaniak-Golik A, Skoneczna A (2015) Mitochondria-nucleus network for genome stability. Free Radic Biol Med 82:73–104. https://doi.org/10.1016/j.freeradbiomed.2015.01.013

Kawashima S, Ogiwara H, Tada S, Harata M, Wintersberger U, Enomoto T, Seki M (2007) The INO80 complex is required for damage-induced recombination. Biochem Biophys Res Commun 355:835–841. https://doi.org/10.1016/j.bbrc.2007.02.036

Kaye JA, Melo JA, Cheung SK, Vaze MB, Haber JE, Toczyski DP (2004) DNA breaks promote genomic instability by impeding proper chromosome segregation. Curr Biol CB 14:2096–2106. https://doi.org/10.1016/j.cub.2004.10.051

Kegel A, Sjöstrand JO, Aström SU (2001) Nej1p, a cell type-specific regulator of nonhomologous end joining in yeast. Curr Biol CB 11:1611–1617

Kemmer D, McHardy LM, Hoon S, Rebérioux D, Giaever G, Nislow C, Roskelley CD, Roberge M (2009) Combining chemical genomics screens in yeast to reveal spectrum of effects of chemical inhibition of sphingolipid biosynthesis. BMC Microbiol 9:9. https://doi.org/10.1186/1471-2180-9-9

Keogh M-C, Kim J-A, Downey M, Fillingham J, Chowdhury D, Harrison JC, Onishi M, Datta N, Galicia S, Emili A, Lieberman J, Shen X, Buratowski S, Haber JE, Durocher D, Greenblatt JF, Krogan NJ (2006) A phosphatase complex that dephosphorylates gammaH2AX regulates DNA damage checkpoint recovery. Nature 439:497–501. https://doi.org/10.1038/nature04384

Klein HL (2001) Spontaneous chromosome loss in *Saccharomyces cerevisiae* is suppressed by DNA damage checkpoint functions. Genetics 159:1501–1509

Kolodner RD, Marsischky GT (1999) Eukaryotic DNA mismatch repair. Curr Opin Genet Dev 9:89–96

Komata M, Bando M, Araki H, Shirahige K (2009) The direct binding of Mrc1, a checkpoint mediator, to Mcm6, a replication helicase, is essential for the replication checkpoint against methyl methanesulfonate-induced stress. Mol Cell Biol 29:5008–5019. https://doi.org/10.1128/MCB.01934-08

Krejci L, Van Komen S, Li Y, Villemain J, Reddy MS, Klein H, Ellenberger T, Sung P (2003) DNA helicase Srs2 disrupts the Rad51 presynaptic filament. Nature 423:305–309. https://doi.org/10.1038/nature01577

Krol K, Brozda I, Skoneczny M, Bretner M, Bretne M, Skoneczna A (2015) A genomic screen revealing the importance of vesicular trafficking pathways in genome maintenance and protection against genotoxic stress in diploid *Saccharomyces cerevisiae* cells. PLoS One 10:e0120702. https://doi.org/10.1371/journal.pone.0120702

Kück U, Hoff B (2010) New tools for the genetic manipulation of filamentous fungi. Appl Microbiol Biotechnol 86:51–62. https://doi.org/10.1007/s00253-009-2416-7

Kültz D, Chakravarty D (2001) Hyperosmolality in the form of elevated NaCl but not urea causes DNA damage in murine kidney cells. Proc Natl Acad Sci USA 98:1999–2004. https://doi.org/10.1073/pnas.98.4.1999

Kumar S, Burgers PM (2013) Lagging strand maturation factor Dna2 is a component of the replication checkpoint initiation machinery. Genes Dev 27:313–321. https://doi.org/10.1101/gad.204750.112

Kunz BA, Henson ES, Roche H, Ramotar D, Nunoshiba T, Demple B (1994) Specificity of the mutator caused by deletion of the yeast structural gene (APN1) for the major apurinic endonuclease. Proc Natl Acad Sci USA 91:8165–8169

Latif C, Harvey SH, O'Connell SJ (2001) Ensuring the stability of the genome: DNA damage checkpoints. Sci World J 1:684–702. https://www.hindawi.com/journals/tswj/2001/914679/abs/. Accessed 31 Dec 2017

Lee SE, Moore JK, Holmes A, Umezu K, Kolodner RD, Haber JE (1998) Saccharomyces Ku70, mre11/rad50 and RPA proteins regulate adaptation to G2/M arrest after DNA damage. Cell 94:399–409

Lee SE, Pellicioli A, Malkova A, Foiani M, Haber JE (2001) The Saccharomyces recombination protein Tid1p is required for adaptation from G2/M arrest induced by a double-strand break. Curr Biol CB 11:1053–1057

Lee SE, Pellicioli A, Vaze MB, Sugawara N, Malkova A, Foiani M, Haber JE (2003a) Yeast Rad52 and Rad51 recombination proteins define a second pathway of DNA damage assessment in response to a single double-strand break. Mol Cell Biol 23:8913–8923

Lee S-J, Schwartz MF, Duong JK, Stern DF (2003b) Rad53 phosphorylation site clusters are important for Rad53 regulation and signaling. Mol Cell Biol 23:6300–6314

Lee K, Zhang Y, Lee SE (2008) *Saccharomyces cerevisiae* ATM orthologue suppresses break-induced chromosome translocations. Nature 454:543–546. https://doi.org/10.1038/nature07054

Lehner K, Mudrak SV, Minesinger BK, Jinks-Robertson S (2012) Frameshift mutagenesis: the roles of primer-template misalignment and the nonhomologous end-joining pathway in *Saccharomyces cerevisiae*. Genetics 190:501–510. https://doi.org/10.1534/genetics.111.134890

Lehoczký P, McHugh PJ, Chovanec M (2007) DNA interstrand cross-link repair in *Saccharomyces cerevisiae*. FEMS Microbiol Rev 31:109–133. https://doi.org/10.1111/j.1574-6976.2006.00046.x

Leroy C, Lee SE, Vaze MB, Ochsenbein F, Ochsenbien F, Guerois R, Haber JE, Marsolier-Kergoat M-C (2003) PP2C phosphatases Ptc2 and Ptc3 are required for DNA checkpoint inactivation after a double-strand break. Mol Cell 11:827–835

Li G-M (2008) Mechanisms and functions of DNA mismatch repair. Cell Res 18:85–98. https://doi.org/10.1038/cr.2007.115

Li S, Smerdon MJ (2004) Dissecting transcription-coupled and global genomic repair in the chromatin of yeast GAL1-10 genes. J Biol Chem 279:14418–14426. https://doi.org/10.1074/jbc.M312004200

Limbo O, Chahwan C, Yamada Y, de Bruin RAM, Wittenberg C, Russell P (2007) Ctp1 is a cell-cycle-regulated protein that functions with Mre11 complex to control double-strand break repair by homologous recombination. Mol Cell 28:134–146. https://doi.org/10.1016/j.molcel.2007.09.009

Lisby M, Rothstein R, Mortensen UH (2001) Rad52 forms DNA repair and recombination centers during S phase. Proc Natl Acad Sci USA 98:8276–8282. https://doi.org/10.1073/pnas.121006298

Lisby M, Mortensen UH, Rothstein R (2003) Colocalization of multiple DNA double-strand breaks at a single Rad52 repair centre. Nat Cell Biol 5:572–577. https://doi.org/10.1038/ncb997

Lisby M, Barlow JH, Burgess RC, Rothstein R (2004) Choreography of the DNA damage response: spatiotemporal relationships among checkpoint and repair proteins. Cell 118:699–713. https://doi.org/10.1016/j.cell.2004.08.015

Litwin I, Bocer T, Dziadkowiec D, Wysocki R (2013) Oxidative stress and replication-independent DNA breakage induced by arsenic in *Saccharomyces cerevisiae*. PLoS Genet 9:e1003640. https://doi.org/10.1371/journal.pgen.1003640

Lum PY, Armour CD, Stepaniants SB, Cavet G, Wolf MK, Butler JS, Hinshaw JC, Garnier P, Prestwich GD, Leonardson A, Garrett-Engele P, Rush CM, Bard M, Schimmack G, Phillips JW, Roberts CJ, Shoemaker DD (2004) Discovering modes of action for therapeutic compounds using a genome-wide screen of yeast heterozygotes. Cell 116:121–137

Lydeard JR, Lipkin-Moore Z, Jain S, Eapen VV, Haber JE (2010) Sgs1 and exo1 redundantly inhibit break-induced replication and de novo telomere addition at broken chromosome ends. PLoS Genet 6:e1000973. https://doi.org/10.1371/journal.pgen.1000973

Mailand N, Bekker-Jensen S, Bartek J, Lukas J (2006) Destruction of Claspin by SCFbetaTrCP restrains Chk1 activation and facilitates recovery from genotoxic stress. Mol Cell 23:307–318. https://doi.org/10.1016/j.molcel.2006.06.016

Majka J, Binz SK, Wold MS, Burgers PMJ (2006) Replication protein A directs loading of the DNA damage checkpoint clamp to 5′-DNA junctions. J Biol Chem 281:27855–27861. https://doi.org/10.1074/jbc.M605176200

Majka J, Niedziela-Majka A, Burgers PMJ (2006a) The checkpoint clamp activates Mec1 kinase during initiation of the DNA damage checkpoint. Mol Cell 24:891–901. https://doi.org/10.1016/j.molcel.2006.11.027

Mamely I, van Vugt MA, Smits VAJ, Semple JI, Lemmens B, Perrakis A, Medema RH, Freire R (2006) Polo-like kinase-1 controls proteasome-dependent degradation of Claspin during checkpoint recovery. Curr Biol CB 16:1950–1955. https://doi.org/10.1016/j.cub.2006.08.026

Mardiros A, Benoun JM, Haughton R, Baxter K, Kelson EP, Fischhaber PL (2011) Rad10-YFP focus induction in response to UV depends on RAD14 in yeast. Acta Histochem 113:409–415. https://doi.org/10.1016/j.acthis.2010.03.005

Maréchal A, Zou L (2015) RPA-coated single-stranded DNA as a platform for post-translational modifications in the DNA damage response. Cell Res 25:9–23. https://doi.org/10.1038/cr.2014.147

Marsischky GT, Lee S, Griffith J, Kolodner RD (1999) *Saccharomyces cerevisiae* MSH2/6 complex interacts with Holliday junctions and facilitates their cleavage by phage resolution enzymes. J Biol Chem 274:7200–7206

Masumoto H, Sugino A, Araki H (2000) Dpb11 controls the association between DNA polymerases alpha and epsilon and the autonomously replicating sequence region of budding yeast. Mol Cell Biol 20:2809–2817

Matsumoto S, Ogino K, Noguchi E, Russell P, Masai H (2005) Hsk1-Dfp1/Him1, the Cdc7-Dbf4 kinase in *Schizosaccharomyces pombe*, associates with Swi1, a component of the replication fork protection complex. J Biol Chem 280:42536–42542. https://doi.org/10.1074/jbc.M510575200

McKinney JS, Sethi S, Tripp JD, Nguyen TN, Sanderson BA, Westmoreland JW, Resnick MA, Lewis LK (2013) A multistep genomic screen identifies new genes required for repair of DNA double-strand breaks in *Saccharomyces cerevisiae*. BMC Genomics 14:251. https://doi.org/10.1186/1471-2164-14-251

Mehta A, Haber JE (2014) Sources of DNA double-strand breaks and models of recombinational DNA repair. Cold Spring Harb Perspect Biol 6:a016428–a016428. https://doi.org/10.1101/cshperspect.a016428

Melo JA, Cohen J, Toczyski DP (2001) Two checkpoint complexes are independently recruited to sites of DNA damage in vivo. Genes Dev 15:2809–2821. https://doi.org/10.1101/gad.903501

Mersaoui SY, Gravel S, Karpov V, Wellinger RJ (2015) DNA damage checkpoint adaptation genes are required for division of cells harbouring eroded telomeres. Microb Cell Graz Austria 2:394–405. https://doi.org/10.15698/mic2015.10.229

Mimitou EP, Symington LS (2008) Sae2, Exo1 and Sgs1 collaborate in DNA double-strand break processing. Nature 455:770–774. https://doi.org/10.1038/nature07312

Mizuguchi G, Shen X, Landry J, Wu W-H, Sen S, Wu C (2004) ATP-driven exchange of histone H2AZ variant catalyzed by SWR1 chromatin remodeling complex. Science 303:343–348. https://doi.org/10.1126/science.1090701

Mohanta TK, Bae H (2015) The diversity of fungal genome. Biol Proc Online 17:8. https://doi.org/10.1186/s12575-015-0020-z

Mojas N, Lopes M, Jiricny J (2007) Mismatch repair-dependent processing of methylation damage gives rise to persistent single-stranded gaps in newly replicated DNA. Genes Dev 21:3342–3355. https://doi.org/10.1101/gad.455407

Motegi A, Kuntz K, Majeed A, Smith S, Myung K (2006) Regulation of gross chromosomal rearrangements by ubiquitin and SUMO ligases in Saccharomyces cerevisiae. Mol Cell Biol 26:1424–1433. https://doi.org/10.1128/MCB.26.4.1424-1433.2006

Myung K, Kolodner RD (2003) Induction of genome instability by DNA damage in Saccharomyces cerevisiae. DNA Repair 2:243–258

Nagai S, Dubrana K, Tsai-Pflugfelder M, Davidson MB, Roberts TM, Brown GW, Varela E, Hediger F, Gasser SM, Krogan NJ (2008) Functional targeting of DNA damage to a nuclear pore-associated SUMO-dependent ubiquitin ligase. Science 322:597–602. https://doi.org/10.1126/science.1162790

Nakada D, Matsumoto K, Sugimoto K (2003) ATM-related Tel1 associates with double-strand breaks through an Xrs2-dependent mechanism. Genes Dev 17:1957–1962. https://doi.org/10.1101/gad.1099003

Nakamura TM, Du L-L, Redon C, Russell P (2004) Histone H2A phosphorylation controls Crb2 recruitment at DNA breaks, maintains checkpoint arrest, and influences DNA repair in fission yeast. Mol Cell Biol 24:6215–6230. https://doi.org/10.1128/MCB.24.14.6215-6230.2004

Navadgi-Patil VM, Burgers PM (2009) The unstructured C-terminal tail of the 9-1-1 clamp subunit Ddc1 activates Mec1/ATR via two distinct mechanisms. Mol Cell 36:743–753. https://doi.org/10.1016/j.molcel.2009.10.014

Navas TA, Sanchez Y, Elledge SJ (1996) RAD9 and DNA polymerase epsilon form parallel sensory branches for transducing the DNA damage checkpoint signal in Saccharomyces cerevisiae. Genes Dev 10:2632–2643

Nayak T, Szewczyk E, Oakley CE, Osmani A, Ukil L, Murray SL, Hynes MJ, Osmani SA, Oakley BR (2006) A versatile and efficient gene-targeting system for Aspergillus nidulans. Genetics 172:1557–1566. https://doi.org/10.1534/genetics.105.052563

Ngo GHP, Balakrishnan L, Dubarry M, Campbell JL, Lydall D (2014) The 9-1-1 checkpoint clamp stimulates DNA resection by Dna2-Sgs1 and Exo1. Nucleic Acids Res 42:10516–10528. https://doi.org/10.1093/nar/gku746

Ni TT, Marsischky GT, Kolodner RD (1999) MSH2 and MSH6 are required for removal of adenine misincorporated opposite 8-oxo-guanine in S. cerevisiae. Mol Cell 4:439–444

Nielsen I, Bentsen IB, Andersen AH, Gasser SM, Bjergbaek L (2013) A Rad53 independent function of Rad9 becomes crucial for genome maintenance in the absence of the Recq helicase Sgs1. PLoS One 8:e81015. https://doi.org/10.1371/journal.pone.0081015

Nikitaki Z, Hellweg CE, Georgakilas AG, Ravanat J-L (2015) Stress-induced DNA damage biomarkers: applications and limitations. Front Chem 3:35. https://doi.org/10.3389/fchem.2015.00035

Nimonkar AV, Dombrowski CC, Siino JS, Stasiak AZ, Stasiak A, Kowalczykowski SC (2012) *Saccharomyces cerevisiae* Dmc1 and Rad51 proteins preferentially function with Tid1 and Rad54 proteins, respectively, to promote DNA strand invasion during genetic recombination. J Biol Chem 287:28727–28737. https://doi.org/10.1074/jbc.M112.373290

Ninomiya Y, Suzuki K, Ishii C, Inoue H (2004) Highly efficient gene replacements in Neurospora strains deficient for nonhomologous end-joining. Proc Natl Acad Sci USA 101:12248–12253. https://doi.org/10.1073/pnas.0402780101

O'Neill BM, Szyjka SJ, Lis ET, Bailey AO, Yates JR, Aparicio OM, Romesberg FE (2007) Pph3-Psy2 is a phosphatase complex required for Rad53 dephosphorylation and replication fork restart during recovery from DNA damage. Proc Natl Acad Sci USA 104:9290–9295. https://doi.org/10.1073/pnas.0703252104

Odds FC, Brown AJ, Gow NA (2004) *Candida albicans* genome sequence: a platform for genomics in the absence of genetics. Genome Biol 5:230. https://doi.org/10.1186/gb-2004-5-7-230

Oza P, Jaspersen SL, Miele A, Dekker J, Peterson CL (2009) Mechanisms that regulate localization of a DNA double-strand break to the nuclear periphery. Genes Dev 23:912–927. https://doi.org/10.1101/gad.1782209

Paciotti V, Clerici M, Lucchini G, Longhese MP (2000) The checkpoint protein Ddc2, functionally related to S. pombe Rad26, interacts with Mec1 and is regulated by Mec1-dependent phosphorylation in budding yeast. Genes Dev 14:2046–2059

Paek AL, Jones H, Kaochar S, Weinert T (2010) The role of replication bypass pathways in dicentric chromosome formation in budding yeast. Genetics 186:1161–1173. https://doi.org/10.1534/genetics.110.122663

Pardo B, Crabbé L, Pasero P (2017) Signaling pathways of replication stress in yeast. FEMS Yeast Res 17. https://doi.org/10.1093/femsyr/fow101

Parsons AB, Brost RL, Ding H, Li Z, Zhang C, Sheikh B, Brown GW, Kane PM, Hughes TR, Boone C (2004) Integration of chemical-genetic and genetic interaction data links bioactive compounds to cellular target pathways. Nat Biotechnol 22:62–69. https://doi.org/10.1038/nbt919

Parsons AB, Lopez A, Givoni IE, Williams DE, Gray CA, Porter J, Chua G, Sopko R, Brost RL, Ho C-H, Wang J, Ketela T, Brenner C, Brill JA, Fernandez GE, Lorenz TC, Payne GS, Ishihara S, Ohya Y, Andrews B, Hughes TR, Frey BJ, Graham TR, Andersen RJ, Boone C (2006) Exploring the mode-of-action of bioactive compounds by chemical-genetic profiling in yeast. Cell 126:611–625. https://doi.org/10.1016/j.cell.2006.06.040

Pavelka N, Rancati G, Zhu J, Bradford WD, Saraf A, Florens L, Sanderson BW, Hattem GL, Li R (2010) Aneuploidy confers quantitative proteome changes and phenotypic variation in budding yeast. Nature 468:321–325. https://doi.org/10.1038/nature09529

Pellicioli A, Lucca C, Liberi G, Marini F, Lopes M, Plevani P, Romano A, Di Fiore PP, Foiani M (1999) Activation of Rad53 kinase in response to DNA damage and its effect in modulating phosphorylation of the lagging strand DNA polymerase. EMBO J 18:6561–6572. https://doi.org/10.1093/emboj/18.22.6561

Peschiaroli A, Dorrello NV, Guardavaccaro D, Venere M, Halazonetis T, Sherman NE, Pagano M (2006) SCFbetaTrCP-mediated degradation of Claspin regulates recovery from the DNA replication checkpoint response. Mol Cell 23:319–329. https://doi.org/10.1016/j.molcel.2006.06.013

Pomerening JR (2009) Positive-feedback loops in cell cycle progression. FEBS Lett 583:3388–3396. https://doi.org/10.1016/j.febslet.2009.10.001

Powell NG, Ferreiro J, Karabetsou N, Mellor J, Waters R (2003) Transcription, nucleosome positioning and protein binding modulate nucleotide excision repair of the *Saccharomyces cerevisiae* MET17 promoter. DNA Repair 2:375–386

Prakash S, Prakash L (2000) Nucleotide excision repair in yeast. Mutat Res 451:13–24

Puddu F, Granata M, Di Nola L, Balestrini A, Piergiovanni G, Lazzaro F, Giannattasio M, Plevani P, Muzi-Falconi M (2008) Phosphorylation of the budding yeast 9-1-1 complex is required for Dpb11 function in the full activation of the UV-induced DNA damage checkpoint. Mol Cell Biol 28:4782–4793. https://doi.org/10.1128/MCB.00330-08

Puddu F, Piergiovanni G, Plevani P, Muzi-Falconi M (2011) Sensing of replication stress and Mec1 activation act through two independent pathways involving the 9-1-1 complex and DNA polymerase ε. PLoS Genet 7:e1002022. https://doi.org/10.1371/journal.pgen.1002022

Puig S, Ramos-Alonso L, Romero AM, Martínez-Pastor MT (2017) The elemental role of iron in DNA synthesis and repair. Met Integr Biometal Sci 9:1483–1500. https://doi.org/10.1039/c7mt00116a

Rancati G, Crispo V, Lucchini G, Piatti S (2005) Mad3/BubR1 phosphorylation during spindle checkpoint activation depends on both Polo and Aurora kinases in budding yeast. Cell Cycle Georget Tex 4:972–980. https://doi.org/10.4161/cc.4.7.1829

Randell JCW, Fan A, Chan C, Francis LI, Heller RC, Galani K, Bell SP (2010) Mec1 is one of multiple kinases that prime the Mcm2-7 helicase for phosphorylation by Cdc7. Mol Cell 40:353–363. https://doi.org/10.1016/j.molcel.2010.10.017

Rawal CC, Riccardo S, Pesenti C, Ferrari M, Marini F, Pellicioli A (2016) Reduced kinase activity of polo kinase Cdc5 affects chromosome stability and DNA damage response in S. cerevisiae. Cell Cycle Georget Tex 15:2906–2919. https://doi.org/10.1080/15384101.2016.1222338

Reed SI (1996) G1/S regulatory mechanisms from yeast to man. Prog Cell Cycle Res 2:15–27

Reed SH, Akiyama M, Stillman B, Friedberg EC (1999) Yeast autonomously replicating sequence binding factor is involved in nucleotide excision repair. Genes Dev 13:3052–3058

Reginato G, Cannavo E, Cejka P (2017) Physiological protein blocks direct the Mre11-Rad50-Xrs2 and Sae2 nuclease complex to initiate DNA end resection. Genes Dev 31:2325–2330. https://doi.org/10.1101/gad.308254.117

Regulus P, Duroux B, Bayle P-A, Favier A, Cadet J, Ravanat J-L (2007) Oxidation of the sugar moiety of DNA by ionizing radiation or bleomycin could induce the formation of a cluster DNA lesion. Proc Natl Acad Sci USA 104:14032–14037. https://doi.org/10.1073/pnas.0706044104

Reha-Krantz LJ, Siddique MSP, Murphy K, Tam A, O'Carroll M, Lou S, Schultz A, Boone C (2011) Drug-sensitive DNA polymerase δ reveals a role for mismatch repair in checkpoint activation in yeast. Genetics 189:1211–1224. https://doi.org/10.1534/genetics.111.131938

Renaud-Young M, Lloyd DC, Chatfield-Reed K, George I, Chua G, Cobb J (2015) The NuA4 complex promotes translesion synthesis (TLS)-mediated DNA damage tolerance. Genetics 199:1065–1076. https://doi.org/10.1534/genetics.115.174490

Rivin CJ, Fangman WL (1980) Cell cycle phase expansion in nitrogen-limited cultures of Saccharomyces cerevisiae. J Cell Biol 85:96–107

Robzyk K, Recht J, Osley MA (2000) Rad6-dependent ubiquitination of histone H2B in yeast. Science 287:501–504

Rossetto D, Truman AW, Kron SJ, Côté J (2010) Epigenetic modifications in double-strand break DNA damage signaling and repair. Clin Cancer Res Off J Am Assoc Cancer Res 16:4543–4552. https://doi.org/10.1158/1078-0432.CCR-10-0513

Rouse J, Jackson SP (2002) Lcd1p recruits Mec1p to DNA lesions in vitro and in vivo. Mol Cell 9:857–869

Sabatinos SA, Forsburg SL (2015) Managing single-stranded DNA during replication stress in fission yeast. Biomolecules 5:2123–2139. https://doi.org/10.3390/biom5032123

Sancar A, Thompson C, Thresher RJ, Araujo F, Mo J, Ozgur S, Vagas E, Dawut L, Selby CP (2000) Photolyase/cryptochrome family blue-light photoreceptors use light energy to repair DNA or set the circadian clock. Cold Spring Harb Symp Quant Biol 65:157–171

Sanchez Y, Desany BA, Jones WJ, Liu Q, Wang B, Elledge SJ (1996) Regulation of RAD53 by the ATM-like kinases MEC1 and TEL1 in yeast cell cycle checkpoint pathways. Science 271:357–360

Sanchez Y, Bachant J, Wang H, Hu F, Liu D, Tetzlaff M, Elledge SJ (1999) Control of the DNA damage checkpoint by chk1 and rad53 protein kinases through distinct mechanisms. Science 286:1166–1171

Sandell LL, Zakian VA (1993) Loss of a yeast telomere: arrest, recovery, and chromosome loss. Cell 75:729–739

Sanders SL, Portoso M, Mata J, Bähler J, Allshire RC, Kouzarides T (2004) Methylation of histone H4 lysine 20 controls recruitment of Crb2 to sites of DNA damage. Cell 119:603–614. https://doi.org/10.1016/j.cell.2004.11.009

Sanvisens N, de Llanos R, Puig S (2013) Function and regulation of yeast ribonucleotide reductase: cell cycle, genotoxic stress, and iron bioavailability. Biomed J 36:51–58. https://doi.org/10.4103/2319-4170.110398

Sarkar S, Kiely R, McHugh PJ (2010) The Ino80 chromatin-remodeling complex restores chromatin structure during UV DNA damage repair. J Cell Biol 191:1061–1068. https://doi.org/10.1083/jcb.201006178

Sattlegger E, Hinnebusch AG (2000) Separate domains in GCN1 for binding protein kinase GCN2 and ribosomes are required for GCN2 activation in amino acid-starved cells. EMBO J 19:6622–6633. https://doi.org/10.1093/emboj/19.23.6622

Sattlegger E, Hinnebusch AG (2005) Polyribosome binding by GCN1 is required for full activation of eukaryotic translation initiation factor 2{alpha} kinase GCN2 during amino acid starvation. J Biol Chem 280:16514–16521. https://doi.org/10.1074/jbc.M414566200

Saugar I, Jimenez-Martin A, Tercero JA (2017) Subnuclear relocalization of structure-specific endonucleases in response to DNA damage. Cell Rep 20(7):1553–1562. https://doi.org/10.1016/j.celrep.2017.07.059

Seeber A, Hegnauer AM, Hustedt N, Deshpande I, Poli J, Eglinger J, Pasero P, Gut H, Shinohara M, Hopfner K-P, Shimada K, Gasser SM (2016) RPA mediates recruitment of MRX to forks and double-strand breaks to hold sister chromatids together. Mol Cell 64:951–966. https://doi.org/10.1016/j.molcel.2016.10.032

Sheltzer JM, Blank HM, Pfau SJ, Tange Y, George BM, Humpton TJ, Brito IL, Hiraoka Y, Niwa O, Amon A (2011) Aneuploidy drives genomic instability in yeast. Science 333:1026–1030. https://doi.org/10.1126/science.1206412

Shibata A, Moiani D, Arvai AS, Perry J, Harding SM, Genois M-M, Maity R, van Rossum-Fikkert S, Kertokalio A, Romoli F, Ismail A, Ismalaj E, Petricci E, Neale MJ, Bristow RG, Masson J-Y, Wyman C, Jeggo PA, Tainer JA (2014) DNA double-strand break repair pathway choice is directed by distinct MRE11 nuclease activities. Mol Cell 53:7–18. https://doi.org/10.1016/j.molcel.2013.11.003

Siler J, Xia B, Wong C, Kath M, Bi X (2017) Cell cycle-dependent positive and negative functions of Fun30 chromatin remodeler in DNA damage response. DNA Repair 50:61–70. https://doi.org/10.1016/j.dnarep.2016.12.009

Skoneczna A, Skoneczny M (2017) Mitotic genome variation in yeast and other fungi. In: Somatic genome variation: in animals, plants, and microorganisms. Wiley, Hoboken, NJ

Skoneczna A, Kaniak A, Skoneczny M (2015) Genetic instability in budding and fission yeast-sources and mechanisms. FEMS Microbiol Rev 39:917–967. https://doi.org/10.1093/femsre/fuv028

Smerdon MJ, Thoma F (1990) Site-specific DNA repair at the nucleosome level in a yeast minichromosome. Cell 61:675–684

Smolka MB, Chen S, Maddox PS, Enserink JM, Albuquerque CP, Wei XX, Desai A, Kolodner RD, Zhou H (2006) An FHA domain-mediated protein interaction network of Rad53 reveals its role in polarized cell growth. J Cell Biol 175:743–753. https://doi.org/10.1083/jcb.200605081

Smolka MB, Albuquerque CP, Chen S, Zhou H (2007) Proteome-wide identification of in vivo targets of DNA damage checkpoint kinases. Proc Natl Acad Sci USA 104:10364–10369. https://doi.org/10.1073/pnas.0701622104

Sogo JM, Lopes M, Foiani M (2002) Fork reversal and ssDNA accumulation at stalled replication forks owing to checkpoint defects. Science 297:599–602. https://doi.org/10.1126/science.1074023

Sousa-Lopes A, Antunes F, Cyrne L, Marinho HS (2004) Decreased cellular permeability to H2O2 protects Saccharomyces cerevisiae cells in stationary phase against oxidative stress. FEBS Lett 578:152–156. https://doi.org/10.1016/j.febslet.2004.10.090

Spampinato CP, Gomez RL, Galles C, Lario LD (2009) From bacteria to plants: a compendium of mismatch repair assays. Mutat Res 682:110–128. https://doi.org/10.1016/j.mrrev.2009.07.001

Sparks JL, Chon H, Cerritelli SM, Kunkel TA, Johansson E, Crouch RJ, Burgers PM (2012) RNase H2-initiated ribonucleotide excision repair. Mol Cell 47:980–986. https://doi.org/10.1016/j.molcel.2012.06.035

Spivak G (2016) Transcription-coupled repair: an update. Arch Toxicol 90:2583–2594. https://doi.org/10.1007/s00204-016-1820-x

Steiner S, Wendland J, Wright MC, Philippsen P (1995) Homologous recombination as the main mechanism for DNA integration and cause of rearrangements in the filamentous ascomycete *Ashbya gossypii*. Genetics 140:973–987

Stirling PC, Bloom MS, Solanki-Patil T, Smith S, Sipahimalani P, Li Z, Kofoed M, Ben-Aroya S, Myung K, Hieter P (2011) The complete spectrum of yeast chromosome instability genes identifies candidate CIN cancer genes and functional roles for ASTRA complex components. PLoS Genet 7:e1002057. https://doi.org/10.1371/journal.pgen.1002057

Ström L, Karlsson C, Lindroos HB, Wedahl S, Katou Y, Shirahige K, Sjögren C (2007) Postreplicative formation of cohesion is required for repair and induced by a single DNA break. Science 317:242–245. https://doi.org/10.1126/science.1140649

Svejstrup JQ (2002) Mechanisms of transcription-coupled DNA repair. Nat Rev Mol Cell Biol 3:21–29. https://doi.org/10.1038/nrm703

Svilar D, Goellner EM, Almeida KH, Sobol RW (2011) Base excision repair and lesion-dependent subpathways for repair of oxidative DNA damage. Antioxid Redox Signal 14:2491–2507. https://doi.org/10.1089/ars.2010.3466

Swartz RK, Rodriguez EC, King MC (2014) A role for nuclear envelope-bridging complexes in homology-directed repair. Mol Biol Cell 25:2461–2471. https://doi.org/10.1091/mbc.E13-10-0569

Syed S, Desler C, Rasmussen LJ, Schmidt KH (2016) A novel Rrm3 function in restricting DNA replication via an Orc5-binding domain is genetically separable from Rrm3 function as an ATPase/Helicase in facilitating fork progression. PLoS Genet 12:e1006451. https://doi.org/10.1371/journal.pgen.1006451

Symington LS (2002) Role of RAD52 epistasis group genes in homologous recombination and double-strand break repair. Microbiol Mol Biol Rev MMBR 66:630–670

Tanaka H, Katou Y, Yagura M, Saitoh K, Itoh T, Araki H, Bando M, Shirahige K (2009) Ctf4 coordinates the progression of helicase and DNA polymerase alpha. Genes Cells Devoted Mol Cell Mech 14:807–820. https://doi.org/10.1111/j.1365-2443.2009.01310.x

Tao R, Xue H, Zhang J, Liu J, Deng H, Chen Y-G (2013) Deacetylase Rpd3 facilitates checkpoint adaptation by preventing Rad53 overactivation. Mol Cell Biol 33:4212–4224. https://doi.org/10.1128/MCB.00618-13

Therizols P, Fairhead C, Cabal GG, Genovesio A, Olivo-Marin J-C, Dujon B, Fabre E (2006) Telomere tethering at the nuclear periphery is essential for efficient DNA double strand break repair in subtelomeric region. J Cell Biol 172:189–199. https://doi.org/10.1083/jcb.200505159

Thoma F (1999) Light and dark in chromatin repair: repair of UV-induced DNA lesions by photolyase and nucleotide excision repair. EMBO J 18:6585–6598. https://doi.org/10.1093/emboj/18.23.6585

Thomas D, Scot AD, Barbey R, Padula M, Boiteux S (1997) Inactivation of OGG1 increases the incidence of G . C-->T . A transversions in *Saccharomyces cerevisiae*: evidence for endogenous oxidative damage to DNA in eukaryotic cells. Mol Gen Genet MGG 254:171–178

Tijsterman M, de Pril R, Tasseron-de Jong JG, Brouwer J (1999) RNA polymerase II transcription suppresses nucleosomal modulation of UV-induced (6-4) photoproduct and cyclobutane pyrimidine dimer repair in yeast. Mol Cell Biol 19:934–940

Tkach JM, Yimit A, Lee AY, Riffle M, Costanzo M, Jaschob D, Hendry JA, Ou J, Moffat J, Boone C, Davis TN, Nislow C, Brown GW (2012) Dissecting DNA damage response pathways by analysing protein localization and abundance changes during DNA replication stress. Nat Cell Biol 14(9):966–976. https://doi.org/10.1038/ncb2549

Toczyski DP, Galgoczy DJ, Hartwell LH (1997) CDC5 and CKII control adaptation to the yeast DNA damage checkpoint. Cell 90:1097–1106

Toh GW-L, Lowndes NF (2003) Role of the *Saccharomyces cerevisiae* Rad9 protein in sensing and responding to DNA damage. Biochem Soc Trans 31:242–246 https://doi.org/10.1042/bst0310242

Tornaletti S, Reines D, Hanawalt PC (1999) Structural characterization of RNA polymerase II complexes arrested by a cyclobutane pyrimidine dimer in the transcribed strand of template DNA. J Biol Chem 274:24124–24130

Torres-Ramos CA, Johnson RE, Prakash L, Prakash S (2000) Evidence for the involvement of nucleotide excision repair in the removal of abasic sites in yeast. Mol Cell Biol 20:3522–3528

Travesa A, Duch A, Quintana DG (2008) Distinct phosphatases mediate the deactivation of the DNA damage checkpoint kinase Rad53. J Biol Chem 283:17123–17130. https://doi.org/10.1074/jbc.M801402200

Travesa A, Kalashnikova TI, de Bruin RAM, Cass SR, Chahwan C, Lee DE, Lowndes NF, Wittenberg C (2013) Repression of G1/S transcription is mediated via interaction of the GTB motifs of Nrm1 and Whi5 with Swi6. Mol Cell Biol 33:1476–1486. https://doi.org/10.1128/MCB.01333-12

Tsabar M, Haber JE (2013) Chromatin modifications and chromatin remodeling during DNA repair in budding yeast. Curr Opin Genet Dev 23:166–173. https://doi.org/10.1016/j.gde.2012.11.015

Tsabar M, Waterman DP, Aguilar F, Katsnelson L, Eapen VV, Memisoglu G, Haber JE (2016) Asf1 facilitates dephosphorylation of Rad53 after DNA double-strand break repair. Genes Dev 30:1211–1224. https://doi.org/10.1101/gad.280685.116

Tseng H-M, Tomkinson AE (2004) Processing and joining of DNA ends coordinated by interactions among Dnl4/Lif1, Pol4, and FEN-1. J Biol Chem 279:47580–47588. https://doi.org/10.1074/jbc.M404492200

Tseng S-F, Gabriel A, Teng S-C (2008) Proofreading activity of DNA polymerase Pol2 mediates 3'-end processing during nonhomologous end joining in yeast. PLoS Genet 4:e1000060. https://doi.org/10.1371/journal.pgen.1000060

Unal E, Heidinger-Pauli JM, Koshland D (2007) DNA double-strand breaks trigger genome-wide sister-chromatid cohesion through Eco1 (Ctf7). Science 317:245–248. https://doi.org/10.1126/science.1140637

Unk I, Haracska L, Johnson RE, Prakash S, Prakash L (2000) Apurinic endonuclease activity of yeast Apn2 protein. J Biol Chem 275:22427–22434. https://doi.org/10.1074/jbc.M002845200

Vaisman A, Woodgate R (2015) Redundancy in ribonucleotide excision repair: competition, compensation, and cooperation. DNA Repair 29:74–82. https://doi.org/10.1016/j.dnarep.2015.02.008

van Attikum H, Gasser SM (2009) Crosstalk between histone modifications during the DNA damage response. Trends Cell Biol 19:207–217. https://doi.org/10.1016/j.tcb.2009.03.001

van Attikum H, Fritsch O, Gasser SM (2007) Distinct roles for SWR1 and INO80 chromatin remodeling complexes at chromosomal double-strand breaks. EMBO J 26:4113–4125. https://doi.org/10.1038/sj.emboj.7601835

Vaze MB, Pellicioli A, Lee SE, Ira G, Liberi G, Arbel-Eden A, Foiani M, Haber JE (2002) Recovery from checkpoint-mediated arrest after repair of a double-strand break requires Srs2 helicase. Mol Cell 10:373–385

Vázquez-Martin C, Rouse J, Cohen PTW (2008) Characterization of the role of a trimeric protein phosphatase complex in recovery from cisplatin-induced versus noncrosslinking DNA damage. FEBS J 275:4211–4221. https://doi.org/10.1111/j.1742-4658.2008.06568.x

Veaute X, Jeusset J, Soustelle C, Kowalczykowski SC, Le Cam E, Fabre F (2003) The Srs2 helicase prevents recombination by disrupting Rad51 nucleoprotein filaments. Nature 423:309–312. https://doi.org/10.1038/nature01585

Vialard JE, Gilbert CS, Green CM, Lowndes NF (1998) The budding yeast Rad9 checkpoint protein is subjected to Mec1/Tel1-dependent hyperphosphorylation and interacts with Rad53 after DNA damage. EMBO J 17:5679–5688. https://doi.org/10.1093/emboj/17.19.5679

Vitale I, Galluzzi L, Castedo M, Kroemer G (2011) Mitotic catastrophe: a mechanism for avoiding genomic instability. Nat Rev Mol Cell Biol 12:385–392. https://doi.org/10.1038/nrm3115

Wang P, Byrum S, Fowler FC, Pal S, Tackett AJ, Tyler JK (2017a) Proteomic identification of histone post-translational modifications and proteins enriched at a DNA double-strand break. Nucleic Acids Res 45:10923–10940. https://doi.org/10.1093/nar/gkx844

Wang W, Daley JM, Kwon Y, Krasner DS, Sung P (2017b) Plasticity of the Mre11-Rad50-Xrs2-Sae2 nuclease ensemble in the processing of DNA-bound obstacles. Genes Dev 31:2331–2336. https://doi.org/10.1101/gad.307900.117

Wanrooij PH, Burgers PM (2015) Yet another job for Dna2: checkpoint activation. DNA Repair 32:17–23. https://doi.org/10.1016/j.dnarep.2015.04.009

Wanrooij PH, Tannous E, Kumar S, Navadgi-Patil VM, Burgers PM (2016) Probing the Mec1ATR checkpoint activation mechanism with small peptides. J Biol Chem 291:393–401. https://doi.org/10.1074/jbc.M115.687145

Waters R, van Eijk P, Reed S (2015) Histone modification and chromatin remodeling during NER. DNA Repair 36:105–113. https://doi.org/10.1016/j.dnarep.2015.09.013

Wellinger RE, Thoma F (1997) Nucleosome structure and positioning modulate nucleotide excision repair in the non-transcribed strand of an active gene. EMBO J 16:5046–5056. https://doi.org/10.1093/emboj/16.16.5046

Wiest NE, Houghtaling S, Sanchez JC, Tomkinson AE, Osley MA (2017) The SWI/SNF ATP-dependent nucleosome remodeler promotes resection initiation at a DNA double-strand break in yeast. Nucleic Acids Res 45:5887–5900. https://doi.org/10.1093/nar/gkx221

Willis NA, Zhou C, Elia AEH, Murray JM, Carr AM, Elledge SJ, Rhind N (2016) Identification of S-phase DNA damage-response targets in fission yeast reveals conservation of damage-response networks. Proc Natl Acad Sci USA 113:E3676–E3685. https://doi.org/10.1073/pnas.1525620113

Wilson TE, Grawunder U, Lieber MR (1997) Yeast DNA ligase IV mediates non-homologous DNA end joining. Nature 388:495–498. https://doi.org/10.1038/41365

Wood V, Gwilliam R, Rajandream M-A, Lyne M, Lyne R, Stewart A, Sgouros J, Peat N, Hayles J, Baker S, Basham D, Bowman S, Brooks K, Brown D, Brown S, Chillingworth T, Churcher C, Collins M, Connor R, Cronin A, Davis P, Feltwell T, Fraser A, Gentles S, Goble A, Hamlin N, Harris D, Hidalgo J, Hodgson G, Holroyd S, Hornsby T, Howarth S, Huckle EJ, Hunt S, Jagels K, James K, Jones L, Jones M, Leather S, McDonald S, McLean J, Mooney P, Moule S, Mungall K, Murphy L, Niblett D, Odell C, Oliver K, O'Neil S, Pearson D, Quail MA, Rabbinowitsch E, Rutherford K, Rutter S, Saunders D, Seeger K, Sharp S, Skelton J, Simmonds M, Squares R, Squares S, Stevens K, Taylor K, Taylor RG, Tivey A, Walsh S, Warren T, Whitehead S, Woodward J, Volckaert G, Aert R, Robben J, Grymonprez B, Weltjens I, Vanstreels E, Rieger M, Schäfer M, Müller-Auer S, Gabel C, Fuchs M, Fritzc C, Holzer E, Moestl D, Hilbert H, Borzym K, Langer I, Beck A, Lehrach H, Reinhardt R, Pohl TM, Eger P, Zimmermann W, Wedler H, Wambutt R, Purnelle B, Goffeau A, Cadieu E, Dréano S, Gloux S, Lelaure V, Mottier S, Galibert F, Aves SJ, Xiang Z, Hunt C, Moore K, Hurst SM, Lucas M, Rochet M, Gaillardin C, Tallada VA, Garzon A, Thode G, Daga RR, Cruzado L, Jimenez J, Sánchez M, del RF, Benito J, Domínguez A, Revuelta JL, Moreno S, Armstrong J, Forsburg SL, Cerrutti L, Lowe T, McCombie WR, Paulsen I, Potashkin J, Shpakovski GV, Ussery D, Barrell BG, Nurse P (2002) The genome sequence of Schizosaccharomyces pombe. Nature 415:871. https://doi.org/10.1038/nature724

Woollard A, Nurse P (1995) G1 regulation and checkpoints operating around START in fission yeast. BioEssays News Rev Mol Cell Dev Biol 17:481–490. https://doi.org/10.1002/bies.950170604

Woolstencroft RN, Beilharz TH, Cook MA, Preiss T, Durocher D, Tyers M (2006) Ccr4 contributes to tolerance of replication stress through control of CRT1 mRNA poly(A) tail length. J Cell Sci 119:5178–5192. https://doi.org/10.1242/jcs.03221

Woudstra EC, Gilbert C, Fellows J, Jansen L, Brouwer J, Erdjument-Bromage H, Tempst P, Svejstrup JQ (2002) A Rad26-Def1 complex coordinates repair and RNA pol II proteolysis in response to DNA damage. Nature 415:929–933. https://doi.org/10.1038/415929a

Wu X, Braithwaite E, Wang Z (1999) DNA ligation during excision repair in yeast cell-free extracts is specifically catalyzed by the CDC9 gene product. Biochemistry (Mosc) 38:2628–2635. https://doi.org/10.1021/bi982592s

Wyatt HDM, West SC (2014) Holliday junction resolvases. Cold Spring Harb Perspect Biol 6: a023192. https://doi.org/10.1101/cshperspect.a023192

Wysocki R, Javaheri A, Allard S, Sha F, Côté J, Kron SJ (2005) Role of Dot1-dependent histone H3 methylation in G1 and S phase DNA damage checkpoint functions of Rad9. Mol Cell Biol 25:8430–8443. https://doi.org/10.1128/MCB.25.19.8430-8443.2005

Yang Y, Gordenin DA, Resnick MA (2010) A single-strand specific lesion drives MMS-induced hyper-mutability at a double-strand break in yeast. DNA Repair 9:914–921. https://doi.org/10.1016/j.dnarep.2010.06.005

Yasutis KM, Kozminski KG (2013) Cell cycle checkpoint regulators reach a zillion. Cell Cycle Georget Tex 12:1501–1509. https://doi.org/10.4161/cc.24637

Yeung M, Durocher D (2011) Srs2 enables checkpoint recovery by promoting disassembly of DNA damage foci from chromatin. DNA Repair 10:1213–1222. https://doi.org/10.1016/j.dnarep.2011.09.005

Yimit A, Riffle M, Brown GW (2015) Genetic regulation of Dna2 localization during the DNA damage response. G3 Bethesda Md 5:1937–1944. https://doi.org/10.1534/g3.115.019208

Yuen KWY, Warren CD, Chen O, Kwok T, Hieter P, Spencer FA (2007) Systematic genome instability screens in yeast and their potential relevance to cancer. Proc Natl Acad Sci USA 104:3925–3930. https://doi.org/10.1073/pnas.0610642104

Zámborszky J, Csikász-Nagy A, Hong CI (2014) Neurospora crassa as a model organism to explore the interconnected network of the cell cycle and the circadian clock. Fungal Genet Biol FG B 71:52–57. https://doi.org/10.1016/j.fgb.2014.08.014

Zhang Y, Hefferin ML, Chen L, Shim EY, Tseng H-M, Kwon Y, Sung P, Lee SE, Tomkinson AE (2007) Role of Dnl4–Lif1 in nonhomologous end-joining repair complex assembly and suppression of homologous recombination. Nat Struct 38 Mol Biol 14:639–646. https://doi.org/10.1038/nsmb1261

Zheng D-Q, Zhang K, Wu X-C, Mieczkowski PA, Petes TD (2016) Global analysis of genomic instability caused by DNA replication stress in Saccharomyces cerevisiae. Proc Natl Acad Sci USA 113:E8114–E8121. https://doi.org/10.1073/pnas.1618129113

Zhou C, Elia AEH, Naylor ML, Dephoure N, Ballif BA, Goel G, Xu Q, Ng A, Chou DM, Xavier RJ, Gygi SP, Elledge SJ (2016) Profiling DNA damage-induced phosphorylation in budding yeast reveals diverse signaling networks. Proc Natl Acad Sci USA 113:E3667–E3675. https://doi.org/10.1073/pnas.1602827113

Zhu Z, Chung W-H, Shim EY, Lee SE, Ira G (2008) Sgs1 helicase and two nucleases Dna2 and Exo1 resect DNA double-strand break ends. Cell 134:981–994. https://doi.org/10.1016/j.cell.2008.08.037

Zou L, Liu D, Elledge SJ (2003) Replication protein A-mediated recruitment and activation of Rad17 complexes. Proc Natl Acad Sci USA 100:13827–13832. https://doi.org/10.1073/pnas.2336100100

Zyrina AN, Sorokin MI, Sokolov SS, Knorre DA, Severin FF (2015) Mitochondrial retrograde signaling inhibits the survival during prolong S/G2 arrest in Saccharomyces cerevisiae. Oncotarget 6:44084–44094. https://doi.org/10.18632/oncotarget.6406

The Nutrient Stress Response in Yeast

4

Vasudha Bharatula and James R. Broach

Abstract

Nutrients such as glucose and glutamine provide yeast not only with substrates for energy production and mass accumulation but also with signals specifying the metabolic and transcriptional program appropriate for the condition in which the yeast cells find themselves. In nutrient replete conditions, nutrient signaling pathways activate growth programs promoting both continuous mass accumulation and the discontinuous process of the cell cycle. Under nutrient limiting conditions, these pathways activate a canonical stress response program and attenuate cell growth in proportion to the extent of nutrient limitation. In this review, we describe recent results elaborating the nature of these nutrient signaling pathways. In addition, we describe the short-term and long-term consequences of nutrient limitation on both the canonical stress response and the regulation of cell growth. We highlight the stochastic nature of the stress response, which permits genetically identical cells in a common environment to pursue distinct survival strategies, maximizing the potential for overall persistence of the clonal population regardless of subsequent environmental conditions. Finally, we examine the mechanistic connection between activation of the stress response and attenuation of cell growth. These recent results provide insight not only into the biology of yeast but also on homologous signaling pathways and stress responses of larger eukaryotes.

V. Bharatula · J. R. Broach (✉)
Department of Biochemistry and Molecular Biology, Penn State College of Medicine, Hershey, PA, USA
e-mail: jbroach@hmc.psu.edu

© Springer Nature Switzerland AG 2018
M. Skoneczny (ed.), *Stress Response Mechanisms in Fungi*,
https://doi.org/10.1007/978-3-030-00683-9_4

4.1   Introduction

Microorganisms vigilantly balance growth versus quiescence in order to maximize survival in the face of an unpredictable and often rapidly changing environment. Changes in the quality and quantity of nutrients, exposure to xenobiotics, shifts in temperature, etc. can trigger physiological, metabolic, developmental, and growth rate changes designed for adaptation to the new conditions. Yeast serves as an excellent model to study evolutionarily conserved processes linking environmental perturbations to modulation of cellular physiology, metabolism, and growth. Many of these environmental changes activate conserved signaling pathways that elicit the environmental stress response—a rapid, canonical reorganization of the transcriptome that prepares cells to weather subsequent stresses. Transcriptional reprograming is accompanied by changes in growth rates leading to transient growth arrest, quiescence, and even death depending on the type, duration, and intensity of the stress.

In this chapter, we discuss changes in gene expression and cell development in conditions of nutrient deprivation. In addition, we describe the signaling pathways actuating those cellular changes and how those changes ultimately effect cellular adaptation and survival. A previous review (Broach 2012) provides a detailed description of the pathways regulating nutrient responses and the consequences of altered nutrient availability. Here we summarize those observations and highlight data that has accumulated since that review. In particular, single-cell studies of yeast have recently demonstrated that signaling pathways exhibit a level of stochasticity that results in heterogeneous responses to stresses even in isogenic cells. This stochasticity in stress signaling creates subpopulations of cells, some of which are able to resume growth quickly when conditions are favorable again and others of which can survive under extended duress. The population-wide variation in response maximizes overall survival in face of an uncertain future. We discuss proposed mechanisms underlying signaling stochasticity and the mutually exclusive nature of stress response versus growth. In particular, we address whether gene expression changes upon stress are the response to or the cause of attenuation of growth rate.

4.2   Nutrient Response Pathways

All cells have the ability to sense changes in external environment and adjust cellular growth rates accordingly. Conserved signaling pathways form the backbone of environment sensing and response mechanisms. For yeast, nutrients serve not only as sources of energy and anabolites but also as signaling molecules that directly regulate transcription and development, which are linked to cellular growth and division. Yeast reside in diverse environments from rotting fruit, insect guts, and even humans, utilizing conserved nutrient sensing and response mechanisms to regulate growth. In unfavorable conditions such as the absence of nutrients, yeast switch from growth mode to a survival mode by activating the response mechanisms to promote distinct developmental programs such as transient growth arrest, quiescence, or sporulation. Ras/protein kinase A (PKA), TORC1 kinase, and the Snf1/AMP-activated protein kinase pathways are essential for sensing nutrient availability and deploying the appropriate developmental programs through modulation of cell metabolism, transcription, translation, and posttranslation processes.

4.2.1 Carbon Responsive Networks

Yeasts grow on a variety of substrates and utilize several types of fermentative and non-fermentative carbon sources for energy production and anabolic processes (Johnston and Carlson 1992). Yeast cells prefer the fermentable sugars glucose or fructose as source of carbon for energy and growth. This preference for glucose over other carbon sources is made possible by allosteric inhibition of catabolic enzymes and transcriptional repression of genes required for respiration and for metabolism of other carbon sources.

Repression of respiration in the presence of glucose is called the Crabtree effect. *S. cerevisiae* catabolize glucose to ethanol even in aerobic conditions, despite the fact that aerobic fermentation yields fewer ATP molecules. One possible reason for such a preference is that production of ethanol in the wild creates an unfavorable environment for other organisms (Thomson et al. 2005). Additionally, by shunting glucose away from the tricarboxylic acid cycle (TCA) and mitochondria for ATP synthesis, fewer reactive oxygen species (ROS) are produced, reducing the incidence of ROS-induced DNA damage during replication (Chen et al. 2007; Silverman et al. 2010). However, perhaps the most compelling explanation for the Crabtree effect is related to its similarity to the Warburg effect observed in cancer cells in which the cell shunts glucose into lactate rather than mitochondrial respiration. While both these processes seem energy inefficient, they facilitate fast ATP production and provide a reservoir of anabolic precursors enabling cells to grow rapidly.

Given the fact that glucose is the preferred source of carbon, removal or addition of glucose leads to large-scale changes in phosphorylation of proteins and transcription of genes (Kresnowati et al. 2006; Wang et al. 2004; Zaman et al. 2009). Overlapping signaling pathways modulate changes in proliferation and gene expression in cells upon glucose starvation or upon addition of glucose to starved cells. The PKA pathway modulates changes in growth, metabolism, and stress responses. The Snf1 pathway is required for utilizing alternative sources of carbon. Regulation of glucose intake through transporters is regulated by Rgt3/Snf3 (see Fig. 4.1).

Ras/Protein Kinase a Pathway A majority of the rapid transcriptional changes in yeast on a carbon source upshift, e.g., shifting from growth on glycerol to growth on glucose, occurs as a consequence of signaling through the PKA pathway. Absence of signaling through PKA abrogates most of these responses (Zaman et al. 2009). PKA is a heterotetrameric complex consisting of two regulatory subunits encoded by *BCY1* and two catalytic subunits encoded by three partially redundant genes *TPK1*, *TPK2*, and *TPK3*. The presence of at least one of these catalytic subunits genes is required for normal growth, though individually these genes are nonessential (Toda et al. 1987). The three catalytic subunits share more that 75% identity in the

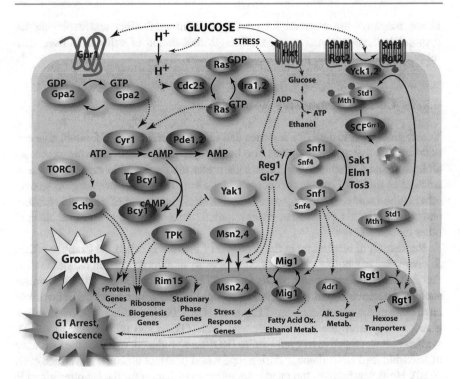

Fig. 4.1 Glucose response pathways. The Ras/PKA (TPK) pathway plays a central role in regulating growth versus quiescence in response to the quality and quantity of the available carbon source, primarily by stimulating mass accumulation and inhibiting the stress response. The major input proceeds through Ras, possibly in response to glucose-stimulated intracellular acidification, with minor input through the G-protein-coupled receptor, Gpr1. The Snf1 and Rgt complexes regulate use of alternate carbon sources, primarily through regulation of a constellation of transcriptional factors. Rgt1 responds to glucose levels through two membrane glucose sensors. Snf1 responds to glucose through the Reg1/Glc7 protein phosphatase 1 and through cellular energy charge, in that binding ADP inhibits removal of the activation loop phosphorylation. Various stresses, including carbon source downshifts, likely impinge on Snf1 and the major stress-responsive transcription factors, Msn2/4, through the Reg1/Glc7 phosphatase. Red dots signify phosphorylation

C-terminal regions of the protein but have heterogeneous N-terminal domains, suggesting that each catalytic subunit is capable of effecting distinct processes along with the redundant functions (Toda et al. 1987). High-throughput studies have demonstrated that all three catalytic subunits have unique targets and promote expression of different downstream genes (Ptacek et al. 2005).

PKA activity is regulated by cAMP, which is produced by adenylyl cyclase and degraded by phosphodiesterases Pde1 and Pde2. Adenylyl cyclase activity is regulated by GTP-bound Ras1 and Ras2 whose level is controlled by a balance between the activities of the guanine nucleotide exchange factor Cdc25 and the GTPase activating proteins, Ira1 and Ira2, that enhance Ras's intrinsic GTPase activity (Colombo et al. 1998). The mechanism by which glucose feeds into this regulatory network to stimulate PKA remains unknown, but cytoplasmic pH has been implicated in the process (Dechant et al. 2010; Isom et al. 2017).

A second kinase, Sch9, a homolog of mammalian S6 kinase, was previously implicated in glucose signaling on the basis of the fact that overexpression of Sch9 suppresses loss-of-function *pka* mutants (Toda et al. 1988). However, subsequent analysis demonstrated that *sch9* mutants are not impaired in eliciting a glucose-mediated transcriptional response (Zaman et al. 2009). Rather, Sch9 is phosphorylated and activated by TORC1 (see below), documenting its central role in nitrogen/amino acid nutrient signaling. The ability of Sch9 overexpression to suppress PKA loss of function likely derives from the fact that both the glucose signaling and nitrogen signaling pathways impinge on highly overlapping sets of genes, namely, those involved in ribosome biogenesis and translation (Lippman and Broach 2009).

The Gα protein, Gpa2, and the associated G-protein-coupled receptor, Gpr1, have also been implicated in glucose signaling. Gpa2 can interact with adenylyl cyclase in vitro (Peeters et al. 2006). Moreover, expression of an activated variant of Gpa2 yields a similar but muted pattern of transcriptional changes to that obtained by induction of an activated variant of Ras2, and those expression changes are dependent on the activity of PKA (Zaman et al. 2009). However, elimination of Gpa2 or Grr1 does not alter glucose-induced transcriptional changes (Zaman et al. 2009). While Gpr1 has been proposed as a glucose receptor activating Gpa2, no direct pharmacologic evidence for glucose as a ligand for the protein has been reported (Lemaire et al. 2004). Moreover, *gpr1Δ* is synthetically lethal with *sch9Δ*, and both mutants yield identical transcriptional changes during growth on non-fermentable carbon sources, and epistasis studies place Gpa2 upstream of Sch9 (Xue et al. 1998; Zaman et al. 2009). Thus, while Gpa2 appears to play a minor role in glucose signaling through PKA, current data suggest a more direct role of Gpa2/Grr1 in signaling through Sch9.

The reduction in cAMP levels and attendant reduction in PKA contribute to the transcriptional and physiological reorganization following transition from a rich to a poor carbon source (Lippman and Broach 2009). In part, these changes are a consequence of the loss of PKA-stimulated activation of growth promoting genes, such as those required for ribosome biogenesis (see below). In part, though, these changes result from loss of PKA-mediated inhibition of the environmental stress response, primarily those mediated by the Msn2/Msn4 transcription factors (Gorner et al. 1998, 2002). These responses are elaborated below. In the inverse process, PKA activation following a carbon source upshift results in substantial transcriptional reprogramming of the cell. Those genes induced by PKA activation include those required for increased translational capacity, thereby enhancing the cell's potential for subsequent mass accumulation. In particular, PKA stimulates RNA polymerase II-mediated ribosomal protein gene transcription and genes required for ribosome biogenesis as well as RNA polymerase I-mediated rRNA transcription and RNA polymerase III-mediated tRNA synthesis (Broach 2012). Those genes repressed by PKA activation predominantly encompass those comprising the environmental stress response (Gasch et al. 2000).

Snf1 and Hap Regulatory Complex The transition from a rich to a poor carbon source also results in suppression of energy-consuming activities and induction of energy-generating processes. These changes are mediated by a Snf1 kinase complex, which alleviates repression of a number of genes required for metabolism of carbon sources other than glucose and activates a number of genes required for gluconeogenesis and fatty acid oxidation. Overall, the Snf1 complex regulates expression of over 400 genes through inactivation of the Mig1 repressor and stimulation of the Adr1 transcriptional activator. The Snf1 complex comprises the catalytic (α) subunit, Snf1, a homolog of the mammalian AMP-activated protein kinase (AMPK); the regulatory β subunit, Snf4; and one of the three γ subunits, Gal38, Sip1, and Sip2. The regulatory subunits modulate Snf1 substrate specificity and subcellular localization. Unlike mammalian AMPK, which is activated predominantly by AMP and thus responds to low energy charge in the cell, yeast Snf1 is activated by phosphorylation of its activation loop, catalyzed by upstream kinases Elm1, Tos3, and Sak1 (Hong et al. 2003; Nath et al. 2003; Sutherland et al. 2003) and inactivated by dephosphorylation by the Glc7-Reg1 protein phosphatase 1 and the Sit4 type 2A phosphatase (Ruiz et al. 2011). However, the activities of these kinases and phosphatases are unchanged by alterations in glucose level. Rather, binding of ADP may induce a conformational change that obstructs access of phosphatases to the activation loop, thus rendering yeast Snf1 responsive also to energy charge (Chandrashekarappa et al. 2011; Mayer et al. 2011).

A recent report suggests that Snf1 activity in response to carbon source may be regulated in part by glucose-induced but reversible protein aggregation. In addition to its role in hexose transporter gene regulation (see below), Std1 is a Snf1 activator protein (Kuchin et al. 2003). In glucose-replete conditions, Std1 forms reversible but nonfunctional aggregates at nuclear-vacuole junctions, which can be reversed by interaction with a protein, Sip5. The protein kinase Vhs1 phosphorylates Sip5 in response to glucose, which prevents the interaction of Sip5 with Std1, allowing aggregation of Std1 and subsequent reduction of Snf1 activity. This model suggests a physiological role for protein aggregation in regulating nutrient responsive networks (Simpson-Lavy et al. 2017).

The Hap2/3/4/5 complex represses genes regulating respiration and oxidative phosphorylation independent of PKA and Snf1 (Zaman et al. 2009). This complex plays a key role in switching from fermentation to respiration following diauxic shift. Levels of Hap4, the activation domain of the complex, increase in response to glucose limitation and induce gene expression (Lascaris et al. 2003). How starvation activates the Hap complex remains unknown.

Rgt and Glucose Transport Glucose also regulates levels of hexose transporters present on the cell membrane (Kaniak et al. 2004; Zaman et al. 2009). Rgt1 along with the co-repressors Mth1 and Std1 repress *HXT1-4* in the presence of glucose (Lakshmanan et al. 2003; Mosley et al. 2003). Glucose sensors Snf3 and Rgt2 respond to external glucose and recruit Mth1 and Std1 to the plasma membrane for degradation, relieving *HXT* genes from repression by Rgt1 (Karhumaa et al. 2010; Snowdon and Johnston 2016; Wu et al. 2006). Snf1 enhances repression activity of Rgt1 through phosphorylation in the absence of glucose. Some hexose transporters are repressed by Mig1, which is also phosphorylated by Snf1 to exclude it from the nucleus. Hence Snf1 can regulate both activation and repression of hexose transporters (Johnston and Kim 2005).

Yak1 and Protein Phosphatase 1 Yak1, a dual specificity kinase, lies in the PKA pathway but acts in opposition to it: while PKA stimulates growth and inhibits the stress response, Yak1 inhibits growth and stimulates the stress response. PKA phosphorylates Yak1, which increases its affinity to 14-3-3 proteins in the cytoplasm and sequesters it from the nucleus (Budovskaya et al. 2005; Lee et al. 2011; Ptacek et al. 2005). Starvation diminishes PKA activity, allowing Yak1 to translocate into the nucleus to influence stability and translation of stress-induced transcripts (Lee et al. 2011; Moriya et al. 2001). Yak1 also directly influences the stress response to starvation by phosphorylation of Msn2, Msn4, and Hsf1 (Lee et al. 2008).

The Glc7 protein phosphatase 1 plays a critical role in glucose sensing. Glc7 promotes glucose repression by attenuating Snf1 activity through its interaction with Reg1, which in turn activates Mig1 (Tu and Carlson 1995). Glc7 is involved in the phosphorylation and degradation of maltose permeases Yck1,2 (Gadura et al. 2006). Msn2, a major stress-responsive transcription factor, is activated by Glc7 through dephosphorylation of its nuclear localizations sequence (Gorner et al. 2002).

4.2.2 Nitrogen-Responsive Networks

Cells require nitrogen for growth and biosynthesis. Limiting nitrogen results in slow growth due to expansion of the G1 phase of the cell cycle (Brauer et al. 2008). While yeast can utilize nitrogen from several sources, glutamine and ammonia are the preferred nitrogen sources. In the presence of a preferred nitrogen source, yeast repress expression of genes required for nitrogen utilization from non-preferred sources, a process designated nitrogen catabolite repression or NCR (Magasanik and Kaiser 2002). NCR also modulates uptake of amino acids by regulating amino acid permeases such as Gap1 at the cell membrane via posttranslational modifications (Magasanik and Kaiser 2002). Availability of a preferred nitrogen source also suppresses autophagy, the recycling of macromolecules and organelles through proteolysis to yield free amino acids (Yang and Klionsky 2009).

Approximately 90 genes are repressed by NCR through the activity of four GATA family transcription factors, Gln3 and Gat1 activators and Dal80 and Gzf3 repressors (Cooper 2002; Magasanik and Kaiser 2002). Subcellular localization of these transcription factors varies with the type of nitrogen source. In the presence of ammonia or glutamine, Gln3 and Gat1 reside in the cytoplasm. Shifting cells to non-preferred nitrogen sources or no nitrogen source causes Gln3 and Gat1 to translocate into the nucleus (Bertram et al. 2000; Cardenas et al. 1999; Li et al. 2017).

Yeast convert nitrogen from non-preferred sources into ammonia and synthesize glutamate by condensing ammonia with α-ketoglutarate. Generation of α-ketoglutarate requires expression of the first three genes in the TCA cycle, which are normally repressed during growth on glucose. Thus, under growth in poor nitrogen and good carbon, activators in the RTG pathway specifically induce expression of these genes for ammonia assimilation. Rtg1 and Rtg3 form a heterodimeric complex, which translocates into the nucleus when nitrogen is depleted to activate target genes (Liu and Butow 1999). Localization of Rgt1 and Rgt3 is mediated through complex interaction with Mks1, Rtg2, and Bmh1/2 (Liu and Butow 2006).

TORC1 The yeast Tor complex I mediates a significant fraction of the nitrogen catabolite repression, in addition to integrating multiple nutrient and stress stimuli to coordinate cell growth and promote cellular homeostasis (see Fig. 4.2). The two yeast Tor proteins are serine/threonine kinases organized into two separate complexes with distinct upstream regulators and downstream substrates and functions. Tor complex 2 (TORC2), consisting of Tor2 and several ancillary proteins, regulates lipid metabolism and glycerol production through modulation of the Akt homolog, Ypk1, to maintain plasma membrane tension homeostasis (Eltschinger and Loewith 2016) but is not involved in response to nitrogen availability. On the other hand, Tor complex I (TORC1), consisting of Tor1 or Tor2 and associated proteins, is inhibited by the macrolide rapamycin and responds to the quality and quantity of nitrogen sources: its activity is diminished upon nitrogen starvation or in the presence of a poor nitrogen source and increased in the presence of a readily metabolized nitrogen source (Binda et al. 2009; De Virgilio and Loewith 2006; Nicastro et al. 2017). Moreover, treatment of cells with rapamycin elicits an NCR-like response. TORC1 regulates the subcellular localization of Gat1 and Gln3 by two distinct mechanisms. Upon removal of a high-quality nitrogen source, TORC1 unfetters and activates a protein phosphatase regulatory subunit, Tap42, which, along with Sit4 and other protein phosphatase 2A-like catalytic subunits, dephosphorylates the transcription factors, releasing them to the nucleus. In contrast, upon transition from a high-quality to a low-quality nitrogen source, Gln3 transitions to the nucleus in a TORC1-dependent but Tap42-independent manner and without obvious dephosphorylation (Li et al. 2017). However, while inhibition of TORC1 by rapamycin mimics the NCR, it does not fully recapitulate it, suggesting an additional regulatory pathway, possibly involving the general amino acid control pathway (Tate et al. 2009, 2017; Tate and Cooper 2003).

Recent observations have clarified, although not fully illuminated, the regulatory circuit that connects nitrogen availability to TORC1 activity (Nicastro et al. 2017). The proteins, Gtr1 and Gtr2, which are members of the Rag family of guanine nucleotide binding proteins, form a heterodimer and regulate TORC1 activity through a guanine nucleotide switch. Gtr1 bound to GTP and Gtr2 bound to GDP stimulate TORC1 activity, while Gtr1 bound to GDP and Gtr2 bound to GTP inhibit TORC1. Thus, regulation of TORC1 involves regulation of the guanine nucleotide states of the Gtr1/Gtr2 complex. The Gtr1/Gtr2 heterodimer is tethered to the vacuolar membrane as part of the Ego complex, which also comprises Ego1, Ego2, and Ego3 (Dubouloz et al. 2005). At the vacuolar membrane, the Ego complex can interact with Vam6, which serves as a guanine nucleotide exchange factor to load GTP onto Gtr1, and with a complex termed SEACIT, comprising Iml1,

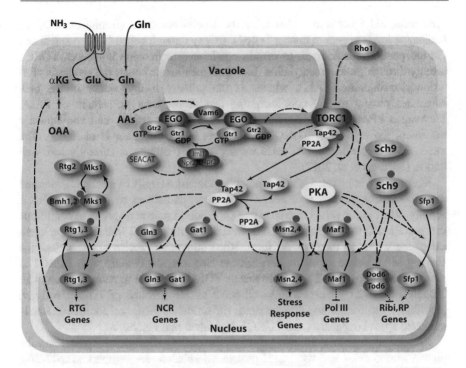

Fig. 4.2 TORC1 and nitrogen regulation. The TORC1 complex adjusts growth as well as expression of genes required for use of alternate nitrogen sources in response to the quality and quantity of available nitrogen sources through regulation of transcriptional activators (blue icons) and repressors (red icons). TORC1 senses nitrogen compound amounts and quality through intracellular amino acid levels, transmitted through a guanine nucleotide switch in the Ego complex, regulates growth primarily through Sch9, and regulates stress and alternative nitrogen source utilization through protein phosphatase 2A. Guanine nucleotide states of the EGO complex are set by competing activities of a guanine nucleotide exchange factor, Vam6, and a GTPase stimulating factor, the SEACIT complex comprising Npr2, Npr3, and Iml1, which itself is inhibited by the SEACAT complex

Npr2, and Npr3 (Neklesa and Davis 2009), which acts as a GTPase activating protein for Gtr1. SEACIT is itself regulated by a complex of five proteins, termed SEACAT, which inhibits SEACIT and thus activates TORC1. In sum, the balance between the SEACIT GTPase activity and the Vam6 guanine nucleotide exchange activity sets the nucleotide status of Gtr1/Gtr2 in the Ego complex, which in turn serves as an immediate upstream regulator of TORC1.

The primary signal(s) for the nitrogen response pathway is the level of intracellular amino acids. For instance, mutations in *GLN1*, which affects glutamine synthetase activity, elicit a transcriptional response similar to TORC1 inhibition (Magasanik and Kaiser 2002). However, how these levels connect to TORC1 activity and whether all amino acids are sensed through the same mechanism are still unresolved. Leucine levels may impact on TORC1 activity through the Gtr complex via the leucyl tRNA synthetase, which binds to Gtr1 in a leucine dependent manner (Bonfils et al. 2012). Glutamine levels appear to impact TORC1 activity through both a Gtr-dependent and Gtr-independent mechanism. In certain conditions, methionine influences TORC1 activity through protein methylation. Increased methionine boosts the synthesis of *S*-adenosyl-methionine (SAM), which, as the primary methyl donor in the cell, stimulates protein methylation, including that of the catalytic subunit of PP2A. Methylated PP2A dephosphorylates Npr2, reducing its activity and consequently increasing the activity of TORC1 (Laxman et al. 2014; Sutter et al. 2013). The means by which other amino acids impact TORC1 activity is currently less well known (Nicastro et al. 2017).

Studies have implicated Rho1, a member of the Rho family of small GTPases and a component of the cell wall integrity pathway, as another regulator of TORC1 activity. Rho1 is converted into its active GTP-bound state in response to conditions that perturb cell wall integrity, such as treatment with the cell wall damaging agent calcofluor white. In its activated form, Rho1 stimulates protein kinase C and the ensuing MAP kinase signaling network but also interacts directly with TORC1 to promote release of Tap42 and to dissociate TORC1 from the membrane (Yan et al. 2012). Thus, Rho1 couples a cell wall stress to transient loss of TORC1 activity, providing a mechanism to integrate stress and nutrient signals and coordinate Rho1-mediated spatial expansion and TORC1-dependent mass increase.

As alluded to above, the two primary downstream effectors of TORC1 are the protein kinase Sch9 and the Tap42-protein phosphatase 2A complex (Tap42-2A), consisting of alternative catalytic subunits Sit4, Pph21, Pph22, and Pph3 and the two regulatory subunits Tap42 and either Rrd1 or Rrd2 (Jiang and Broach 1999; Zheng and Jiang 2005). Active TORC1 phosphorylates and activates Sch9, whereas TORC1 phosphorylates and sequesters Tap42 to restrain its activity (Urban et al. 2007). Inactivation of TORC1 results in transient release of Tap42, which then targets the phosphatase catalytic subunits to specific proteins. Phosphoproteomic and transcriptomic studies of TORC1 signaling revealed that Sch9 and Tap42-2A branches of TORC1 signaling have overlapping as well as distinct functions. For instance, as noted above, Tap42 is required for dephosphorylation of Gln3 and Gat1 (Beck and Hall 1999; Cardenas et al. 1999; Tate et al. 2009) and expression of genes in the RTG pathway upon rapamycin treatment or nitrogen depletion (Duvel et al. 2003). Sch9, on the other hand, predominantly serves to stimulate ribosome biogenesis through activation of RNA polymerase I transcription of rDNA, RNA polymerase III transcription of tRNA genes, and RNA polymerase II transcription of ribosome protein and ribosome biogenesis genes. Sch9 also functions to inhibit macroautophagy. Some transcriptional and phosphoproteomic changes are independent of both Sch9 and Tap42, suggesting the possible existence of additional TORC1 downstream effectors (Breitkreutz et al. 2010; Huber et al. 2009).

TORC1 and PKA are the major nutrient sensing pathways and converge on a variety of critical growth and developmental processes. Both PKA and Sch9 impinge on biogenesis of translational capacity through regulation of transcription of genes for ribosomal proteins, ribosomal biogenesis, and ribosomal RNAs and tRNAs. Similarly, both these pathways regulate autophagy and quiescence. However, the two different pathways effect these processes in distinct but complementary processes (Broach 2012; Lippman and Broach 2009) that have distinct logical properties. Their regulation of translational components essentially constitute an AND gate such that both pathways have to be active in order to promote accumulation of translational capacity. This insures that absence of either an adequate carbon source or an adequate nitrogen source restricts further accumulation of translational capacity. In contrast, their regulation of autophagy and quiescence constitutes an OR gate such that diminished signaling through either pathway activates autophagy in the short term and promotes quiescence in the long term (Pedruzzi et al. 2003; Yorimitsu et al. 2007). Thus, the two pathways could simply constitute conduits for two different nutrient sources impinging on the same central processes. However, an alternative or additional component to this overlapping regulatory process could be the time frame in which, or stimulus to which, regulation is achieved. In particular, PKA may predominantly facilitate rapid re-equilibration following nutrient transition, whereas TORC1 may promote homeostasis over extended time. In support of the latter proposition, TORC1 appears to be subject to various forms of negative feedback (Eltschinger and Loewith 2016). For instance, inhibition of translation, which liberates amino acids, stimulates TORC1 and in turn stimulates translation. Similarly, activation of macroautophagy, which also generates intracellular amino acids, stimulates TORC1, which inhibits macroautophagy. Thus, TORC1 is well positioned to restore internal homeostasis following minor perturbations in the cellular processes, whereas PKA is poised and sensitized to respond to a variety of external environmental changes.

4.3 Response to Nutrient Limitation

Nutrient deprivation has a profound effect on the metabolism, physiology, and developmental programs of a cell both on a short-term and long-term fashion. These have been described in detail for *Saccharomyces* previously (Broach 2012), and here we focus primarily on recent observation regarding the stress response and the related effect on cellular growth.

4.3.1 Regulation of Cellular Growth

Cell Cycle Regulation Saccharomyces cerevisiae reproduces by budding. New mass and organelles produced by continuous synthesis in the mother cell are deposited in a bud that grows to particular size, after which it separates into an individual cell. Concurrently, the discontinuous process of genome duplication and cell division progress as distinct stages of the cell cycle. The budding cycle in yeast provides a morphological indicator of the different stages of the cell cycle (Hartwell 1974), an observation that served as the basis of several elegant studies using genetic mutants to determine the molecular mechanisms of cell cycle regulation in yeast (Hartwell et al. 1970) (see Fig. 4.3).

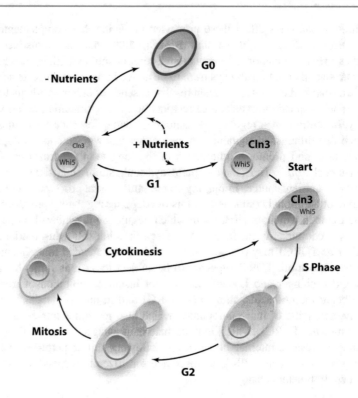

Fig. 4.3 The yeast growth and cell cycle. The discontinuous process of cell division, comprising DNA replication, chromosome segregation, and cytokinesis, is coupled to the growth cycle by requiring that the daughter cell achieve a certain "size" before initiating the cell division cycle. This occurs at "Start," when Whi5 exits the nucleus as a consequence of the increasing accumulation through growth during G1 of Cyclin 3, which titrates out the fixed amount of Whi5 previously deposited in the nucleus. In the absence of nutrients, newly born cells exit the cell cycle and enter one of a number of quiescent states, collectively referred to as G0

The continuous process of mass accumulation and the discontinuous process of genome duplication are coupled at the G1 stage of the cell cycle (Johnson and Skotheim 2013). Coupling of the two processes results from the fact that the discontinuous, or cell division, cycle generally requires less time than that needed for doubling the mass of the cell. To coordinate these two facets of cell growth at each cell cycle, initiation of the cell division cycle occurs in G1, at a point referred to as "Start," only after cells have attained an appropriate "size," which is essentially equivalent to that of the initial mass of the previous cell. Start can be defined as the point at which a cell becomes committed to completing the next cell division cycle even upon withdrawal of all nutrients or exposure to growth inhibitory signals such as mating pheromone. Functionally, Start can be defined in *Saccharomyces* as the point in the cell cycle at which the transcriptional repressor, Whi5, exits the nucleus (Costanzo et al. 2004). Nuclear exit results from Whi5 phosphorylation by the cyclin-dependent kinase, Cdc28, in complex with the G1 cyclin, Cln3, and results in transcriptional activation of a cohort of genes that comprise the initial set of the cascading series of genes responsible for executing the sequential stages of the cell cycle.

Nutrients impinge on execution of Start in two respects (Jorgensen and Tyers 2004). First, nutrients are required for attaining the appropriate size through promotion of mass accumulation. Nutrient regulation of mass accumulation is discussed below. The mechanism by which a cell measures its own size has been elusive but has recently been proposed to involve a dilution process by which Whi5 is titrated by Cdc28/Cln3 (Schmoller and Skotheim 2015; Schmoller et al. 2015). Specifically, the model rests on the observation that a constant amount of Whi5 resides in the cell during G1, since its synthesis occurs only during the G2/M phase of the cell cycle. In contrast, Cdc28/Cln3 are continually synthesized and thus are present in the cell at a constant concentration, albeit at an increasing amount as the cell mass increases. Start occurs when the amount of Cdc28/Cln3, which is proportional to cell size, functionally exceeds that of Whi5 and thereby inactivates it through phosphorylation and removal from the nucleus.

Nutrients also impinge on the Start by setting the actual size threshold (Jorgensen and Tyers 2004). For instance, cells growing on limiting nitrogen initiate cell division at a smaller size than cell growing on a rich nitrogen source. The mechanism by which nutrients set the size threshold is unknown but may be related to the recent observation that the size of the cell at Start is highly correlated with the rate of mass accumulation in G1 (Ferrezuelo et al. 2012). Thus, cells in a poor nutritional environment accumulate mass slowly and thus initiate Start at a smaller size than cells in a nutrient-rich environment. How this observation integrates into the titration model above requires further exploration. How nutrition controls mass accumulation rate also requires further study, although the rate of mass accumulation is likely related to the cell-wide translational rate, which is a product of the number of translational units, i.e., active ribosomes, in the cell and the efficiency of translation of each individual translational unit. As noted below, nutrient-stimulated signaling pathways impinge on both the production of ribosomes and the rate of ongoing translation. Moreover, mutants affecting specific nutrient signaling pathways have profound effects on cell size specification (Jorgensen and Tyers 2004; Turner et al. 2012).

START and the yeast metabolic cycle (YMC) also appear to be connected (Burnetti et al. 2016). YMC is defined by oscillations in the use of dissolved oxygen by synchronous and asynchronous cells, observed primarily at high culture density. Cells oscillate between low and high oxygen consumption (LOC and HOC) states. LOC cells accumulate storage carbohydrates, which are metabolized in HOC to release energy and promote cell biosynthesis. About one quarter of yeast genes respond to the quality and quantity of carbon sources (Brauer et al. 2008). These genes promote ribogenesis, mitochondrial function, and protein translation and are expressed periodically during YMC. YMC is shorter than the cell division cycle (CDC) but is coupled to it. DNA replication occurs only once during the YMC but not in all cells during all YMCs (Burnetti et al. 2016). Earlier studies indicated that DNA replication occurred only during LOC, prompting the hypothesis that temporal segregation of CDC and YMC was required to prevent reactive oxygen species from damaging DNA during replication (Chen et al. 2007; Klevecz et al. 2004; Silverman et al. 2010; Tu et al. 2005). More recent studies have observed DNA replication in both LOC and HOC states (Gasch et al. 2017; Slavov and Botstein 2011). While CDC and YMC seem to be connected in some fashion, the exact nature of this relationship with respect to nutrient signaling remains unknown.

Nutrient Regulation of Mass Accumulation Cell growth is highly dependent on ribogenesis, and the signaling pathways discussed above play a key role in regulating ribosome biogenesis. Ribosomes contain 4 RNA molecules and 79 ribosomal proteins encoded by 138 genes. Over 230 proteins, called the ribosomal biogenesis (RiBi) proteins, are involved in synthesis and assembly of functional ribosomes. Nutrients regulate ribosomal RNA synthesis, ribosomal protein, and Ribi protein expression as well as tRNA synthesis by modulating the activities of RNA polymerases Pol I, Pol II, and Pol III (Broach 2012). As an example, RiBi gene expression is activated by the Sfp1 transcription factor and repressed by Stb3, Dot6, and Tod6 transcription repressors through recruitment of Rpd3L histone deacetylase (Huber et al. 2011; Humphrey et al. 2004). Localization of Sfp1 to the nucleus responds to the nutritional state of the cell via the PKA and TORC1 pathways (Jorgensen et al. 2004; Lempiainen et al. 2009; Marion et al. 2004; Singh and Tyers 2009). PKA and Sch9 also phosphorylate Dot6 and Tod6 to relieve repression of target genes (Lippman and Broach 2009) and also phosphorylate Stb3 to evict it from the nucleus (Huber et al. 2011; Liko et al. 2010). Through different mechanisms, the PKA and TORC1 also collaborate on regulation of ribosomal protein gene expression and rRNA and tRNA synthesis, thereby connecting nutrient availability to expansion of the cell's biosynthetic capacity.

The stress response pathway also regulates cell growth, both acutely and long term. Starvation activates the environmental stress response (ESR) pathway (Gasch et al. 2000). Activation of ESR leads to repression of Ribi and RP genes possibly by influencing localization of Sfp1 (Jorgensen et al. 2004). Transcription factors Msn2/4 mediate the majority of gene expression changes during stress (see below). Phosphorylation of Msn2/4 by PKA prevents Msn2/4 from localizing in the nucleus while phosphorylation by Snf1 promotes its nuclear export (Gorner et al. 2002). Msn2/4 is dephosphorylated by PP2A and PP1, whose actions are likely responsible for acute responses of Msn2/4 to nutrient withdrawal (De Wever et al. 2005; Petrenko et al. 2013). Yak1 also phosphorylates Msn2 and activates it, an action that plays a key role in long-term adaptation to nutrient deprivation (Malcher et al. 2011). In sum, the primary nutrient sensing pathways—PKA and TORC1—impinge on cell growth both as positive regulators of components of mass accumulation and as negative regulators of the stress response. The fact that deletion of *MSN2* or *YAK1* reverses the growth defect resulting from the absence of active PKA (Garrett and Broach 1989; Smith et al. 1998) may suggest that inhibition of the stress response is the critical growth-promoting feature of the nutrient signaling pathways.

4.3.2 Acute Nutrient Stress Response

All cells perceive and respond to environmental stresses, and that response is critical to a cell's survival under continued adverse conditions. In yeast, a wide variety of stresses, including acute nutrient downshifts, induce ESR, which is highly dependent on the activity of Msn2/4 (Gasch et al. 2000). While this response is robust and conserved, it does not appear to provide protection against immediate insults (Lu et al. 2009). This is not surprising, since the time lag associated with transcription and translation of new gene products would preclude a cell's ability to mount an immediate response to stress through transcriptional activation. Nonetheless, Msn2/4-dependent stress-stimulated induction of gene expression is required for an acquired stress resistance or adaptive response, in which a mild initial stress provides protection of cells against a subsequent lethal exposure to the same or a different stress (Berry and Gasch 2008). This adaptive response is not unique to yeast cells but pervades all cells in which it has been examined (Charng et al. 2006; Chinnusamy et al. 2004; Durrant and Dong 2004; Matsumoto et al. 2007; Zhao et al. 2007). Thus, cells use transcriptional activation in response to an initial stress as a means of preparing for long-term adversity.

Stresses affect gene expression not only at the transcriptional level but also at the translation level. Glucose depletion results in a rapid cessation of translation initiation, dependent on signaling through the PKA pathway (Ashe et al. 2000). Similarly, inhibition of TORC1 through rapamycin treatment or starvation for nitrogen also results in rapid loss of translation initiation as a result of increased phosphorylation of the translation initiation factor eIF2α (Cherkasova and Hinnebusch 2003; Urban et al. 2007) and reduced stability of eIF4G (Berset et al. 1998). As a separate effect on translation, stress increases the abundance of cytoplasmic P-bodies, ribonucleoprotein aggregates containing non-translating mRNAs, and enzymes important for decapping and mRNA degradation (Balagopal and Parker 2009). While P-bodies may be involved in mRNA turnover, they appear to play a more significant role in mRNA sequestration and storage for subsequent use. Consistent with this hypothesis, mutants that fail to accumulate P-bodies during stationary phase exhibit reduced survival (Shah et al. 2013). P-body formation is regulated at least in part by the PKA pathway through direct phosphorylation of its core constituent, Pat1: inhibition of the pathway results in increased formation of P-bodies and activation of the pathway blocks formation induced by a variety of stresses (Ramachandran et al. 2011). A final noteworthy effect of stress on translation is repression of specific subsets of transcripts via the Pumilio family of translational repressors. A prime example is Puf3, which binds to the 3' UTR of a number of mitochondrial protein mRNAs and in nutrient-replete conditions promotes their degradation. However, on glucose depletion, Puf3 becomes phosphorylated in a PKA- and Sch9-dependent fashion and relocates to polysomes and promotes translation of its associated transcripts (Lee and Tu 2015). In sum, these studies document that a variety of stress-mediated translational effects complement the stress-induced transcriptional responses that protect the cell from environmental adversity.

As noted above, stresses induce a canonical transcriptional response through the transcriptional factors, Msn2 and Msn4. Stress modulates these factors' activity through cytoplasmic to nuclear relocalization but do so in an unusual way: stresses induce increased frequency of bursts of short-lived, recurrent periods of Msn2 nuclear localization with more intense stresses eliciting a greater frequency of bursts (Hao et al. 2013; Hao and O'Shea 2012; Petrenko et al. 2013) (see Fig. 4.4). While such bursting behavior is not a common feature of transcriptional factors in response to external stimuli, several yeast transcription factors do exhibit it, primarily those factors responsive to nutrients, including calcium-responsive Crz1, low glucose-responsive Mig1/2 and Nrg1/2, nitrogen-responsive Rtg1, and the nutrient-responsive repressors Dot6/Tod6 (Dalal et al. 2014). In addition, different levels of stress can elicit different patterns of transcription factor localization. For instance, mild nutrient downshifts yield a bursting response pattern of Msn2 while complete nutrient withdrawal yields its sustained nuclear relocalization.

One question raised by these observations is how the cycling patterns are generated. How are recurrent bursts generated and how do most of the molecules in an individual cell behave in concert? Garmendia-Torres et al. (2007) initially speculated that such cycles of Msn2 could be established by linking its nuclear localization to an intracellular PKA oscillator generated by a negative feedback loop in which stress inhibited PKA activation and loss of PKA activity stimulated reactivation of adenylyl cyclase. This model has been subsequently expanded to include protein phosphatase 1 and Snf1 in a larger feedback loop of PKA inhibition and reactivation (Jiang et al. 2017; Petrenko et al. 2013). Nonetheless, the general conclusion is that intracellular cycles of low and high PKA activity in a cell render the cellular environment instructive for Msn2 nuclear entry and subsequent exit.

A second question posed by cycling behavior of a transcription factor is the functional benefit of such behavior. One hypothesis is that different patterns of transcription factor localization can yield activation of different sets of genes, dependent on pattern. Hansen and O'Shea (2013) provided evidence that short bursts of Msn2 nuclear entry attendant on mild stress exposure activated a group of stress-responsive genes, termed fast responders, while prolonged nuclear entry upon severe stress activated the fast responders as well as an additional set of genes defined as slow responders. Elfving et al. (2014) extended that concept to suggest that stress-responsive genes exhibited a continuum of responsiveness such that the more intense the stress, the greater number of genes would be marshalled for a response. Finally, Lin et al. (2015) proposed that bursting behavior allowed for combinatorial regulation in which different transcription factors with different bursting patterns would overlap in certain situations to accomplish one task while different set of factors might overlap under different stimuli. Consistent with this hypothesis, Gasch et al. (2017) noted that Dot6 and Msn2 often colocalize to the nucleus following a salt stress stimulus but not under unstressed conditions. Certainly, this is still an unresolved issue, and additional data is required to determine the functional utility of transcription factor bursting behavior.

A second feature of the acute stress response is that genetically identical cells subject to an identical stress can behave quite differently, with some cells mounting a

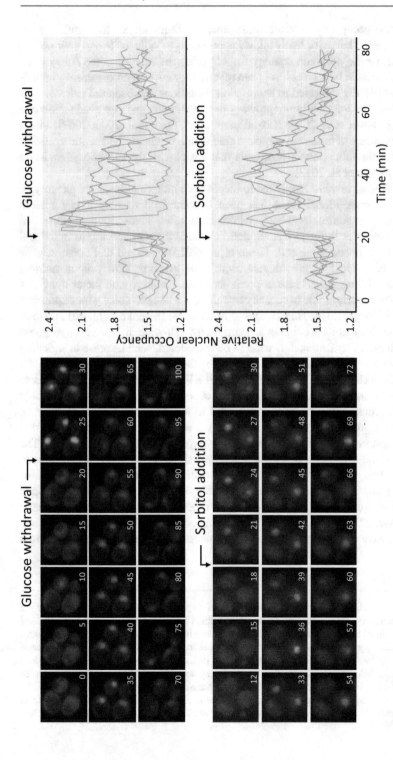

Fig. 4.4 Stochastic behavior of the Msn2 transcription factor in response to stress. Cells expressing Msn2-GFP were immobilized in a microfluidic chamber, bathed in standard growth media and observed by fluorescence microscopy over time prior to and following withdrawal of glucose (upper panels) or addition of sorbitol (lower panels). Images of the same field taken at 5 min (upper) or 3 min (lower) intervals are shown on the left. As evident in the sequential images of the

robust nuclear occupancy of Msn2 while others show no nuclear localization at all. This idiosyncratic behavior likely allows populations of cells to "hedge their bet" as to what will be the optimum strategy for surviving the ensuing stress (Levy et al. 2012). Idiosyncratic behaviors have been reported for other microorganisms, including *Bacillus subtilis* subjected to sporulation conditions and bacterial cultures with subpopulations of transiently nongrowing persistent cells that can evade antibiotic treatment (Balaban et al. 2004; Kussell et al. 2005; Suel et al. 2006, 2007). Moreover, the resistance of tumors to complete eradication by chemotherapeutic treatment may result from subpopulations of cells that are transiently and idiosyncratically quiescent (Moore et al. 2012).

Several studies in yeast have tried to identify the molecular basis for phenotypic variability. Transcriptional noise has been hypothesized to play a key role in heterogeneous responses to stress on individual cells. Transcriptional noise can arise from promoter intrinsic factors, that is, events that are specific to individual promoters even in the same cell. Keren et al. (2015) observed that noise in gene expression is coupled to growth rate, with cells growing slowly upon nutrient starvation exhibiting an increase in promoter noise. Transcription factor dynamics and variation in nucleosome occupancy can also contribute to noisy gene expression (Brown and Boeger 2014; Hansen and O'Shea 2013; Petrenko et al. 2013). However, such intrinsic noise would not account for the cell-wide differences in response to stress. Rather, extrinsic factors need to be considered, such as noise in signaling pathways themselves, which can be amplified by the small number of molecules of a transcription factor within a cell to generate cell-to-cell variation in responses to stress. Moreover, as distinct from persisters in a bacterial population, yeast cells exhibit a continuum of behaviors with regard to the stress response. For instance, yeast colonies growing on replete media show a continuum of growth rates with a correlated continuum of trehalose synthetase expression (Levy et al. 2012). The resistance of those colonies to a subsequent heat shock is inversely proportional to the growth rate of the colony. Thus, yeast cells do not exhibit an all or none bet-hedging process to mitigate an unpredictable future but rather present a diverse but ever-changing population of individuals to meet a variety of different potential future stresses. This indicates that the noise associated with stress response must induce a variety of epigenetic states that provide a continuum of potential stress responses.

Fig. 4.4 (continued) sorbitol-treated sample, two cells responded robustly to the stress while the third cell in the image (in the upper left hand corner) failed to respond at all. This is an example of "bet hedging" in which genetically identical cells subjected to the same environmental stress behave differently, allowing the population as a whole to explore multiple survival strategies in the face of an uncertain future. The right-hand panels are traces of the nuclear occupancy of Msn2 in four individual cells followed for 80 min during which they are subjected to the indicated stress. In response to both acute stress treatments, Msn2 enters the nucleus en masse in most cells and then exits shortly after, following which it exhibits bursts of nuclear entry and exit. Modeling suggests that this behavior is driven by intracellular oscillations of PKA activity, which controls the nuclear import of Msn2

4.3.3   Long-Term Nutrient Stress Response: Stationary Phase and Quiescence

Different developmental programs can be accessed following extended nutrient starvation depending on cell type. Diploid cells grown on a non-fermentable carbon source in the absence of nitrogen can undergo meiosis and sporulation. Similarly, certain strains of *Saccharomyces* subjected to limited starvation alter their morphology and budding pattern to generate pseudohyphae. How nutrient signaling pathways inform and coordinate these developmental programs has been reviewed earlier (Broach 2012). In the absence of these specific developmental program, yeast cells exit the cell cycle upon nutrient starvation and enter an ill-defined G0 state or quiescence.

The most extensively studied form of quiescence is stationary phase, which yeast cells grown in liquid medium access following depletion of glucose and then subsequent depletion through aerobic respiration of the ethanol produced during the fermentation stage. Once they attain stationary phase, i.e., no further growth, cells remain viable and can remain dormant for long durations (Gray et al. 2004; Smets et al. 2010; Werner-Washburne et al. 1993). The transitions of cells from growth to stationary phase require readjustment of growth rates and metabolism coordinated by nutrient signaling pathways (De Virgilio 2012). Transition to diauxic phase is accompanied by transcriptional induction of genes regulating respiration, fatty-acid oxidation, glyoxylate reactions, and ROS scavenging. Stationary phase cells exhibit distinguishing features such as thickened cell wall, accumulation of glycogen and trehalose, decreased transcription and translation, condensed chromatin, and upregulation of the stress response pathway. These quiescent cells can resume growth by reentering the cell cycle when replenished with nutrients.

The Rim15 pathway plays an important role in regulating entry to stationary. Rim15 belongs to the PAS kinase or Greatwall family and was first identified in regulation of meiosis in diploid cells by arresting cells and inducing expression of early meiotic genes. In haploid cells, Rim15 is essential for entering G0 phase of the cell cycle (Cameroni et al. 2004). Rim15 mutants arrest at G1 in stationary phase and never progress to quiescence. Reciprocally, overexpression of Rim15 in log phase cells can reprogram cells to exhibit quiescent characteristics. Nutrients impinge on Rim15 by signaling through PKA, TORC1 and the Pho80-Pho85 pathway, whose activities lead to its sequestration in the cytoplasm (Pedruzzi et al. 2003; Reinders et al. 1998; Wanke et al. 2005). Inhibition of any of the nutrient signaling pathway results in activation of Rim15, which in turn regulates the transcription factors Msn2/4 and Gis1 to facilitate diauxic shift and quiescence (Roosen et al. 2005).

Yeast cells can also become quiescent upon acute starvation for any of the major nutrients—carbon, nitrogen, or phosphate—and each starvation can induce a quiescent state within 24 h that exhibits both common and distinct characteristics (Klosinska et al. 2011). Starvation for any of the three nutrients elicits all of the physiological properties associated with quiescence, such as resistance to heat shock and oxidative stress. Moreover, the starvations result in a common transcriptional program, which is in large part a direct extrapolation of the changes that occur during slow growth. In contrast, the metabolic changes that occur upon starvation and the genetic requirements for surviving starvation differ significantly depending on the nutrient for which the cell is starved. Mitochondrial function is important for survival upon carbon but not nitrogen starvation. Moreover, *snf1* and *ira2* mutants, which exhibit signaling patterns mimicking glucose abundance, fail to become quiescent under glucose starvation but access quiescence normally under nitrogen or phosphate starvation. Autophagy and signaling through the TORC1 pathway are essential for survival upon nitrogen but not carbon starvation (Gresham et al. 2011; Klosinska et al. 2011; Lavoie and Whiteway 2008). Requirements for survival under phosphate starvation overlap those needed for carbon and nitrogen starvation except for TORC1 signaling. Rim15, which functions downstream of TORC1, is required for quiescence and meiosis initiation during nitrogen starvation (Klosinska et al. 2011; Pedruzzi et al. 2000). Hyperactive PKA activity in a *bcy1* mutant renders cells sensitive to any type of starvation (De Virgilio 2012), while hypoactive Ras2 causes a delay in recovery from carbon starvation (Jiang et al. 1998). Thus, nutrient signaling pathways are essential for entry, maintenance, as well as exit from quiescence.

Besides changes in metabolic and transcriptional programs, quiescent yeast cells undergo significant changes in cellular and genome architecture. During the G1 phase of exponential growth, yeast chromosomes assume a "Rabl" configuration in which the centromeres congregate near the spindle pole body through short microtubules with the chromosome arms extending outward into the nucleoplasm, terminating at telomeres anchored to the nuclear membrane (Bystricky et al. 2004; Guacci et al. 1997). Upon transition into quiescence under glucose starvation, yeast nuclei undergo significant reorganization in which an extended microtubule array is assembled, along which the declustered centromeres become distributed (Laporte et al. 2013). Moreover, chromatin undergoes a condensin-dependent compaction, and telomeres exhibit hyperclusting (Laporte et al. 2016; Rutledge et al. 2015; Schafer et al. 2008). The functional significance of these changes is not fully resolved.

4.3.4 The Relationship Between Stress and Cell Growth

Stress and cell growth are reciprocally interconnected. Acute stresses such as osmotic shock, nutrient deprivation, and exposure to oxidative agents all lead to transient cell cycle arrest. One mechanism for this arrest is the Msn2/4-dependent induction of Cip1, a stress-specific G1 cyclin-CDK inhibitor (Chang et al. 2017). In a separate process, inhibition of TORC1 or nitrogen depletion induces G1 arrest through phosphorylation and stabilization of the CDK inhibitor Sic1 via a Rim15-dependent inhibition of PP2A (Moreno-Torres et al. 2015). Further, osmotic stress yields cell cycle arrest at multiple stages of the cell cycle through several distinct mechanisms interdicting the cell cycle machinery (Radmaneshfar et al. 2013). Finally, as a more indirect mechanism, nuclear localization of Msn2 and of the ribosome biogenesis repressors, Dot6 and Tod6, are highly correlated following acute stress, consistent with their nuclear import being regulated in concert by upstream signaling pathways (Gasch et al. 2017). This leads to a coordinate activation of the environmental stress response and suppression of the growth capacity of the cell.

While the reports above demonstrate that stress induces cell cycle arrest, other studies have shown that slower growing cells activate the stress response pathway, rendering them resistant to a subsequent lethal heat shock. This is true for cells growing in chemostat cultures under limiting nutrients (Lu et al. 2009) or for subclones of cells growing in a benign environment (Levy et al. 2012). In this latter situation, heat shock resistance, trehalose accumulation, and slow growth were all correlated, suggesting the existence of heterogeneous subclones of cells within a genetically uniform population with distinct epigenetic state(s). In fact, the slow-growing and fast-growing microcolonies in such a genetically identical population exhibited distinct transcriptional profiles (Garcia-Martinez et al. 2016). These observations highlight the question of the causal relationships among stress response, nutrient signaling, and cell growth.

Yeast cells can adjust cell doubling time over a tenfold range in proportion to nutrient availability, primarily through expansion or contraction of the G1 phase of the cell cycle (Brauer et al. 2008). This is true for limitation for any of the primary nutrients—carbon, nitrogen, phosphate, or sulfur—and for auxotrophic cells limited for the required amino acid. Moreover, the cellular growth rate is highly correlated with a specific transcriptional signature, primarily comprised of ESR and ribosomal biogenesis genes, such that the growth rate of the cell can be precisely predicted from the level of expression of the genes in that signature set, regardless of the perturbation responsible for the reduced or enhanced growth rate (Airoldi et al. 2009).

This observation raises the question of whether gene expression is set by the growth rate of the cell in a feedback mechanism or whether environmental sensing and gene expression changes modulate growth rates in a feed-forward mechanism (Levy and Barkai 2009). The feed-forward mechanism in theory would limit growth regulation to only those environments that cells have been exposed. The feedback mechanism, on the other hand, would equip cells to adapt to new environmental conditions. A possible mechanism for feedback regulation of gene expression by growth rate was suggested by Freddolino et al. (2018), who proposed fitness-derived transcriptional tuning of thousands of genes without directly sensing external changes as a mechanism for cellular response to stress. However, it is not clear that this mechanism would work on a time frame consistent with the observed transcriptional adaptation to environmental changes. Rather, other data are more consistent with a feed-forward mechanism. Namely, the gene expression levels of members of the growth rate-associated signature set are established by the nutrient signaling pathways rather than the growth rate of the cell or the levels of internal metabolites. For instance, artificial activation of the Ras/PKA pathway during growth on non-fermentable carbon source yields the rapid growth transcriptional signature (Zaman et al. 2009). Similarly, growth of an *adh1* mutant in the presence of glucose exhibit the rapid growth rate signature despite the fact that such mutants cannot ferment glucose and thus grow slowly (Levy et al. 2007). This suggests that yeast cells set their growth rate on the basis of their perception of the nutrient state of the cell, rather than their consumption of nutrients. Accordingly, a disconnect between perception and actuality is potentially lethal, as is the case for subjecting cells with activated Ras/PKA signaling to growth on a suboptimal carbon source.

4.4   Conclusions

We discussed the extensive network of signaling pathways involved in cellular response to starvation, stationary phase, and hypoxia in yeast and other fungi. Many of these signaling proteins are conserved in higher eukaryotes and provide clues on the regulation of growth in complex organisms as well. Regulating cellular growth in adverse conditions is a fundamental challenge faced by all cells and yeast serves as a great model to continue exploring several complex unanswered questions to further our knowledge in this field.

References

Airoldi EM, Huttenhower C, Gresham D, Lu C, Caudy AA, Dunham MJ, Broach JR, Botstein D, Troyanskaya OG (2009) Predicting cellular growth from gene expression signatures. PLoS Comput Biol 5(1):e1000257

Ashe MP, De Long SK, Sachs AB (2000) Glucose depletion rapidly inhibits translation initiation in yeast. Mol Biol Cell 11:833–848

Balaban NQ, Merrin J, Chait R, Kowalik L, Leibler S (2004) Bacterial persistence as a phenotypic switch. Science 305:1622–1625

Balagopal V, Parker R (2009) Polysomes, P bodies and stress granules: states and fates of eukaryotic mRNAs. Curr Opin Cell Biol 21:403–408

Beck T, Hall MN (1999) The TOR signalling pathway controls nuclear localization of nutrient-regulated transcription factors. Nature 402:689–692

Berry DB, Gasch AP (2008) Stress-activated genomic expression changes serve a preparative role for impending stress in yeast. Mol Biol Cell 19:4580–4587

Berset C, Trachsel H, Altmann M (1998) The TOR (target of rapamycin) signal transduction pathway regulates the stability of translation initiation factor eIF4G in the yeast *Saccharomyces cerevisiae*. Proc Natl Acad Sci USA 95:4264–4269

Bertram PG, Choi JH, Carvalho J, Ai W, Zeng C, Chan TF, Zheng XF (2000) Tripartite regulation of Gln3p by TOR, Ure2p, and phosphatases. J Biol Chem 275:35727–35733

Binda M, Peli-Gulli MP, Bonfils G, Panchaud N, Urban J, Sturgill TW, Loewith R, De Virgilio C (2009) The Vam6 GEF controls TORC1 by activating the EGO complex. Mol Cell 35:563–573

Bonfils G, Jaquenoud M, Bontron S, Ostrowicz C, Ungermann C, De Virgilio C (2012) Leucyl-tRNA synthetase controls TORC1 via the EGO complex. Mol Cell 46:105–110

Brauer MJ, Huttenhower C, Airoldi EM, Rosenstein R, Matese JC, Gresham D, Boer VM, Troyanskaya OG, Botstein D (2008) Coordination of growth rate, cell cycle, stress response, and metabolic activity in yeast. Mol Biol Cell 19:352–367

Breitkreutz A, Choi H, Sharom JR, Boucher L, Neduva V, Larsen B, Lin ZY, Breitkreutz BJ, Stark C, Liu G et al (2010) A global protein kinase and phosphatase interaction network in yeast. Science 328:1043–1046

Broach JR (2012) Nutritional control of growth and development in yeast. Genetics 192:73–105

Brown CR, Boeger H (2014) Nucleosomal promoter variation generates gene expression noise. Proc Natl Acad Sci USA 111:17893–17898

Budovskaya YV, Stephan JS, Deminoff SJ, Herman PK (2005) An evolutionary proteomics approach identifies substrates of the cAMP-dependent protein kinase. Proc Natl Acad Sci USA 102:13933–13938

Burnetti AJ, Aydin M, Buchler NE (2016) Cell cycle start is coupled to entry into the yeast metabolic cycle across diverse strains and growth rates. Mol Biol Cell 27:64–74

Bystricky K, Heun P, Gehlen L, Langowski J, Gasser SM (2004) Long-range compaction and flexibility of interphase chromatin in budding yeast analyzed by high-resolution imaging techniques. Proc Natl Acad Sci USA 101:16495–16500

Cameroni E, Hulo N, Roosen J, Winderickx J, De Virgilio C (2004) The novel yeast PAS kinase Rim 15 orchestrates G0-associated antioxidant defense mechanisms. Cell Cycle 3:462–468

Cardenas ME, Cutler NS, Lorenz MC, Di Como CJ, Heitman J (1999) The TOR signaling cascade regulates gene expression in response to nutrients. Genes Dev 13:3271–3279

Chandrashekarappa DG, McCartney RR, Schmidt MC (2011) Subunit and domain requirements for adenylate-mediated protection of Snf1 kinase activation loop from dephosphorylation. J Biol Chem 286:44532–44541

Chang YL, Tseng SF, Huang YC, Shen ZJ, Hsu PH, Hsieh MH, Yang CW, Tognetti S, Canal B, Subirana L et al (2017) Yeast Cip1 is activated by environmental stress to inhibit Cdk1-G1 cyclins via Mcm1 and Msn2/4. Nat Commun 8:56

Charng YY, Liu HC, Liu NY, Hsu FC, Ko SS (2006) Arabidopsis Hsa32, a novel heat shock protein, is essential for acquired thermotolerance during long recovery after acclimation. Plant Physiol 140:1297–1305

Chen Z, Odstrcil EA, Tu BP, McKnight SL (2007) Restriction of DNA replication to the reductive phase of the metabolic cycle protects genome integrity. Science 316:1916–1919

Cherkasova VA, Hinnebusch AG (2003) Translational control by TOR and TAP42 through dephosphorylation of eIF2alpha kinase GCN2. Genes Dev 17:859–872

Chinnusamy V, Schumaker K, Zhu JK (2004) Molecular genetic perspectives on cross-talk and specificity in abiotic stress signalling in plants. J Exp Bot 55:225–236

Colombo S, Ma P, Cauwenberg L, Winderickx J, Crauwels M, Teunissen A, Nauwelaers D, de Winde JH, Gorwa MF, Colavizza D et al (1998) Involvement of distinct G-proteins, Gpa2 and Ras, in glucose- and intracellular acidification-induced cAMP signalling in the yeast *Saccharomyces cerevisiae*. Embo J 17:3326–3341

Cooper TG (2002) Transmitting the signal of excess nitrogen in *Saccharomyces cerevisiae* from the Tor proteins to the GATA factors: connecting the dots. FEMS Microbiol Rev 26:223–238

Costanzo M, Nishikawa JL, Tang X, Millman JS, Schub O, Breitkreuz K, Dewar D, Rupes I, Andrews B, Tyers M (2004) CDK activity antagonizes Whi5, an inhibitor of G1/S transcription in yeast. Cell 117:899–913

Dalal CK, Cai L, Lin Y, Rahbar K, Elowitz MB (2014) Pulsatile dynamics in the yeast proteome. Curr Biol 24:2189–2194

De Virgilio C (2012) The essence of yeast quiescence. FEMS Microbiol Rev 36:306–339

De Virgilio C, Loewith R (2006) The TOR signalling network from yeast to man. Int J Biochem Cell Biol 38:1476–1481

De Wever V, Reiter W, Ballarini A, Ammerer G, Brocard C (2005) A dual role for PP1 in shaping the Msn2-dependent transcriptional response to glucose starvation. Embo J 24:4115–4123

Dechant R, Binda M, Lee SS, Pelet S, Winderickx J, Peter M (2010) Cytosolic pH is a second messenger for glucose and regulates the PKA pathway through V-ATPase. EMBO J 29:2515–2526

Dubouloz F, Deloche O, Wanke V, Cameroni E, De Virgilio C (2005) The TOR and EGO protein complexes orchestrate microautophagy in yeast. Mol Cell 19:15–26

Durrant WE, Dong X (2004) Systemic acquired resistance. Annu Rev Phytopathol 42:185–209

Duvel K, Santhanam A, Garrett S, Schneper L, Broach JR (2003) Multiple roles of Tap42 in mediating rapamycin-induced transcriptional changes in yeast. Mol Cell 11:1467–1478

Elfving N, Chereji RV, Bharatula V, Bjorklund S, Morozov AV, Broach JR (2014) A dynamic interplay of nucleosome and Msn2 binding regulates kinetics of gene activation and repression following stress. Nucleic Acids Res 42:5468–5482

Eltschinger S, Loewith R (2016) TOR complexes and the maintenance of cellular homeostasis. Trends Cell Biol 26:148–159

Ferrezuelo F, Colomina N, Palmisano A, Gari E, Gallego C, Csikasz-Nagy A, Aldea M (2012) The critical size is set at a single-cell level by growth rate to attain homeostasis and adaptation. Nat Commun 3:1012

Freddolino PL, Yang J, Momen-Roknabadi A, Tavazoie S (2018) Stochastic tuning of gene expression enables cellular adaptation in the absence of pre-existing regulatory circuitry. eLife 7:e31867. https://doi.org/10.7554/eLife.31867

Gadura N, Robinson LC, Michels CA (2006) Glc7-Reg1 phosphatase signals to Yck1,2 casein kinase 1 to regulate transport activity and glucose-induced inactivation of *Saccharomyces* maltose permease. Genetics 172:1427–1439

Garcia-Martinez J, Delgado-Ramos L, Ayala G, Pelechano V, Medina DA, Carrasco F, Gonzalez R, Andres-Leon E, Steinmetz L, Warringer J et al (2016) The cellular growth rate controls overall mRNA turnover, and modulates either transcription or degradation rates of particular gene regulons. Nucleic Acids Res 44:3643–3658

Garmendia-Torres C, Goldbeter A, Jacquet M (2007) Nucleocytoplasmic oscillations of the yeast transcription factor Msn2: evidence for periodic PKA activation. Curr Biol 17:1044–1049

Garrett S, Broach J (1989) Loss of Ras activity in *Saccharomyces cerevisiae* is suppressed by disruptions of a new kinase gene, YAKI, whose product may act downstream of the cAMP-dependent protein kinase. Genes Dev 3:1336–1348

Gasch AP, Spellman PT, Kao CM, Carmel-Harel O, Eisen MB, Storz G, Botstein D, Brown PO (2000) Genomic expression programs in the response of yeast cells to environmental changes. Mol Biol Cell 11:4241–4257

Gasch AP, Yu FB, Hose J, Escalante LE, Place M, Bacher R, Kanbar J, Ciobanu D, Sandor L, Grigoriev IV et al (2017) Single-cell RNA sequencing reveals intrinsic and extrinsic regulatory heterogeneity in yeast responding to stress. PLoS Biol 15:e2004050

Gorner W, Durchschlag E, Martinez-Pastor MT, Estruch F, Ammerer G, Hamilton B, Ruis H, Schuller C (1998) Nuclear localization of the C2H2 zinc finger protein Msn2p is regulated by stress and protein kinase A activity. Genes Dev 12:586–597

Gorner W, Durchschlag E, Wolf J, Brown EL, Ammerer G, Ruis H, Schuller C (2002) Acute glucose starvation activates the nuclear localization signal of a stress-specific yeast transcription factor. EMBO J 21:135–144

Gray JV, Petsko GA, Johnston GC, Ringe D, Singer RA, Werner-Washburne M (2004) "Sleeping beauty": quiescence in *Saccharomyces cerevisiae*. Microbiol Mol Biol Rev 68:187–206

Gresham D, Boer VM, Caudy A, Ziv N, Brandt NJ, Storey JD, Botstein D (2011) System-level analysis of genes and functions affecting survival during nutrient starvation in *Saccharomyces cerevisiae*. Genetics 187:299–317

Guacci V, Hogan E, Koshland D (1997) Centromere position in budding yeast: evidence for anaphase A. Mol Biol Cell 8:957–972

Hansen AS, O'Shea EK (2013) Promoter decoding of transcription factor dynamics involves a trade-off between noise and control of gene expression. Mol Syst Biol 9:704

Hao N, O'Shea EK (2012) Signal-dependent dynamics of transcription factor translocation controls gene expression. Nat Struct Mol Biol 19:31–39

Hao N, Budnik BA, Gunawardena J, O'Shea EK (2013) Tunable signal processing through modular control of transcription factor translocation. Science 339:460–464

Hartwell LH (1974) *Saccharomyces cerevisiae* cell cycle. Bacteriol Rev 38:164–198

Hartwell LH, Culotti J, Reid B (1970) Genetic control of the cell-division cycle in yeast. I. Detection of mutants. Proc Natl Acad Sci USA 66:352–359

Hong SP, Leiper FC, Woods A, Carling D, Carlson M (2003) Activation of yeast Snf1 and mammalian AMP-activated protein kinase by upstream kinases. Proc Natl Acad Sci USA 100:8839–8843

Huber A, Bodenmiller B, Uotila A, Stahl M, Wanka S, Gerrits B, Aebersold R, Loewith R (2009) Characterization of the rapamycin-sensitive phosphoproteome reveals that Sch9 is a central coordinator of protein synthesis. Genes Dev 23(16):1929–1943

Huber A, French SL, Tekotte H, Yerlikaya S, Stahl M, Perepelkina MP, Tyers M, Rougemont J, Beyer AL, Loewith R (2011) Sch9 regulates ribosome biogenesis via Stb3, Dot6 and Tod6 and the histone deacetylase complex RPD3L. EMBO J 30(15):3052–3064

Humphrey EL, Shamji AF, Bernstein BE, Schreiber SL (2004) Rpd3p relocation mediates a transcriptional response to rapamycin in yeast. Chem Biol 11:295–299

Isom DG, Page SC, Collins LB, Kapolka NJ, Taghon GJ, Dohlman HG (2017) Coordinated regulation of intracellular pH by two glucose sensing pathways in yeast. J Biol Chem 293 (7):2318–2329

Jiang Y, Broach JR (1999) Tor proteins and protein phosphatase 2A reciprocally regulate Tap42 in controlling cell growth in yeast. EMBO J 18:2782–2792

Jiang Y, Davis C, Broach JR (1998) Efficient transition to growth on fermentable carbon sources in Saccharomyces cerevisiae requires signaling through the Ras pathway. EMBO J 17:6942–6951

Jiang Y, AkhavanAghdam Z, Tsimring LS, Hao N (2017) Coupled feedback loops control the stimulus-dependent dynamics of the yeast transcription factor Msn2. J Biol Chem 292:12366–12372

Johnson A, Skotheim JM (2013) Start and the restriction point. Curr Opin Cell Biol 25:717–723

Johnston M, Carlson M (1992) Carbon regulation in Saccharomyces. In: Broach JR, Pringle JR, Jones EW (eds) Molecular and cellular biology of the yeast Saccharomyces. Cold Spring Harbor Laboratory Press, Cold Spring Harbor, NY

Johnston M, Kim JH (2005) Glucose as a hormone: receptor-mediated glucose sensing in the yeast Saccharomyces cerevisiae. Biochem Soc Trans 33:247–252

Jorgensen P, Tyers M (2004) How cells coordinate growth and division. Curr Biol 14:R1014–R1027

Jorgensen P, Rupes I, Sharom JR, Schneper L, Broach JR, Tyers M (2004) A dynamic transcriptional network communicates growth potential to ribosome synthesis and critical cell size. Genes Dev 18:2491–2505

Kaniak A, Xue Z, Macool D, Kim JH, Johnston M (2004) Regulatory network connecting two glucose signal transduction pathways in Saccharomyces cerevisiae. Eukaryot Cell 3:221–231

Karhumaa K, Wu B, Kielland-Brandt MC (2010) Conditions with high intracellular glucose inhibit sensing through glucose sensor Snf3 in Saccharomyces cerevisiae. J Cell Biochem 110:920–925

Keren L, van Dijk D, Weingarten-Gabbay S, Davidi D, Jona G, Weinberger A, Milo R, Segal E (2015) Noise in gene expression is coupled to growth rate. Genome Res 25:1893–1902

Klevecz RR, Bolen J, Forrest G, Murray DB (2004) A genomewide oscillation in transcription gates DNA replication and cell cycle. Proc Natl Acad Sci USA 101:1200–1205

Klosinska MM, Crutchfield CA, Bradley PH, Rabinowitz JD, Broach JR (2011) Yeast cells can access distinct quiescent states. Genes Dev 25:336–349

Kresnowati MT, van Winden WA, Almering MJ, ten Pierick A, Ras C, Knijnenburg TA, Daran-Lapujade P, Pronk JT, Heijnen JJ, Daran JM (2006) When transcriptome meets metabolome: fast cellular responses of yeast to sudden relief of glucose limitation. Mol Syst Biol 2:49

Kuchin S, Vyas VK, Kanter E, Hong SP, Carlson M (2003) Std1p (Msn3p) positively regulates the Snf1 kinase in Saccharomyces cerevisiae. Genetics 163:507–514

Kussell E, Kishony R, Balaban NQ, Leibler S (2005) Bacterial persistence: a model of survival in changing environments. Genetics 169:1807–1814

Lakshmanan J, Mosley AL, Ozcan S (2003) Repression of transcription by Rgt1 in the absence of glucose requires Std1 and Mth1. Curr Genet 44:19–25

Laporte D, Courtout F, Salin B, Ceschin J, Sagot I (2013) An array of nuclear microtubules reorganizes the budding yeast nucleus during quiescence. J Cell Biol 203:585–594

Laporte D, Courtout F, Tollis S, Sagot I (2016) Quiescent Saccharomyces cerevisiae forms telomere hyperclusters at the nuclear membrane vicinity through a multifaceted mechanism involving Esc1, the Sir complex, and chromatin condensation. Mol Biol Cell 27:1875–1884

Lascaris R, Bussemaker HJ, Boorsma A, Piper M, van der Spek H, Grivell L, Blom J (2003) Hap4p overexpression in glucose-grown *Saccharomyces cerevisiae* induces cells to enter a novel metabolic state. Genome Biol 4:R3

Lavoie H, Whiteway M (2008) Increased respiration in the sch9Delta mutant is required for increasing chronological life span but not replicative life span. Eukaryot Cell 7:1127–1135

Laxman S, Sutter BM, Tu BP (2014) Methionine is a signal of amino acid sufficiency that inhibits autophagy through the methylation of PP2A. Autophagy 10:386–387

Lee CD, Tu BP (2015) Glucose-regulated phosphorylation of the PUF protein Puf3 regulates the translational fate of its bound mRNAs and association with RNA granules. Cell Rep 11:1638–1650

Lee P, Cho BR, Joo HS, Hahn JS (2008) Yeast Yak1 kinase, a bridge between PKA and stress-responsive transcription factors, Hsf1 and Msn2/Msn4. Mol Microbiol 70:882–895

Lee P, Paik SM, Shin CS, Huh WK, Hahn JS (2011) Regulation of yeast Yak1 kinase by PKA and autophosphorylation-dependent 14-3-3 binding. Mol Microbiol 79:633–646

Lemaire K, Van de Velde S, Van Dijck P, Thevelein JM (2004) Glucose and sucrose act as agonist and mannose as antagonist ligands of the G protein-coupled receptor Gpr1 in the yeast *Saccharomyces cerevisiae*. Mol Cell 16:293–299

Lempiainen H, Uotila A, Urban J, Dohnal I, Ammerer G, Loewith R, Shore D (2009) Sfp1 interaction with TORC1 and Mrs6 reveals feedback regulation on TOR signaling. Mol Cell 33:704–716

Levy S, Barkai N (2009) Coordination of gene expression with growth rate: a feedback or a feed-forward strategy? FEBS Lett 583:3974–3978

Levy S, Ihmels J, Carmi M, Weinberger A, Friedlander G, Barkai N (2007) Strategy of transcription regulation in the budding yeast. PLoS One 2:e250

Levy SF, Ziv N, Siegal ML (2012) Bet hedging in yeast by heterogeneous, age-correlated expression of a stress protectant. PLoS Biol 10:e1001325

Li J, Yan G, Liu S, Jiang T, Zhong M, Yuan W, Chen S, Zheng Y, Jiang Y, Jiang Y (2017) Target of rapamycin complex 1 and Tap42-associated phosphatases are required for sensing changes in nitrogen conditions in the yeast *Saccharomyces cerevisiae*. Mol Microbiol 106:938–948

Liko D, Conway MK, Grunwald DS, Heideman W (2010) Stb3 plays a role in the glucose-induced transition from quiescence to growth in *Saccharomyces cerevisiae*. Genetics 185(3):797–810

Lin Y, Sohn CH, Dalal CK, Cai L, Elowitz MB (2015) Combinatorial gene regulation by modulation of relative pulse timing. Nature 527:54–58

Lippman SI, Broach JR (2009) Protein kinase A and TORC1 activate genes for ribosomal biogenesis by inactivating repressors encoded by Dot6 and its homolog Tod6. Proc Natl Acad Sci USA 106:19928–19933

Liu Z, Butow RA (1999) A transcriptional switch in the expression of yeast tricarboxylic acid cycle genes in response to a reduction or loss of respiratory function. Mol Cell Biol 19:6720–6728

Liu Z, Butow RA (2006) Mitochondrial retrograde signaling. Annu Rev Genet 40:159–185

Lu C, Brauer MJ, Botstein D (2009) Slow growth induces heat-shock resistance in normal and respiratory-deficient yeast. Mol Biol Cell 20:891–903

Magasanik B, Kaiser CA (2002) Nitrogen regulation in *Saccharomyces cerevisiae*. Gene 290:1–18

Malcher M, Schladebeck S, Mosch HU (2011) The Yak1 protein kinase lies at the center of a regulatory cascade affecting adhesive growth and stress resistance in *Saccharomyces cerevisiae*. Genetics 187:717–730

Marion RM, Regev A, Segal E, Barash Y, Koller D, Friedman N, O'Shea EK (2004) Sfp1 is a stress- and nutrient-sensitive regulator of ribosomal protein gene expression. Proc Natl Acad Sci USA 101:14315–14322

Matsumoto H, Hamada N, Takahashi A, Kobayashi Y, Ohnishi T (2007) Vanguards of paradigm shift in radiation biology: radiation-induced adaptive and bystander responses. J Radiat Res (Tokyo) 48:97–106

Mayer FV, Heath R, Underwood E, Sanders MJ, Carmena D, McCartney RR, Leiper FC, Xiao B, Jing C, Walker PA et al (2011) ADP regulates SNF1, the *Saccharomyces cerevisiae* homolog of AMP-activated protein kinase. Cell Metab 14:707–714

Moore N, Houghton J, Lyle S (2012) Slow-cycling therapy-resistant cancer cells. Stem Cells Dev 21:1822–1830

Moreno-Torres M, Jaquenoud M, De Virgilio C (2015) TORC1 controls G1-S cell cycle transition in yeast via Mpk1 and the greatwall kinase pathway. Nat Commun 6:8256

Moriya H, Shimizu-Yoshida Y, Omori A, Iwashita S, Katoh M, Sakai A (2001) Yak1p, a DYRK family kinase, translocates to the nucleus and phosphorylates yeast Pop2p in response to a glucose signal. Genes Dev 15:1217–1228

Mosley AL, Lakshmanan J, Aryal BK, Ozcan S (2003) Glucose-mediated phosphorylation converts the transcription factor Rgt1 from a repressor to an activator. J Biol Chem 278:10322–10327

Nath N, McCartney RR, Schmidt MC (2003) Yeast Pak1 kinase associates with and activates Snf1. Mol Cell Biol 23:3909–3917

Neklesa TK, Davis RW (2009) A genome-wide screen for regulators of TORC1 in response to amino acid starvation reveals a conserved Npr2/3 complex. PLoS Genet 5:e1000515

Nicastro R, Sardu A, Panchaud N, De Virgilio C (2017) The architecture of the rag GTPase signaling network. Biomolecules 7(3):48

Pedruzzi I, Burckert N, Egger P, De Virgilio C (2000) *Saccharomyces cerevisiae* Ras/cAMP pathway controls post-diauxic shift element-dependent transcription through the zinc finger protein Gis1. Embo J 19:2569–2579

Pedruzzi I, Dubouloz F, Cameroni E, Wanke V, Roosen J, Winderickx J, De Virgilio C (2003) TOR and PKA signaling pathways converge on the protein kinase Rim15 to control entry into G0. Mol Cell 12:1607–1613

Peeters T, Louwet W, Gelade R, Nauwelaers D, Thevelein JM, Versele M (2006) Kelch-repeat proteins interacting with the Galpha protein Gpa2 bypass adenylate cyclase for direct regulation of protein kinase A in yeast. Proc Natl Acad Sci USA 103:13034–13039

Petrenko N, Chereji RV, McClean MN, Morozov AV, Broach JR (2013) Noise and interlocking signaling pathways promote distinct transcription factor dynamics in response to different stresses. Mol Biol Cell 24:2045–2057

Ptacek J, Devgan G, Michaud G, Zhu H, Zhu X, Fasolo J, Guo H, Jona G, Breitkreutz A, Sopko R et al (2005) Global analysis of protein phosphorylation in yeast. Nature 438:679–684

Radmaneshfar E, Kaloriti D, Gustin MC, Gow NA, Brown AJ, Grebogi C, Romano MC, Thiel M (2013) From START to FINISH: the influence of osmotic stress on the cell cycle. PLoS One 8:e68067

Ramachandran V, Shah KH, Herman PK (2011) The cAMP-dependent protein kinase signaling pathway is a key regulator of P body foci formation. Mol Cell 43:973–981

Reinders A, Burckert N, Boller T, Wiemken A, De Virgilio C (1998) *Saccharomyces cerevisiae* cAMP-dependent protein kinase controls entry into stationary phase through the Rim15p protein kinase. Genes Dev 12:2943–2955

Roosen J, Engelen K, Marchal K, Mathys J, Griffioen G, Cameroni E, Thevelein JM, De Virgilio C, De Moor B, Winderickx J (2005) PKA and Sch9 control a molecular switch important for the proper adaptation to nutrient availability. Mol Microbiol 55:862–880

Ruiz A, Xu X, Carlson M (2011) Roles of two protein phosphatases, Reg1-Glc7 and Sit4, and glycogen synthesis in regulation of SNF1 protein kinase. Proc Natl Acad Sci USA 108:6349–6354

Rutledge MT, Russo M, Belton JM, Dekker J, Broach JR (2015) The yeast genome undergoes significant topological reorganization in quiescence. Nucleic Acids Res 43:8299–8313

Schafer G, McEvoy CR, Patterton HG (2008) The *Saccharomyces cerevisiae* linker histone Hho1p is essential for chromatin compaction in stationary phase and is displaced by transcription. Proc Natl Acad Sci USA 105:14838–14843

Schmoller KM, Skotheim JM (2015) The biosynthetic basis of cell size control. Trends Cell Biol 25:793–802

Schmoller KM, Turner JJ, Koivomagi M, Skotheim JM (2015) Dilution of the cell cycle inhibitor Whi5 controls budding-yeast cell size. Nature 526:268–272

Shah KH, Zhang B, Ramachandran V, Herman PK (2013) Processing body and stress granule assembly occur by independent and differentially regulated pathways in *Saccharomyces cerevisiae*. Genetics 193:109–123

Silverman SJ, Petti AA, Slavov N, Parsons L, Briehof R, Thiberge SY, Zenklusen D, Gandhi SJ, Larson DR, Singer RH et al (2010) Metabolic cycling in single yeast cells from unsynchronized steady-state populations limited on glucose or phosphate. Proc Natl Acad Sci USA 107:6946–6951

Simpson-Lavy K, Xu T, Johnston M, Kupiec M (2017) The Std1 activator of the Snf1/AMPK kinase controls glucose response in yeast by a regulated protein aggregation. Mol Cell 68 (6):1120–1133

Singh J, Tyers M (2009) A Rab escort protein integrates the secretion system with TOR signaling and ribosome biogenesis. Genes Dev 23:1944–1958

Slavov N, Botstein D (2011) Coupling among growth rate response, metabolic cycle, and cell division cycle in yeast. Mol Biol Cell 22:1997–2009

Smets B, Ghillebert R, De Snijder P, Binda M, Swinnen E, De Virgilio C, Winderickx J (2010) Life in the midst of scarcity: adaptations to nutrient availability in *Saccharomyces cerevisiae*. Curr Genet 56:1–32

Smith A, Ward MP, Garrett S (1998) Yeast PKA represses Msn2p/Msn4p-dependent gene expression to regulate growth, stress response and glycogen accumulation. EMBO J 17:3556–3564

Snowdon C, Johnston M (2016) A novel role for yeast casein kinases in glucose sensing and signaling. Mol Biol Cell 27:3369–3375

Suel GM, Garcia-Ojalvo J, Liberman LM, Elowitz MB (2006) An excitable gene regulatory circuit induces transient cellular differentiation. Nature 440:545–550

Suel GM, Kulkarni RP, Dworkin J, Garcia-Ojalvo J, Elowitz MB (2007) Tunability and noise dependence in differentiation dynamics. Science 315:1716–1719

Sutherland CM, Hawley SA, McCartney RR, Leech A, Stark MJ, Schmidt MC, Hardie DG (2003) Elm1p is one of three upstream kinases for the *Saccharomyces cerevisiae* SNF1 complex. Curr Biol 13:1299–1305

Sutter BM, Wu X, Laxman S, Tu BP (2013) Methionine inhibits autophagy and promotes growth by inducing the SAM-responsive methylation of PP2A. Cell 154:403–415

Tate JJ, Cooper TG (2003) Tor1/2 regulation of retrograde gene expression in *Saccharomyces cerevisiae* derives indirectly as a consequence of alterations in ammonia metabolism. J Biol Chem 278:36924–36933

Tate JJ, Georis I, Feller A, Dubois E, Cooper TG (2009) Rapamycin-induced Gln3 dephosphorylation is insufficient for nuclear localization: Sit4 and PP2A phosphatases are regulated and function differently. J Biol Chem 284:2522–2534

Tate JJ, Buford D, Rai R, Cooper TG (2017) General amino acid control and 14-3-3 proteins Bmh1/2 are required for nitrogen catabolite repression-sensitive regulation of Gln3 and Gat1 localization. Genetics 205:633–655

Thomson JM, Gaucher EA, Burgan MF, De Kee DW, Li T, Aris JP, Benner SA (2005) Resurrecting ancestral alcohol dehydrogenases from yeast. Nat Genet 37:630–635

Toda T, Cameron S, Sass P, Zoller M, Wigler M (1987) Three different genes in *S. cerevisiae* encode the catalytic subunits of the cAMP-dependent protein kinase. Cell 50:277–287

Toda T, Cameron S, Sass P, Wigler M (1988) SCH9, a gene of *Saccharomyces cerevisiae* that encodes a protein distinct from, but functionally and structurally related to, cAMP-dependent protein kinase catalytic subunits. Genes Dev 2:517–527

Tu J, Carlson M (1995) REG1 binds to protein phosphatase type 1 and regulates glucose repression in *Saccharomyces cerevisiae*. EMBO J 14:5939–5946

Tu BP, Kudlicki A, Rowicka M, McKnight SL (2005) Logic of the yeast metabolic cycle: temporal compartmentalization of cellular processes. Science 310:1152–1158

Turner JJ, Ewald JC, Skotheim JM (2012) Cell size control in yeast. Curr Biol 22:R350–R359

Urban J, Soulard A, Huber A, Lippman S, Mukhopadhyay D, Deloche O, Wanke V, Anrather D, Ammerer G, Riezman H et al (2007) Sch9 is a major target of TORC1 in *Saccharomyces cerevisiae*. Mol Cell 26:663–674

Wang Y, Pierce M, Schneper L, Guldal CG, Zhang X, Tavazoie S, Broach JR (2004) Ras and Gpa2 mediate one branch of a redundant glucose signaling pathway in yeast. PLoS Biol 2:E128

Wanke V, Pedruzzi I, Cameroni E, Dubouloz F, De Virgilio C (2005) Regulation of G0 entry by the Pho80-Pho85 cyclin-CDK complex. EMBO J 24:4271–4278

Werner-Washburne M, Braun E, Johnston GC, Singer RA (1993) Stationary phase in the yeast *Saccharomyces cerevisiae*. Microbiol Rev 57:383–401

Wu B, Ottow K, Poulsen P, Gaber RF, Albers E, Kielland-Brandt MC (2006) Competitive intra- and extracellular nutrient sensing by the transporter homologue Ssy1p. J Cell Biol 173:327–331

Xue Y, Batlle M, Hirsch JP (1998) *GPR1* encodes a putative G protein-coupled receptor that associates with the Gpa2p Galpha subunit and functions in a Ras-independent pathway. EMBO J 17:1996–2007

Yan G, Lai Y, Jiang Y (2012) The TOR complex 1 is a direct target of Rho1 GTPase. Mol Cell 45:743–753

Yang Z, Klionsky DJ (2009) An overview of the molecular mechanism of autophagy. Curr Top Microbiol Immunol 335:1–32

Yorimitsu T, Zaman S, Broach JR, Klionsky DJ (2007) Protein kinase A and Sch9 cooperatively regulate induction of autophagy in *Saccharomyces cerevisiae*. Mol Biol Cell 18:4180–4189

Zaman S, Lippman SI, Schneper L, Slonim N, Broach JR (2009) Glucose regulates transcription in yeast through a network of signaling pathways. Mol Syst Biol 5:245

Zhao H, Wang JQ, Shimohata T, Sun G, Yenari MA, Sapolsky RM, Steinberg GK (2007) Conditions of protection by hypothermia and effects on apoptotic pathways in a rat model of permanent middle cerebral artery occlusion. J Neurosurg 107:636–641

Zheng Y, Jiang Y (2005) The yeast phosphotyrosyl phosphatase activator is part of the Tap42-phosphatase complexes. Mol Biol Cell 16:2119–2127

Response and Cytoprotective Mechanisms Against Proteotoxic Stress in Yeast and Fungi

5

Yukio Kimata, Thi Mai Phuong Nguyen, and Kenji Kohno

Abstract

Misfolded and/or aggregated proteins are not only functionless but also hazardous for cells. The budding yeast *Saccharomyces cerevisiae* has been a significant model organism for exploring cellular strategies against protein misfolding. There are various intracellular machineries for sequestering protein aggregates away from other cellular compartments or for disaggregating them. In addition, the ubiquitin-proteasome system functions in degradation of misfolded proteins accumulated in the cytosol, nuclei, and the endoplasmic reticulum. The heat shock response and unfolded protein response are gene expression programs for the induction of factors that contribute to these protein quality control systems in response to an accumulation of misfolded proteins. While, in general, observations obtained through yeast studies can be applied to other eukaryotic species, yeast- and fungal-specific factors or phenomena are also intriguing not only as matters of basic cell biology but also for clinical (pathogenic yeast and fungi) and industrial (fermentation) applications.

5.1   Introduction

Genetic mutations or heterologous gene expression often causes misfolding of product proteins in cells. Moreover, various stress stimuli, such as exposure to high temperature or certain chemicals, are known to impair proper folding of cellular proteins, leading to their aggregation. It is widely accepted that protein misfolding is hazardous

Y. Kimata (✉) · T. M. P. Nguyen
Graduate School of Biological Sciences, Nara Institute of Science and Technology, Ikoma, Japan
e-mail: kimata@bs.naist.jp

K. Kohno
Institute for Research Initiatives, Nara Institute of Science and Technology, Ikoma, Japan

© Springer Nature Switzerland AG 2018
M. Skoneczny (ed.), *Stress Response Mechanisms in Fungi*,
https://doi.org/10.1007/978-3-030-00683-9_5

161

to cells for a variety of reasons, including the general fact that misfolded proteins cannot exhibit their proper functions. Misfolding often leads to a massive insufficiency of the protein. As described in Sects. 5.4–5.9, the protein quality control systems work to capture misfolded proteins, which are then degraded and/or prevented from being transported to their final destinations. While beneficial for cells in most cases, the protein quality control systems are thought to often aggravate human diseases. For example, some mutant forms of the cystic fibrosis transmembrane conductance regulator (CFTR) protein are not transported to the plasma membrane and are quickly digested via endoplasmic reticulum (ER)-associated degradation (ERAD) in cells from patients with cystic fibrosis (Farinha and Canato 2017). Moreover, misfolded proteins often aggregate, are per se toxic for cells, and cause various neurodegenerative diseases. The disadvantages of misfolded protein expression in budding yeast *Saccharomyces cerevisiae* cells have been quantitatively described by Geiler-Samerotte et al. (2011). It is widely believed that protein aggregates can physically interact with and damage other cellular components.

Yeast cells are an important model organism for cell biological studies of the protein quality control systems and cellular responses against protein misfolding and aggregation. This is primarily because, in many cases, these cellular functions are evolutionarily conserved in eukaryotic species. Thus, a number of noteworthy insights were obtained through studies on yeasts, which are easy-to-use unicellular eukaryotic organisms. For instance, yeasts are used as model organisms for studies on the bioeffects of toxic chemicals (Sect. 5.3) and on human diseases (Sect. 5.2) in which protein misfolding is involved. Asymmetric cell division and senescence are also biological issues that can be investigated via yeast studies. Most yeast species including *S. cerevisiae* proliferate through budding to yield mother and daughter cells. Importantly, mother cells age, as demonstrated, for example, by the observation that they produce only a limited number of daughter cells (Sinclair et al. 1998). The relationship between protein misfolding/aggregation and asymmetric cell division is also described in this Sect. 5.4.

It is also intriguing that there exist various yeast- and fungal-specific factors and phenomena related to protein misfolding/aggregation. For instance, pathogenic yeast and fungi are likely to utilize their protein quality surveillance systems in the infection progress. It also should be noted that, as touched in Sect. 5.3, stress stimuli to which yeast cells are exposed during their use in food production and ethanol fermentation often result in misfolding and aggregation of cellular proteins.

5.2   Protein Aggregation and Prion Formation in Yeast and Fungal Cells

S. cerevisiae cells heterologously expressing aggregate-prone proteins that cause human neurodegenerative disorders have been used as disease models (Coughlan and Brodsky 2005). Such neurodegenerative disorders include the polyglutamine (polyQ) expansion diseases such as Huntington's disease, which are caused by expanded cytosine-adenine-guanine (CAG) repeats encoding a long polyQ tract in

Fig. 5.1 Growth and propagation of yeast prions. Through contact with prion seeds or fibers, normally folded proteins are converted into the prion form and stacked for prion growth. Splitting of prion fibers generates new prion seeds, which cause prion propagation

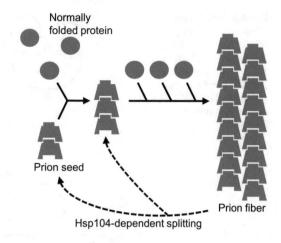

the huntingtin protein (Fan et al. 2014). According to Meriin et al. (2002), GFP carrying a polyQ expansion aggregates, resulting in growth retardation in a polyQ length-dependent manner when expressed in *S. cerevisiae* cells. Furthermore, α-synuclein, which aggregates in neurons to cause Parkinson's disease, also aggregates in *S. cerevisiae* cells when highly expressed (Outeiro and Lindquist 2003). Powerful yeast genetic techniques have since been employed for genome-wide comprehensive screens for factors that are related to α-synuclein toxicity (Willingham et al. 2003; Liang et al. 2008; Zabrocki et al. 2008; Yeger-Lotem et al. 2009). A number of proteins including the polyQ proteins and α-synuclein are known to aggregate as amyloids, which are highly ordered cross-β-sheet stacks of denatured monomers, while other proteins form more amorphous aggregates. It should also be noted that it is possible that soluble prefibrillar amyloid oligomers are more cytotoxic than insoluble rigid amyloid fibers (Sakono and Zako 2010).

Small fragments of amyloid fibers often work as seeds and grow to be new amyloid fibers through the incorporation of "normally folded" endogenous proteins. Notably, these kinds of amyloid oligomers can act as prions, which are infectious agents that do not carry genome DNA or RNA. Yeast prions are aggregates of endogenous proteins that are transmitted through cell division and grow in progeny cells even in the absence of genetic mutations of the corresponding genes (Fig. 5.1; Glover et al. 1997; Li and Kowal 2012; Wickner 2016). One of the intriguing features of yeast prions is that they serve as epigenetic inheritable units that are inherited in a non-Mendelian manner (Fig. 5.2). As listed in Garcia and Jarosz (2014), a number of yeast or fungal proteins can form prions resulting in notable phenotypes. For example, prion formation of Sup35, which is the class-II translation-termination factor, in *S. cerevisiae* cells, leads to loss-of-function of this protein. The non-Mendelian inheritable unit linked to the Sup35 prion formation is named [*PSI*⁺] and causes frequent readthrough of the translation-termination stop codons. Kawai-Noma et al. (2010) demonstrated that Sup35 prions actually form a fibrillar structure in *S. cerevisiae* cells.

Fig. 5.2 Non-Mendelian inheritance of yeast prions. Mating between [*PRION⁺*] and [*prion⁻*] haploids of *S. cerevisiae* generates a [*PRION⁺*] diploid, which yields four [*PRION⁺*] haploid spores after meiosis

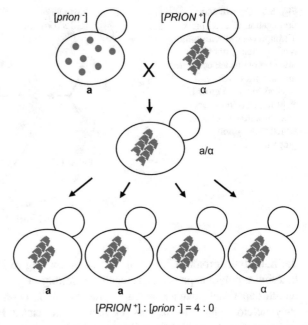

As described above, protein misfolding and aggregation are believed to be hazardous to cells. For instance, McGlinchey et al. (2011) demonstrated the harmfulness of yeast prions. However, it is also possible that yeast prion formation is an intrinsic ability of certain proteins and can be beneficial in certain situations (True and Lindquist 2000; True et al. 2004; Shorter and Lindquist 2005; Suzuki et al. 2012). For example, prion conversion of Mod5 protein affects the sterol biosynthetic pathway, leading to cellular resistance against certain antifungal agents. According to Suzuki et al. (2012), selective pressure from antifungal drugs facilitates the de novo appearance of the Mod5 prion state for yeast cell survival. Furthermore, the HETs protein of filamentous fungus *Podospora anserina* gains its function in heterokaryon incompatibility when it is prionized (Coustou et al. 1997; Wickner 2016). Recent studies have also demonstrated intriguing examples of harmful and beneficial aspects of noninheritable protein aggregates. Age-dependent aggregation of the Whi3 protein has been reported to lead to a loss of pheromone sensitivity of old *S. cerevisiae* cells (Schlissel et al. 2017). In contrast, stress-induced reversible aggregation and inactivation of the Cdc19 protein is likely to be a cellular strategy to cope with starvation stress (Saad et al. 2017).

Protein misfolding and aggregation occur not only in the cytosol and nuclei but also in membranous cellular compartments such as the ER. The ER is a flat- or tubular-shaped membrane-bound sac in and on which secretory and transmembrane proteins are folded. As shown in Fig. 5.3, fluorescent proteins that localize to the ER show a double-ring-like distribution pattern in *S. cerevisiae* cells. The outer ring represents the cortical ER, which proximately locates to the plasma membrane, whereas the inner ring is identical to the nuclear envelope, which also functions as

Fig. 5.3 The endoplasmic reticulum (ER) of *S. cerevisiae* cells. (**a**) Cells producing the ER version of GFP (eroGFP; Merksamer et al. 2008) were imaged using a fluorescent microscope. (**b**) The ER of a cell is schematically illustrated. Green lines represent membranes, while the ER luminal space is colored yellow. Note that, as well as the ER in higher eukaryotic cells, the cortical ER in *S. cerevisiae* cells exists as a network of interconnected tubules (Prinz et al. 2000), which is not represented in this picture. (**c**) A cell producing both eroGFP and mCherry-tagged CFTR

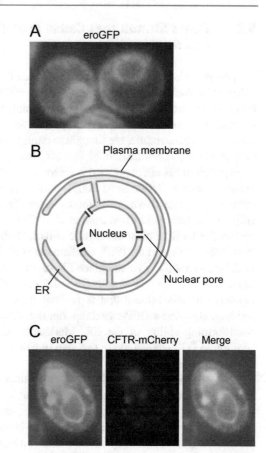

an ER (Fig. 5.3). In general, correctly folded ER client proteins are carried from the ER to the Golgi apparatus via COPII-mediated vesicular transport or direct contact of the two organelles (Sato and Nakano 2007; Kurokawa et al. 2014), while misfolded proteins are retained and accumulate in the ER. According to Umebayashi et al. (1997), a mutant version of an aspartic proteinase from a filamentous fungus *Rhizopus niveus* forms aggregates, which can be observed using transmission electron microscopy, in the ER lumen, and causes deformation of ER shape when heterologously expressed in *S. cerevisiae* cells. Figure 5.3 shows a punctate distribution of mammalian CFTR on the *S. cerevisiae* ER, as originally presented by Huyer et al. (2004).

5.3 Stress Stimuli that Cause Protein Misfolding and Aggregation

Nascent peptides translocated into the ER are often modified by asparagine residue-linked (N-linked) glycosylation (Aebi 2013) and/or disulfide-bond linkage between two cysteine residues. Therefore, stress stimuli that impair these posttranslational modifications predominantly disturb protein folding in the ER, while other stresses, such as high temperature, are thought to damage protein-folding status in cells rather non-selectively. Dysfunction of the ER that is tightly linked to misfolding and aggregation of ER client proteins is referred to as ER stress and triggers the unfolded protein response (UPR; Mori 2009), which, as described in Sect. 5.9, is a stress-induced cytoprotective transcriptome change. The N-linked glycosylation-inhibiting antibiotic tunicamycin (Takatsuki et al. 1975) is a potent ER stressor and strongly evokes the UPR in most eukaryotic cells including yeasts and fungi (Normington et al. 1989; Kohno et al. 1993; Guillemette et al. 2007). Thiol-reducing agents such as dithiothreitol (DTT) also induce ER stress and cause ER accumulation of protein aggregates in *S. cerevisiae* cells (Promlek et al. 2011) because of the disturbance of oxidative protein folding that is facilitated by disulfide-bond linkage of cysteine residues. By using a GFP-based reporter that allows for assessments of the disulfide bond-forming ability of the ER, Merksamer et al. (2008) demonstrated that in addition to DTT, tunicamycin can also impair the disulfide bond-forming ability of the *S. cerevisiae* ER.

As described in Sect. 5.8, the heat shock response (HSR) that is mediated by the transcription-factor protein Hsf1 is believed, at least partly, to be a result of impaired protein folding in the cytosol and nuclei, while the UPR is triggered upon accumulation of misfolded proteins in the ER. The best-known example of a stress stimulus that induces the HSR is caused by an acute shift to a higher-than-normal temperature. It has also been widely accepted that the HSR is involved in ethanol tolerance of *S. cerevisiae* cells (Ma and Liu 2010a, b; Teixeira et al. 2011). On the other hand, the UPR is also induced potently in ethanol-stressed *S. cerevisiae* cells (Miyagawa et al. 2014). Moreover, the UPR machinery contributes to ethanol tolerance of *S. cerevisiae* cells, and ethanol stress facilitates protein aggregation in the *S. cerevisiae* ER (Miyagawa et al. 2014). Ethanol stress is thus likely to cause protein misfolding and aggregation throughout cells.

While some environment-polluting heavy metals are known to work as protein denaturants (Tamas et al. 2014), they seem to harm yeast cells by distinct toxicity mechanisms. Cadmium (Cd^{2+}), but not toxic concentrations of mercury (Hg^{2+}), lead (Pb^{2+}), arsenic (As^{3+}), or chromium ($CrO4^{2-}$), induces a potent UPR when added into *S. cerevisiae* cultures (Gardarin et al. 2010; Le et al. 2016). Le et al. (2016) demonstrated that cadmium ions impair oxidative protein folding in the ER, leading to aggregation of ER client proteins. In contrast, Jacobson et al. (2017) demonstrated an aggregation of cytosolic proteins by cadmium-ion stress. According to Jacobson et al. (2012), arsenite has a potent ability to aggregate proteins, which explains the toxicity mechanism of this metalloid in yeast cells. Arsenite, hydrogen peroxide (oxidative stress) and azetidine-2-carboxylic acid (an amino acid analogue) are likely to facilitate aggregation of similar kinds of proteins in *S. cerevisiae* cells (Weids et al. 2016). This observation suggests that the intrinsic nature of proteins determines their propensity to aggregate.

5.4   Spatial Sequestration of Misfolded/Aggregated Proteins

According to Parsell et al. (1994), transmission electron microscopy of severely heat-stressed *S. cerevisiae* cells demonstrated the existence of many electron-dense foci dispersed throughout the cytosol and nuclei. In other words, this type of protein aggregate is not distributed to a specific cellular compartment(s).

On the other hand, Kaganovich et al. (2008) reported two distinct non-membranous cellular compartments in which cytosolic protein aggregates are deposited. One is the insoluble protein deposit (IPOD), which is located adjacent to the vacuole. The IPOD is thought to be the site where rigidly and non-reversibly aggregated proteins, such as amyloid aggregates, accumulate. According to Tyedmers et al. (2010a), prions undergo their maturation process in the IPOD. In contrast, proteins that are aggregated in the juxtanuclear quality control compartment (JUNQ), which is located near the nuclear membrane, seem to be more mobile. As described in Sect. 5.6, a major (and probably the best-known) pathway for the disposal of cellular misfolded proteins is via their ubiquitination and subsequent degradation by proteasomes, which are large proteinaceous particles located in the cytosol and nuclei. Because aggregation-prone proteins accumulate in the JUNQ when ubiquitinated, it is likely that the JUNQ is the site in which proteasome substrate proteins are transiently stored before their degradation.

It should be noted that the JUNQ is clearly observable in cells with proteasome dysfunction (Kaganovich et al. 2008). Escusa-Toret et al. (2013) reported that, under more natural conditions, misfolded cytosolic proteins gather on the cortical ER of *S. cerevisiae* cells to form mobile puncta, which are named the Q-bodies, also known as cytoQ (Miller et al. 2015a), before being digested. The JUNQ may be an end-point product that is formed through the coalescence of Q-bodies as a result of impaired clearance. Moreover, Miller et al. (2015b) reported a protein aggregate-accumulation site in the nucleus, which is named the intranuclear quality control compartment (INQ), and may be identical to the JUNQ. Protein transport via the nuclear pores is required for the deposition of cytosolic misfolded proteins in the INQ.

Importantly, deposition of misfolded and/or aggregated proteins into a large cellular compartment(s) such as the IPOD, JUNQ, INQ, and/or Q-bodies requires cellular factors and may be a cellular strategy to sequester potentially toxic species from other cellular components (Malinovska et al. 2012; Escusa-Toret et al. 2013; Miller et al. 2015a). Foci formed by CFTR aggregation on the ER are named the ER-associated compartment (ERAC; Huyer et al. 2004) and may also be beneficial for cells. Kakoi et al. (2013) proposed the involvement of the COPII machinery for ERAC formation.

Upon division of *S. cerevisiae* cells, damaged or aggregated cytosolic proteins appear not to be distributed to daughter cells (Aguilaniu et al. 2003; Liu et al. 2010; Zhou et al. 2014). According to Liu et al. (2010), actin cables are involved in the intracellular transport of protein aggregates during their asymmetric distribution, which results in "young and healthy" progeny. Meanwhile, Zhou et al. (2014) proposed that protein aggregates are initially bound on the ER surface and then transferred to the mitochondria surface for their retention in the mother cells. As described by Clay et al. (2014), misfolded ER client proteins are also retained in mother cells. Damaged proteins are also asymmetrically distributed upon binary cell division of fission yeast *Schizosaccharomyces pombe* (Erjavec et al. 2008).

5.5   Disaggregation of Cytosolic and Nuclear Proteins

Heat shock proteins (Hsps) originally referred to a group of proteins whose cellular expression levels are elevated when cells are exposed to heat stress. Many Hsps, for example, *S. cerevisiae* Hsp104, are transcriptionally induced by the HSR that is mediated by Hsf1 and serve as molecular chaperones.

As described above, protein aggregates can be observed in the cytosol and nuclei of heat-stressed *S. cerevisiae* cells via transmission electron microscopy. According to Parsell et al. (1994), disappearance of the protein aggregates during the post-stress recovery phase is markedly retarded when cells carry a mutation in the *HSP104* gene. In agreement with this observation, mutations in *HSP104* considerably diminish the thermotolerance of *S. cerevisiae* cells (Sanchez and Lindquist 1990). Hsp104 belongs to the AAA+ (ATPases associated with various cellular activities) ATPase superfamily (Ogura and Wilkinson 2001; Grimminger-Marquardt and Lashuel 2010). Briefly, AAA+ ATPases self-oligomerize to form a ring-shaped structure, into which linear macromolecules are threaded using energy obtained through ATP hydrolysis, and thus work to dissociate and disentangle macromolecular complexes. Consistent with this concept, Hsp104 acts as a disaggregase for the re-solubilization of protein aggregates, while it does not seem to prevent de novo protein aggregation (Parsell et al. 1994; Glover and Lindquist 1998; Tyedmers et al. 2010b). The JUNQ and IPOD contain Hsp104 (Kaganovich et al. 2008), possibly indicating ongoing protein disaggregation in these compartments. Unlike fungi, bacteria, or plants, metazoan cells do not carry Hsp104 orthologues.

Intriguingly, both loss and overproduction of Hsp104 lead to the elimination of certain *S. cerevisiae* prions (Patino et al. 1996; Grimminger-Marquardt and Lashuel 2010). This observation strongly suggests an involvement of Hps104 in prion disaggregation in yeast cells. As for the reason why loss of Hsp104 abolishes yeast prion propagation, Hsp104 is likely to function to split prion fibers, yielding prion seeds.

Another family of molecular chaperones, Hsp70, is the most ubiquitous chaperone family and is highly conserved throughout prokaryotic and eukaryotic organisms (Sharma and Masison 2009). The Hsp70 family of chaperones are composed of N-terminal nucleotide-binding and C-terminal substrate-binding

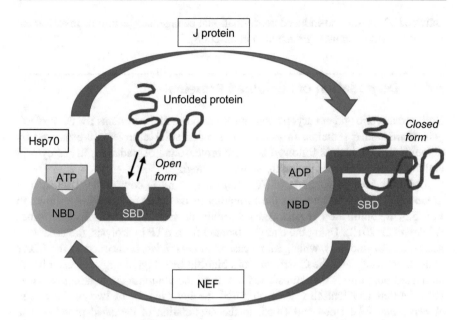

Fig. 5.4 ATPase cycle of the Hsp70 family chaperones. ATP association with the nucleotide-binding domain (NBD) of Hsp70 results in its substrate-binding domain (SBD) to be in the "open form," which quickly captures and releases unfolded substrate proteins. Meanwhile, the ADP-bound form of Hsp70 results in the "closed-form" of the SBD, in which the unfolded substrate protein is tightly trapped. J proteins promote ATP hydrolysis by Hsp70, while the nucleotide exchange factors (NEFs) stimulate ADP/ATP exchange of Hsp70

domains. As illustrated in Fig. 5.4, they capture and release unfolded protein substrates dependent on the ATP cycle, which is stimulated by various cofactors including J proteins. J proteins are defined as proteins that carry the highly conserved J domain, which is responsible for binding to the Hsp70 family of chaperones. Nuclear and cytosolic *S. cerevisiae* chaperones in the Hsp70 family can be divided into subfamilies (Peisker et al. 2010), and the most essential and major subfamily is composed of four structurally similar proteins Ssa1, 2, 3, and 4. Meanwhile, the members of the Ssb subfamily of chaperones are associated with ribosomes and are thought to specialize in the quality control of nascent peptides (Peisker et al. 2010). An in vitro reconstitution study by Glover and Lindquist (1998) showed that Hsp104 disaggregates and reactivates aggregated client proteins in concert with the cytosolic Hsp70/J-protein system, Ssa1/Ydj1 in *S. cerevisiae*.

In contrast to Hsp104 and the Hsp70 family chaperones, small Hsps (sHsps) exhibit an ATP-independent chaperone activity. *S. cerevisiae* sHsps, Hsp42 and Hsp26, function to prevent protein aggregation and to re-solubilize aggregated proteins via multiple mechanisms (Tyedmers et al. 2010b; Mogk and Bukau 2017). In contrast, Specht et al. (2011) reported that Hsp42 also contributes to aggregation and Q-body deposition of misfolded proteins. Interestingly, Hsp26 is activated via its thermo-induced conformational change, suggesting its involvement in yeast thermotolerance (Franzmann et al. 2008).

5.6   Degradation of Misfolded Proteins

Both prokaryotic and eukaryotic cells have various cellular systems for the proteolysis of misfolded proteins. In general, misfolded soluble proteins are marked by ubiquitination, which is followed by their proteasomal degradation, in the cytosol and nuclei of eukaryotes including yeast and fungi.

Ubiquitin is a polypeptide of 76 amino acids and is covalently attached to a number of cellular proteins via the formation of an isopeptide bond. As shown in Fig. 5.5, ubiquitination is performed through the E1-E2-E3 cascade of enzymes (Finley et al. 2012). Using the energy obtained from ATP hydrolysis, the ubiquitin-activating enzyme (E1), which, in the case of *S. cerevisiae*, is encoded by the *UBA1* gene, links itself with the C terminus of ubiquitin by a high-energy thioester bond. Activated ubiquitin is then transferred to one of the ubiquitin-conjugating enzymes (E2). Seufert and Jentsch (1990) proposed the involvement of two of the eleven *S. cerevisiae* E2s, Ubc4 and Ubc5, in the degradation of damaged proteins. The ubiquitin ligases (E3) are responsible for the transfer of ubiquitin from E2s to substrate proteins. In a reflection of a wide variety of functions of ubiquitination (Finley et al. 2012), there exist a large number of E3s (60–100 in *S. cerevisiae*). In

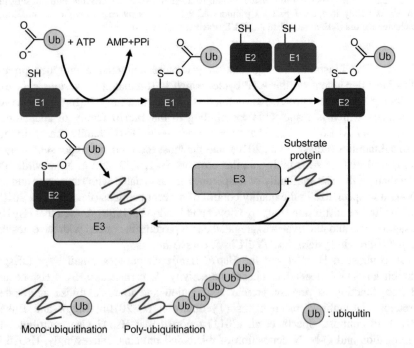

Fig. 5.5 The E1-E2-E3 cascade for protein ubiquitination

most cases studied so far, the carboxyl group of the C-terminal glycine residue on ubiquitin is conjugated to the ε-amino group of a lysine residue(s) on a substrate protein (Finley et al. 2012). Since ubiquitin itself carries multiple lysine residues, a ubiquitin molecule can ubiquitinate another ubiquitin, resulting in the formation of a poly-ubiquitin chain on a substrate protein (Fig. 5.5). While canonically, the Lys48-linked poly-ubiquitin chain serves as a signal for proteasomal degradation, Xu et al. (2009) proposed the existence of abundant poly-ubiquitin chains linked through other lysine residues and their involvement in target protein degradation. In particular, E2 Ubc6 predominantly causes the Lys11 linkage of ubiquitin molecules on substrate proteins to facilitate the ERAD, which is described in Sect. 5.7.

The E3s are thought to determine substrate specificity of ubiquitination, since they have the ability to associate with substrate proteins. While Ubr1 and its paralogue Ubr2 are likely to serve as the E3s for ubiquitination of misfolded cytosolic proteins in *S. cerevisiae* cells (Eisele and Wolf 2008; Nillegoda et al. 2010), Gardner et al. (2005) proposed that the nuclear-localized E3 San1 of *S. cerevisiae* promotes the ubiquitination of abnormal nuclear proteins together with E2s Ubc1 and Cdc34. The San1 protein has N-terminal and C-terminal intrinsically disordered regions, which can capture a wide variety of misfolded proteins (Rosenbaum et al. 2011). According to Fredrickson et al. (2011), San1 recognizes hydrophobic segments of a substrate protein. This observation explains the molecular basis for the distinction between correctly folded and misfolded proteins, since in general, misfolding of a soluble and globular protein leads to exposure of its hydrophobic segments, which are buried inside the protein when it is correctly folded. Heck et al. (2010) and Prasad et al. (2010) reported that cytosolic misfolded proteins are carried into the nuclei dependent on the Hsp70/J-protein chaperone system for their San1-mediated degradation. Another E3, Hul5, also appears to function in the poly-ubiquitination of misfolded cytosolic proteins, especially in heat-stressed *S. cerevisiae* cells (Fang et al. 2011).

The canonical proteasome is a large proteinaceous complex consisting of a 20S core particle, and one or two 19S regulatory particles (Tanaka 2009). The core particle is a cylinder-like structure, and the proteolytic site faces its internal cavity. The 19S regulatory particle caps the cylinder-shaped 20S core and delivers degradation substrate proteins, which, in most case, are poly-ubiquitinated, into the cylinder.

Mono-ubiquitination or Lys63-linked poly-ubiquitination, which, for example, is carried out by the E3 Rsp5 in *S. cerevisiae* cells, does not directly result in substrate proteins being targeted for proteasomal degradation. Rsp5-mediated ubiquitination of plasma membrane-located transmembrane proteins promotes their endocytosis, which is followed by vacuolar degradation (Kaliszewski and Zoladek 2008). According to Zhao et al. (2013) and Keener and Babst (2013), this phenomenon contributes to the clearance of misfolded and/or damaged cell-surface proteins. Moreover, Lu et al. (2014) proposed that aggregation-prone polyQ proteins are marked by Rsp5-mediated ubiquitination and then digested by autophagy, which, unlike proteasomal proteolysis, can eliminate large intracellular particles. In contrast, it is also likely that Rsp5 poly-ubiquitinates misfolded cytosolic proteins to mark them for proteasomal degradation in heat-stressed *S. cerevisiae* cells (Fang et al. 2014). In this case, poly-ubiquitin chains produced by Rsp5 are "edited" to be a signal for proteasomal degradation (Fang et al. 2016).

Since mitochondria originated from endosymbiosis of aerobic bacteria into ancient eukaryotic cells, they bear various bacteria-like features. For instance, the mitochondria carry several AAA+ family proteases including the Lon protease, which, in the case of *S. cerevisiae*, is encoded by the *PIM1* gene. The *S. cerevisiae* Lon protease appears to contribute to clearance of misfolded proteins accumulated in the mitochondria, and *PIM1*-gene knockout leads to mitochondrial dysfunction (Suzuki et al. 1994; Van Dyck et al. 1994; Wagner et al. 1994). Intriguingly, Ruan et al. (2017) reported that, in heat-stressed *S. cerevisiae* cells, misfolded cytosolic proteins are transported into the mitochondria, where they are degraded by Lon.

5.7 Protein Quality Control in the ER and ERAD

As described above, the ER is a cellular membranous sac. In the luminal area of the ER, soluble proteins are folded before being transported to the Golgi apparatus and then finally delivered to the cell surface (protein secretion) or other membranous compartments including the vacuoles. Moreover, transmembrane proteins are integrated into the membrane and folded on the ER. As a protein quality control system, the ER retains misfolded proteins and often digests them via the ERAD pathway. BiP and calnexin are the two most conserved and well-characterized molecular chaperones which function in these ER activities.

Calnexin is a calcium-binding transmembrane lectin that localizes to the ER and contributes to the quality control of N-glycosylated proteins (Helenius and Aebi 2004; Lederkremer 2009). When glucose is removed from the N-linked oligosaccharide chain of a client protein, it can escape from calnexin-dependent ER retention. However, if such a protein is still misfolded, it is re-glycosylated by UDP-Glc: glycoprotein glucosyltransferase (UGGT), an enzyme in the ER that can sense the folding status of substrate proteins, and then reassociates with calnexin. This system, namely, the calnexin cycle, is believed to exist in various eukaryotes including *S. pombe*. Fanchiotti et al. (1998) indicated that the UGGT is required for survival of heavily stressed *S. pombe* cells. Meanwhile, calnexin is not just a calnexin-cycle component, since the calnexin gene *CNX1*, but not the UGGT gene, is an essential *S. pombe* gene (Jannatipour and Rokeach 1995; Parlati et al. 1995a). Consistent with this insight, *S. cerevisiae* does not have the calnexin-cycle system (Fernandez et al. 1994), even though it has the calnexin gene *CNE1* (Parlati et al. 1995b). Disruption of *S. cerevisiae CNE1*, unlike that of *S. pombe CNX1*, results in only weak phenotypes, suggesting that the importance of calnexin differs among yeast and fungal species. Zhang et al. (2017) demonstrated the involvement of calnexin in calcium-ion homeostasis in the filamentous fungus *Aspergillus nidulans*.

The immunoglobulin heavy-chain-binding protein (BiP), which is named for its association with unassembled immunoglobulin heavy chains in mammalian antibody-producing cells, is an ER-located Hsp70 family molecular chaperone, which is conserved throughout eukaryotic species (Behnke et al. 2015). The

S. cerevisiae BiP gene is named *KAR2*, because a mutation in this gene impairs *kar*yogamy, which is the fusion of two haploid nuclei upon sexual mating of cells (Rose et al. 1989). In general, J proteins confer specific roles to the Hsp70 family of molecular chaperones. *S. cerevisiae* BiP has been demonstrated to have three J protein partners including Sec63, which facilitates the translocation of nascent peptides across the ER membrane (Rapoport et al. 2017). Meanwhile, other ER-located J proteins, Scj1 and Jem1, have been reported to function in the folding and quality control of ER client proteins together with BiP (Simons et al. 1995; Silberstein et al. 1998; Nishikawa et al. 2001). In agreement with the multiplicity of BiP's role, some *kar2* mutant alleles result in severe translocation deficiencies, while other *kar2* mutant alleles predominantly impair protein folding in the ER (Kimata et al. 2003, 2004). As shown in Umebayashi et al. (1997) and Huyer et al. (2004), BiP is preferentially co-localized and associated with protein aggregates that formed in the ER.

Pioneer *S. cerevisiae* studies published in the 1990s revealed that some mutants of ER client proteins were digested before being transported to the Golgi apparatus (Finger et al. 1993) and that cytosolic factor(s) and ATP are required for this phenomenon called ERAD (McCracken and Brodsky 1996). As reviewed in Thibault and Ng (2012), while the molecular machinery promoting ERAD has been uncovered through a number of studies in *S. cerevisiae*, the mechanisms are mostly conserved throughout eukaryotic species. One landmark finding was that misfolded proteins that accumulated in the ER lumen were poly-ubiquitinated and degraded in a proteasome-dependent manner (Hiller et al. 1996).

Szathmary et al. (2005) indicated the involvement of the ER luminal protein Yos9 in the degradation of glycosylated ERAD substrate proteins (Fig. 5.6). The ability of Yos9 to associate with a model ERAD substrate strongly suggests a role of this protein as a sensor of misfolded proteins. BiP and its J protein partners Scj1 and Jem1 are involved in the ERAD, possibly indirectly by maintaining the solubility of substrate proteins in the ER lumen (Plemper et al. 1997; Nishikawa et al. 2001). Meanwhile, Denic et al. (2006) proposed a role of BiP as a misfolded-protein sensory complex together with Yos9. Although a lectin, Yos9 is likely to function also in the degradation of an unglycosylated ERAD substrate (Benitez et al. 2011).

To be digested by the proteasome, ERAD substrate proteins located in the ER lumen must be transported across the ER membrane to the cytosol (Fig. 5.6). The ER-located E3 Hrd1 self-oligomerizes and serves as the most important membrane component in the ERAD (Carvalho et al. 2010). According to a cryo-EM structural analysis by Schoebel et al. (2017), Hrd1 forms an aqueous channel, which is likely to act as the polypeptide-conducting gate for the ERAD. The ER transmembrane protein Der1 also contributes to this retro-translocation step (Mehnert et al. 2014). Hrd1 is also responsible for the poly-ubiquitination of ERAD substrates, for which Ubc1, Ubc6, and Ubc7 act as the E2 (Biederer et al. 1997; Bays et al. 2001). After being poly-ubiquitinated, ERAD substrates are dissociated from the Hrd1 complex at the cytosolic side of the ER membrane by the Cdc48 AAA+-family ATPase complex (Fig. 5.6; Stein et al. 2014).

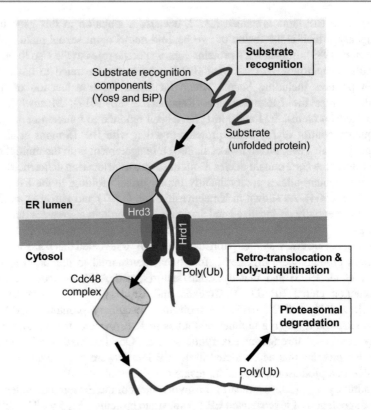

Fig. 5.6 The endoplasmic reticulum (ER)-associated degradation pathway for digestion of misfolded luminal proteins (ERAD-L). Yos9 and BiP form a complex for the surveillance of misfolded proteins that accumulate in the ER. After being captured by the BiP/Yos9 complex, misfolded proteins are transferred to the retro-translocation complex, which is composed of Hrd1, Hrd3, and other proteins including Der1. Hrd1 serves both as the retro-translocation channel and the ubiquitin (Ub) ligase (E3). The Cdc48 complex, which contains Udf1 and Npl4, pulls misfolded substrate proteins to the cytosolic side of the ER membrane for their proteasomal degradation

The ERAD described thus far has been on the digestion of misfolded proteins that accumulate in the ER lumen and is therefore called ERAD-L (L for Luminal). On the other hand, misfolded transmembrane proteins are also subjected to proteasomal degradation. Based on the factor requirements, two other subtypes of the ERAD, namely, ERAD-C (C for Cytosolic) and ERAD-M (M for Membrane), have been proposed as degradation pathways for transmembrane proteins carrying misfolded regions, respectively, on their cytosolic and transmembrane domains (Vashist and Ng 2004; Carvalho et al. 2006). Unlike ERAD-L, ERAD-M does not require Der1. Instead of Hrd1, another transmembrane E3, namely, Doa10, is employed in ERAD-C. A transmembrane protein damaged in its luminal domain is thought to be digested via the ERAD-L pathway.

As described above, the nuclear envelope is a part of the ER. The outer nuclear membrane faces the cytosol and seems to be functionally equivalent to the cortical

ER membrane (for instance, in attachment of ribosomes). However, the inner nuclear membrane is thought to have quite different properties, and notably, the ERAD across the inner membrane is carried out through a distinct mechanism in which the Asi complex serves as the E3 (Foresti et al. 2014; Khmelinskii et al. 2014).

5.8   HSR

It is rational that, as a strategy for cellular survival, proteins contributing to the protein quality control systems are transcriptionally induced in response to stress stimuli that damage the cellular protein-folding status. In brief, two regulatory pathways, which are controlled, respectively, by the transcription factors Msn2 (and its paralogue Msn4) and Hsf1, are involved in gene induction in response to heat stress in *S. cerevisiae* cells (Teixeira et al. 2011; Verghese et al. 2012; Morano et al. 2012). In addition to high temperature, various other stress stimuli, including nutrient limitation and oxidative stress, trigger the Msn2/Msn4-mediated transcriptional change, which is thus thought to be a main branch of the general stress response or the environmental stress response (ESR). According to Gasch et al. (2000), approximately 180 genes are transcriptionally upregulated through the ESR dependent on Msn2/Msn4. Meanwhile, as described below, the HSR that is controlled by Hsf1 appears to be more directly linked to a disturbance of protein-folding integrity and cellular defense against it.

Msn2/Msn4 and Hsf1 recognize different promoter sequences and thus, according to protein expression profiling of unstressed and heat-stressed *S. cerevisiae* cells (Boy-Marcotte et al. 1999), upregulate the expression of different proteins, i.e., Msn2/4 upregulates antioxidants and enzymes involved in carbon metabolism, and Hsf1 upregulates molecular chaperones and their cofactors. It should be noted that heat stress also leads to other undesirable outcomes besides protein misfolding, such as the production of reactive oxygen species (Moraitis and Curran 2004), with which the Msn2-/Msn4-mediated ESR may cope. Meanwhile, expression of Hsp26 and Hsp104 is driven both by Msn2/Msn4 and Hsf1 (Amoros and Estruch 2001).

Unlike fungus-specific Msn2 homologues, Hsf1 is conserved throughout eukaryotic species. While vertebrate and plant cells carry multiple Hsf1 paralogue genes, *S. cerevisiae* cells express a single Hsf1 gene, which is essential for cellular growth even under non-stress conditions. According to Solis et al. (2016), 18 genes, which include those encoding Hsp104, Hsp42, cytosolic/nuclear Hsp70 paralogues (Ssa1 and Ssa2), and their cofactors, are strictly under the control of Hsf1 in *S. cerevisiae* cells. The critical nature of Hsf1 under non-stress conditions can be explained by the basal-level transcription of some of these genes (Solis et al. 2016).

The regulatory mechanism of Hsf1 also highlights a tight relationship between the HSR and protein misfolding in the cytosol and nuclei. Hsf1 is activated and binds to target DNA as trimers. It is widely accepted that, in unstressed mammalian cells, molecular chaperones (Hsp70 and Hsp90) associate with Hsf1, causing trimer dissociation and cytosolic localization of Hsf1 (Dayalan Naidu and Dinkova-Kostova 2017). According to the "chaperone titration" model, proteotoxic stress, including heat shock, massively elevates the cellular levels of misfolded proteins, which deprives Hsf1 of molecular chaperones. While this model explains the activation mechanism of Hsf1 upon heat stress, it should be noted that, reflecting its activity under non-stress conditions, Hsf1 is located in the nuclei and binds to target DNA sequences constitutively in *S. cerevisiae* cells (McDaniel et al. 1989; Hahn et al. 2004). Nevertheless, it is also evident that, even in the case of *S. cerevisiae*, proteotoxic stress upregulates Hsf1. Zheng et al. (2016) demonstrated association of the cytosolic/ nuclear Hsp70 chaperone(s) with Hsf1, which is abolished in response to heat stress in *S. cerevisiae* cells. This phenomenon is likely to serve as the core mechanism regulating the activity of Hsf1, suggesting that the "chaperone titration" model is applicable also to the *S. cerevisiae* HSR, though possibly in a different way from that of mammalian cells. Since the cytosolic/nuclear Hsp70 chaperone(s) is massively induced by the HSR, this regulatory mechanism of Hsf1 can be considered to be a negative feedback loop. Furthermore, Zheng et al. (2016) also proposed that phosphorylation of Hsf1 facilitates recruitment of the basal transcription machinery to the promoters of the Hsf1 target genes, which appears to contribute to a sustained HSR during heat stress.

The human pathogenic yeast *Candida albicans* requires the activation of Hsf1 for virulence (Nicholls et al. 2011). According to Leach et al. (2016), Hsf1 positively controls not only molecular chaperone genes but also genes involved in pathogenicity in *C. albicans* cells. It is likely that infection of warm-bodied animals per se is a "heat shock" for the pathogenic yeast cells and that they use the "heat shock" signal both for cellular protection against proteotoxic stress and for virulent progression.

5.9   The Unfolded Protein Response (UPR)

Another gene expression program which copes with proteotoxic stress is the UPR, which, unlike the HSR, appears to be specialized for the maintenance of ER homeostasis. Three sequential pioneer studies showed that, in addition to the Hsf1-responsive sequence, the promoter region of the *S. cerevisiae* BiP gene *KAR2* carries an element that contributes to the gene induction in response to various stress stimuli that cumulatively impair protein-folding status in the ER (Normington et al. 1989; Mori et al. 1992; Kohno et al. 1993). Importantly, this observation indicates that the HSR and UPR are distinct cellular events.

Through genetic screens for *S. cerevisiae* mutant cells in which the UPR is not triggered, the ER-located type I transmembrane Ire1 was identified as a key factor for the UPR (Cox et al. 1993; Mori et al. 1993). Upon ER stress, the Ire1-dependent UPR machinery transcriptionally induces a large number of genes including those encoding factors for protein translocation across the ER membrane, COPII vesicle components and enzymes for lipid biosynthesis, as well as molecular chaperones and protein modification enzymes that contribute to protein folding in the ER (Travers et al. 2000; Kimata et al. 2006). According to Casagrande et al. (2000) and Friedlander et al. (2000), expression of some of the ERAD machinery components is also positively controlled by the UPR machinery. Therefore, the UPR is likely to

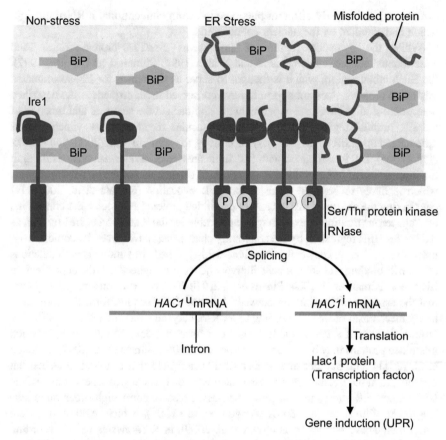

Fig. 5.7 Activation steps and function of *S. cerevisiae* Ire1. Upon ER stress, Ire1 self-oligomerizes and is activated through the release of BiP and the direct association of misfolded proteins. The N-terminal intrinsically disordered stretch of Ire1 also negatively contributes to the regulation of Ire1 oligomerization

be a cellular system devoted to coping with misfolded proteins that accumulate in the ER and to restoring total ER functions under ER stress conditions.

Two other factors, Hac1 and Rlg1/Trl1, have also been demonstrated to be indispensable for intracellular signaling in the UPR in *S. cerevisiae* cells (Cox and Walter 1996; Mori et al. 1996; Sidrauski et al. 1996). The transcript from the *HAC1* gene (*HAC1*[u], u for uninduced) has an intron and cannot be translated into an active transcription factor (Fig. 5.7). However, under ER stress conditions, Ire1 serves as an endoribonuclease to splice the *HAC1*[u] mRNA, yielding the *HAC1*[i] mRNA (i for induced), the translation product of which, namely, the Hac1 protein, is responsible for the transcriptional induction in the UPR (Fig. 5.7; Cox and Walter 1996; Sidrauski and Walter 1997). The tRNA ligase Rlg1/Trl1 functions in exon ligation in this splicing reaction (Sidrauski et al. 1996). It should be noted that Ire1- and Rlg1/Trl1-dependent splicing of the *HAC1*[u] mRNA, which is carried out in the cytosol, is a completely different phenomenon from conventional mRNA splicing, which is dependent on the nuclear spliceosome.

While the cytosolic region of Ire1 bears Ser/Thr protein kinase and endoribonuclease activities (Shamu and Walter 1996; Sidrauski and Walter 1997), the ER luminal region, which is thought to serve as the sensor for ER-accumulated misfolded proteins, does not appear to have conserved motif sequences. As described here, some studies have thus experimentally approached the structure and function of the ER luminal region, elucidating the molecular mechanism by which Ire1 is upregulated upon ER stress (Fig. 5.7). According to Okamura et al. (2000) and Kimata et al. (2003), BiP is associated with Ire1 in unstressed *S. cerevisiae* cells, resulting in the suppression of Ire1 activity. ER stress causes dissociation of the Ire1-BiP complex, which is likely to be a prerequisite for UPR evocation (Kimata et al. 2003). By considering the ability of BiP to capture misfolded proteins as a molecular chaperone, the "chaperone titration" model may be applicable for the regulation of Ire1 by BiP, as well as for HSR regulation by the Ssa family chaperone(s). However, it should also be noted that in *S. cerevisiae* cells, as well as wild-type Ire1, Ire1 mutants with deletions of the BiP-binding site are not constitutively active but trigger the UPR dependent on ER stress (Kimata et al. 2004; Pincus et al. 2010). This observation strongly suggests that the association/dissociation between Ire1 and BiP does not form the core regulatory mechanism of Ire1. According to an X-ray crystal structural analysis of the ER luminal domain of *S. cerevisiae* Ire1, its dimer forms a groove-like structure, in which misfolded proteins may be captured (Credle et al. 2005). Kimata et al. (2007), Promlek et al. (2011) and Gardner and Walter (2011) indicated that unfolded proteins that accumulate in the ER are actually associated with Ire1, leading to an evocation of the UPR. Upon ER stress, Ire1 becomes self-associated to form high-order oligomers (Fig. 5.8; Kimata et al. 2007; Aragon et al. 2009), which exhibit a potent endoribonuclease activity (Korennykh et al. 2009), in *S. cerevisiae* cells. According to Gardner and Walter (2011), the association between ER-accumulated misfolded proteins and Ire1 leads to the bundling of the latter.

The kinase domain of Ire1 functions in its autophosphorylation and capture of ADP, both of which are required for the conformational change of Ire1, which then exhibits potent endoribonuclease activity (Shamu and Walter 1996, Lee et al. 2008; Korennykh et al. 2009; Rubio et al. 2011). Dephosphorylation of Ire1 contributes to the attenuation of Ire1 activity during the recovery phase of ER-stressed *S. cerevisiae* cells (Chawla et al. 2011; Rubio et al. 2011). Reassociation of BiP with Ire1 is also involved in the repression of Ire1 after peak induction of the UPR (Pincus et al. 2010; Ishiwata-Kimata et al. 2013). Since BiP is transcriptionally induced by the UPR, this phenomenon is likely to be a negative feedback regulatory mechanism. The luminal domain of yeast and fungal Ire1 orthologues, but not those of metazoan or plant cells, carry N-terminal intrinsically disordered stretches, which also contribute to the repression of Ire1 activity (Fig. 5.7; Mathuranyanon et al. 2015).

Unlike the mRNAs of metazoan and plant transcription factor that are targets of Ire1-dependent splicing (metazoan XBP1 and plant (*Arabidopsis*) bZIP60), translation of the yeast and fungal *HAC1* mRNAs appears to be blocked unless they are spliced by Ire1. In the case of the *S. cerevisiae* *HAC1*[u] mRNA, an intramolecular

Unstressed cells

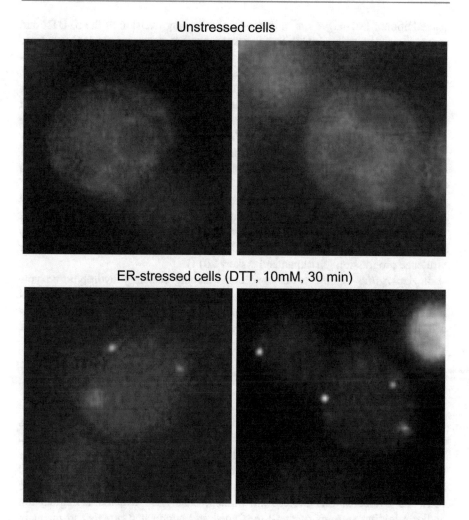

ER-stressed cells (DTT, 10mM, 30 min)

Fig. 5.8 High-order oligomerization of Ire1 in *S. cerevisiae* cells. Cells producing mNeonGreen-tagged Ire1 were imaged via fluorescence microscopy. When stressed by DTT, the Ire1 molecules gather to form dot-like foci

base pairing between its 5′-untranslated region (UTR) and intron attenuates its translation (Ruegsegger et al. 2001). Moreover, a leak-translation product from the *HAC1*ⁱ mRNA is quickly degraded by the proteasome (Di Santo et al. 2016). On the other hand, translation of the *hacA* mRNAs (the *HAC1* orthologue) of filamentous fungi *Aspergillus niger*, *A. nidulans*, and *Trichoderma reesei* is inhibited by their upstream short open reading frames located on the 5′-UTRs and/or intramolecular base pairing between the 5′-UTRs and coding regions (Saloheimo et al. 2003; Mulder and Nikolaev 2009). The *hacA* mRNAs of these filamentous fungi carry one additional Ire1-target site in the 5'-UTR, allowing excision of the 5'-UTR and escape from translational inhibition in response to ER stress and activation of Ire1.

In ER-stressed metazoan and plant cells, Ire1 orthologues degrade mRNAs encoding ER client proteins (Hollien and Weissman 2006; Mishiba et al. 2013). This phenomenon is known as the regulated IRE1-dependent decay of mRNA (RIDD) and thought to reduce protein load into the ER. *S. pombe* cells but not *S. cerevisiae* cells are likely to perform the RIDD (Kimmig et al. 2012). It should be noted that *S. pombe* does not carry the *HAC1* orthologue gene in its genome. For induction of the expression level of BiP under ER stress conditions in *S. pombe*, Ire1 cleaves the 3'-UTR of BiP mRNA, which carries an unstabilizing signal (Kimmig et al. 2012).

Based on phenotypic differences between the knockout mutants of the *hacA* and *ireA* (the *IRE1* orthologue) genes, Feng et al. (2011) argued that *Aspergillus fumigatus* Ire1A has a downstream target(s) other than the *hacA* mRNA. Possibly through the promotion of protein secretion, pathogenic fungi apply the UPR for virulence progression (Krishnan and Askew 2014).

In *Aspergillus oryzae*, under condition inducing secretory hydrolytic enzyme production, repression of *ireA* completely blocked cell growth, but this defect was restored by the expression of intronless *hacA*. Thus UPR is required for *A. oryzae* cell growth to alleviate ER stress induced by excessive production of hydrolytic enzymes (Tanaka et al. 2015).

5.10   Conclusion

Even under normal physiological conditions, misfolded and aggregated proteins would be produced with low frequency but at a constant rate in all living cells. Furthermore, some genetic mutations and environmental stresses, as represented by heat shock and chemicals, result in the formation of unfolded, misfolded, and aggregated proteins at a higher rate, leading to increased proteotoxicity to the cells. To overcome this proteotoxic burden, cells have evolved a degradation and/or refolding systems of these misfolded and aggregated proteins to maintain proteostasis in cells, which includes ubiquitin-proteasome and molecular chaperone systems. The most effective and sophisticated response systems are HSR in the cytosol/nucleus and UPR in the ER. HSR activates essential transcription factors to induce various molecular chaperones to mitigate proteotoxicity in cells, whereas UPR activates transcription factors to induce not only molecular chaperones but folding enzymes and components of degradation machinery of misfolded proteins. HSR is evolutionarily conserved in all living cells, whereas UPR is developed in eukaryotes, because the signals from ER luminal events should be transferred to cytosol/nucleus overcoming membrane barrier. Increasing evidence has been shown to suggest that each organelle has its own UPR system. Although protein aggregation has been believed to be toxic to the cells, cells have developed the other different system for alleviating proteotoxic effects, represented by JUNQ and IPOD, which sequester aggregated proteins from other cellular compartments. These studies have been mainly analyzed by using yeast *S. cerevisiae*, and similar systems are also conserved in fungi and other higher eukaryotes. To understand the molecular mechanisms of proteostasis in yeast and fungi more precisely is important for not only basic and clinical science but also for fermentation or large-scale production of important proteins by using yeast and fungi.

References

Aebi M (2013) N-linked protein glycosylation in the ER. Biochim Biophys Acta 1833 (11):2430–2437

Aguilaniu H, Gustafsson L, Rigoulet M et al (2003) Asymmetric inheritance of oxidatively damaged proteins during cytokinesis. Science 299(5613):1751–1753

Amoros M, Estruch F (2001) Hsf1p and Msn2/4p cooperate in the expression of *Saccharomyces cerevisiae* genes HSP26 and HSP104 in a gene- and stress type-dependent manner. Mol Microbiol 39(6):1523–1532

Aragon T, van Anken E, Pincus D et al (2009) Messenger RNA targeting to endoplasmic reticulum stress signalling sites. Nature 457(7230):736–740

Bays NW, Gardner RG, Seelig LP et al (2001) Hrd1p/Der3p is a membrane-anchored ubiquitin ligase required for ER-associated degradation. Nat Cell Biol 3(1):24–29

Behnke J, Feige MJ, Hendershot LM (2015) BiP and its nucleotide exchange factors Grp170 and Sil1: mechanisms of action and biological functions. J Mol Biol 427(7):1589–1608

Benitez EM, Stolz A, Wolf DH (2011) Yos9, a control protein for misfolded glycosylated and non-glycosylated proteins in ERAD. FEBS Lett 585(19):3015–3019

Biederer T, Volkwein C, Sommer T (1997) Role of Cue1p in ubiquitination and degradation at the ER surface. Science 278(5344):1806–1809

Boy-Marcotte E, Lagniel G, Perrot M et al (1999) The heat shock response in yeast: differential regulations and contributions of the Msn2p/Msn4p and Hsf1p regulons. Mol Microbiol 33 (2):274–283

Carvalho P, Goder V, Rapoport TA (2006) Distinct ubiquitin-ligase complexes define convergent pathways for the degradation of ER proteins. Cell 126(2):361–373

Carvalho P, Stanley AM, Rapoport TA (2010) Retrotranslocation of a misfolded luminal ER protein by the ubiquitin-ligase Hrd1p. Cell 143(4):579–591

Casagrande R, Stern P, Diehn M et al (2000) Degradation of proteins from the ER of *S. cerevisiae* requires an intact unfolded protein response pathway. Mol Cell 5(4):729–735

Chawla A, Chakrabarti S, Ghosh G et al (2011) Attenuation of yeast UPR is essential for survival and is mediated by IRE1 kinase. J Cell Biol 193(1):41–50

Clay L, Caudron F, Denoth-Lippuner A et al (2014) A sphingolipid-dependent diffusion barrier confines ER stress to the yeast mother cell. Elife 3:e01883

Coughlan CM, Brodsky JL (2005) Use of yeast as a model system to investigate protein conformational diseases. Mol Biotechnol 30(2):171–180

Coustou V, Deleu C, Saupe S et al (1997) The protein product of the het-s heterokaryon incompatibility gene of the fungus *Podospora anserina* behaves as a prion analog. Proc Natl Acad Sci USA 94 (18):9773–9778

Cox JS, Walter P (1996) A novel mechanism for regulating activity of a transcription factor that controls the unfolded protein response. Cell 87(3):391–404

Cox JS, Shamu CE, Walter P (1993) Transcriptional induction of genes encoding endoplasmic reticulum resident proteins requires a transmembrane protein kinase. Cell 73(6):1197–1206

Credle JJ, Finer-Moore JS, Papa FR et al (2005) On the mechanism of sensing unfolded protein in the endoplasmic reticulum. Proc Natl Acad Sci USA 102(52):18773–18784

Dayalan Naidu S, Dinkova-Kostova AT (2017) Regulation of the mammalian heat shock factor 1. FEBS J 284(11):1606–1627

Denic V, Quan EM, Weissman JS (2006) A luminal surveillance complex that selects misfolded glycoproteins for ER-associated degradation. Cell 126(2):349–359

Di Santo R, Aboulhouda S, Weinberg DE (2016) The fail-safe mechanism of post-transcriptional silencing of unspliced HAC1 mRNA. Elife 5:e20069

Eisele F, Wolf DH (2008) Degradation of misfolded protein in the cytoplasm is mediated by the ubiquitin ligase Ubr1. FEBS Lett 582(30):4143–4146

Erjavec N, Cvijovic M, Klipp E et al (2008) Selective benefits of damage partitioning in unicellular systems and its effects on aging. Proc Natl Acad Sci USA 105(48):18764–18769

Escusa-Toret S, Vonk WI, Frydman J (2013) Spatial sequestration of misfolded proteins by a dynamic chaperone pathway enhances cellular fitness during stress. Nat Cell Biol 15 (10):1231–1243

Fan HC, Ho LI, Chi CS et al (2014) Polyglutamine (PolyQ) diseases: genetics to treatments. Cell Transplant 23(4–5):441–458

Fanchiotti S, Fernandez F, D'Alessio C et al (1998) The UDP-Glc:Glycoprotein glucosyltransferase is essential for *Schizosaccharomyces pombe* viability under conditions of extreme endoplasmic reticulum stress. J Cell Biol 143(3):625–635

Fang NN, Ng AH, Measday V et al (2011) Hul5 HECT ubiquitin ligase plays a major role in the ubiquitylation and turnover of cytosolic misfolded proteins. Nat Cell Biol 13(11):1344–1352

Fang NN, Chan GT, Zhu M et al (2014) Rsp5/Nedd4 is the main ubiquitin ligase that targets cytosolic misfolded proteins following heat stress. Nat Cell Biol 16(12):1227–1237

Fang NN, Zhu M, Rose A et al (2016) Deubiquitinase activity is required for the proteasomal degradation of misfolded cytosolic proteins upon heat-stress. Nat Commun 7:12907

Farinha CM, Canato S (2017) From the endoplasmic reticulum to the plasma membrane: mechanisms of CFTR folding and trafficking. Cell Mol Life Sci 74(1):39–55

Feng X, Krishnan K, Richie DL et al (2011) HacA-independent functions of the ER stress sensor IreA synergize with the canonical UPR to influence virulence traits in *Aspergillus fumigatus*. PLoS Pathog 7(10):e1002330

Fernandez FS, Trombetta SE, Hellman U et al (1994) Purification to homogeneity of UDP-glucose: glycoprotein glucosyltransferase from *Schizosaccharomyces pombe* and apparent absence of the enzyme fro *Saccharomyces cerevisiae*. J Biol Chem 269(48):30701–30706

Finger A, Knop M, Wolf DH (1993) Analysis of two mutated vacuolar proteins reveals a degradation pathway in the endoplasmic reticulum or a related compartment of yeast. Eur J Biochem 218(2):565–574

Finley D, Ulrich HD, Sommer T et al (2012) The ubiquitin-proteasome system of *Saccharomyces cerevisiae*. Genetics 192(2):319–360

Foresti O, Rodriguez-Vaello V, Funaya C et al (2014) Quality control of inner nuclear membrane proteins by the Asi complex. Science 346(6210):751–755

Franzmann TM, Menhorn P, Walter S et al (2008) Activation of the chaperone Hsp26 is controlled by the rearrangement of its thermosensor domain. Mol Cell 29(2):207–216

Fredrickson EK, Rosenbaum JC, Locke MN et al (2011) Exposed hydrophobicity is a key determinant of nuclear quality control degradation. Mol Biol Cell 22(13):2384–2395

Friedlander R, Jarosch E, Urban J et al (2000) A regulatory link between ER-associated protein degradation and the unfolded-protein response. Nat Cell Biol 2(7):379–384

Garcia DM, Jarosz DF (2014) Rebels with a cause: molecular features and physiological consequences of yeast prions. FEMS Yeast Res 14(1):136–147

Gardarin A, Chedin S, Lagniel G et al (2010) Endoplasmic reticulum is a major target of cadmium toxicity in yeast. Mol Microbiol 76(4):1034–1048

Gardner BM, Walter P (2011) Unfolded proteins are Ire1-activating ligands that directly induce the unfolded protein response. Science 333(6051):1891–1894

Gardner RG, Nelson ZW, Gottschling DE (2005) Degradation-mediated protein quality control in the nucleus. Cell 120(6):803–815

Gasch AP, Spellman PT, Kao CM et al (2000) Genomic expression programs in the response of yeast cells to environmental changes. Mol Biol Cell 11(12):4241–4257

Geiler-Samerotte KA, Dion MF, Budnik BA et al (2011) Misfolded proteins impose a dosage-dependent fitness cost and trigger a cytosolic unfolded protein response in yeast. Proc Natl Acad Sci USA 108(2):680–685

Glover JR, Lindquist S (1998) Hsp104, Hsp70, and Hsp40: a novel chaperone system that rescues previously aggregated proteins. Cell 94(1):73–82

Glover JR, Kowal AS, Schirmer EC et al (1997) Self-seeded fibers formed by Sup35, the protein determinant of [PSI+], a heritable prion-like factor of S. cerevisiae. Cell 89(5):811–819

Grimminger-Marquardt V, Lashuel HA (2010) Structure and function of the molecular chaperone Hsp104 from yeast. Biopolymers 93(3):252–276

Guillemette T, van Peij N, Goosen T et al (2007) Genomic analysis of the secretion stress response in the enzyme-producing cell factory Aspergillus niger. BMC Genomics 8:158

Hahn JS, Hu Z, Thiele DJ et al (2004) Genome-wide analysis of the biology of stress responses through heat shock transcription factor. Mol Cell Biol 24(12):5249–5256

Heck JW, Cheung SK, Hampton RY (2010) Cytoplasmic protein quality control degradation mediated by parallel actions of the E3 ubiquitin ligases Ubr1 and San1. Proc Natl Acad Sci USA 107(3):1106–1111

Helenius A, Aebi M (2004) Roles of N-linked glycans in the endoplasmic reticulum. Annu Rev Biochem 73:1019–1049

Hiller MM, Finger A, Schweiger M et al (1996) ER degradation of a misfolded luminal protein by the cytosolic ubiquitin-proteasome pathway. Science 273(5282):1725–1728

Hollien J, Weissman JS (2006) Decay of endoplasmic reticulum-localized mRNAs during the unfolded protein response. Science 313(5783):104–107

Huyer G, Longsworth GL, Mason DL et al (2004) A striking quality control subcompartment in Saccharomyces cerevisiae: the endoplasmic reticulum-associated compartment. Mol Biol Cell 15(2):908–921

Ishiwata-Kimata Y, Promlek T, Kohno K et al (2013) BiP-bound and nonclustered mode of Ire1 evokes a weak but sustained unfolded protein response. Genes Cells 18(4):288–301

Jacobson T, Navarrete C, Sharma SK et al (2012) Arsenite interferes with protein folding and triggers formation of protein aggregates in yeast. J Cell Sci 125(Pt 21):5073–5083

Jacobson T, Priya S, Sharma SK et al (2017) Cadmium causes misfolding and aggregation of cytosolic proteins in yeast. Mol Cell Biol 37(17):e00490–e00416

Jannatipour M, Rokeach LA (1995) The Schizosaccharomyces pombe homologue of the chaperone calnexin is essential for viability. J Biol Chem 270(9):4845–4853

Kakoi S, Yorimitsu T, Sato K (2013) COPII machinery cooperates with ER-localized Hsp40 to sequester misfolded membrane proteins into ER-associated compartments. Mol Biol Cell 24 (5):633–642

Kaganovich D, Kopito R, Frydman J (2008) Misfolded proteins partition between two distinct quality control compartments. Nature 454(7208):1088–1095

Kaliszewski P, Zoladek T (2008) The role of Rsp5 ubiquitin ligase in regulation of diverse processes in yeast cells. Acta Biochim Pol 55(4):649–662

Kawai-Noma S, Pack CG, Kojidani T et al (2010) In vivo evidence for the fibrillar structures of Sup35 prions in yeast cells. J Cell Biol 190(2):223–231

Keener JM, Babst M (2013) Quality control and substrate-dependent downregulation of the nutrient transporter Fur4. Traffic 14(4):412–427

Khmelinskii A, Blaszczak E, Pantazopoulou M et al (2014) Protein quality control at the inner nuclear membrane. Nature 516(7531):410–413

Kimata Y, Kimata YI, Shimizu Y et al (2003) Genetic evidence for a role of BiP/Kar2 that regulates Ire1 in response to accumulation of unfolded proteins. Mol Biol Cell 14(6):2559–2569

Kimata Y, Oikawa D, Shimizu Y et al (2004) A role for BiP as an adjustor for the endoplasmic reticulum stress-sensing protein Ire1. J Cell Biol 167(3):445–456

Kimata Y, Ishiwata-Kimata Y, Yamada S et al (2006) Yeast unfolded protein response pathway regulates expression of genes for anti-oxidative stress and for cell surface proteins. Genes Cells 11(1):59–69

Kimata Y, Ishiwata-Kimata Y, Ito T et al (2007) Two regulatory steps of ER-stress sensor Ire1 involving its cluster formation and interaction with unfolded proteins. J Cell Biol 179(1):75–86

Kimmig P, Diaz M, Zheng J et al (2012) The unfolded protein response in fission yeast modulates stability of select mRNAs to maintain protein homeostasis. Elife 1:e00048

Kohno K, Normington K, Sambrook J et al (1993) The promoter region of the yeast KAR2 (BiP) gene contains a regulatory domain that responds to the presence of unfolded proteins in the endoplasmic reticulum. Mol Cell Biol 13(2):877–890

Korennykh AV, Egea PF, Korostelev AA et al (2009) The unfolded protein response signals through high-order assembly of Ire1. Nature 457(7230):687–693

Krishnan K, Askew DS (2014) Endoplasmic reticulum stress and fungal pathogenesis. Fungal Biol Rev 28(2–3):29–35

Kurokawa K, Okamoto M, Nakano A (2014) Contact of cis-Golgi with ER exit sites executes cargo capture and delivery from the ER. Nat Commun 5:3653

Le QG, Ishiwata-Kimata Y, Kohno K et al (2016) Cadmium impairs protein folding in the endoplasmic reticulum and induces the unfolded protein response. FEMS Yeast Res 16(5): fow049

Leach MD, Farrer RA, Tan K et al (2016) Hsf1 and Hsp90 orchestrate temperature-dependent global transcriptional remodelling and chromatin architecture in *Candida albicans*. Nat Commun 7:11704

Lederkremer GZ (2009) Glycoprotein folding, quality control and ER-associated degradation. Curr Opin Struct Biol 19(5):515–523

Lee KP, Dey M, Neculai D et al (2008) Structure of the dual enzyme Ire1 reveals the basis for catalysis and regulation in nonconventional RNA splicing. Cell 132(1):89–100

Li L, Kowal AS (2012) Environmental regulation of prions in yeast. PLoS Pathog 8(11):e1002973

Liang J, Clark-Dixon C, Wang S et al (2008) Novel suppressors of alpha-synuclein toxicity identified using yeast. Hum Mol Genet 17(23):3784–3795

Liu B, Larsson L, Caballero A et al (2010) The polarisome is required for segregation and retrograde transport of protein aggregates. Cell 140(2):257–267

Lu K, Psakhye I, Jentsch S (2014) Autophagic clearance of polyQ proteins mediated by ubiquitin-Atg8 adaptors of the conserved CUET protein family. Cell 158(3):549–563

Ma M, Liu LZ (2010a) Quantitative transcription dynamic analysis reveals candidate genes and key regulators for ethanol tolerance in *Saccharomyces cerevisiae*. BMC Microbiol 10:169

Ma M, Liu ZL (2010b) Mechanisms of ethanol tolerance in *Saccharomyces cerevisiae*. Appl Microbiol Biotechnol 87(3):829–845

McCracken AA, Brodsky JL (1996) Assembly of ER-associated protein degradation in vitro: dependence on cytosol, calnexin, and ATP. J Cell Biol 132(3):291–298

Malinovska L, Kroschwald S, Munder MC et al (2012) Molecular chaperones and stress-inducible protein-sorting factors coordinate the spatiotemporal distribution of protein aggregates. Mol Biol Cell 23(16):3041–3056

Mathuranyanon R, Tsukamoto T, Takeuchi A et al (2015) Tight regulation of the unfolded protein sensor Ire1 by its intramolecularly antagonizing subdomain. J Cell Sci 128(9):1762–1772

McDaniel D, Caplan AJ, Lee MS et al (1989) Basal-level expression of the yeast HSP82 gene requires a heat shock regulatory element. Mol Cell Biol 9(11):4789–4798

McGlinchey RP, Kryndushkin D, Wickner RB (2011) Suicidal [PSI+] is a lethal yeast prion. Proc Natl Acad Sci USA 108(13):5337–5341

Mehnert M, Sommer T, Jarosch E (2014) Der1 promotes movement of misfolded proteins through the endoplasmic reticulum membrane. Nat Cell Biol 16(1):77–86

Meriin AB, Zhang X, He X et al (2002) Huntington toxicity in yeast model depends on polyglutamine aggregation mediated by a prion-like protein Rnq1. J Cell Biol 157(6):997–1004

Merksamer PI, Trusina A, Papa FR (2008) Real-time redox measurements during endoplasmic reticulum stress reveal interlinked protein folding functions. Cell 135(5):933–947

Miller SB, Mogk A, Bukau B (2015a) Spatially organized aggregation of misfolded proteins as cellular stress defense strategy. J Mol Biol 427(7):1564–1574

Miller SB, Ho CT, Winkler J et al (2015b) Compartment-specific aggregases direct distinct nuclear and cytoplasmic aggregate deposition. EMBO J 34(6):778–797

Mishiba K, Nagashima Y, Suzuki E et al (2013) Defects in IRE1 enhance cell death and fail to degrade mRNAs encoding secretory pathway proteins in the *Arabidopsis* unfolded protein response. Proc Natl Acad Sci USA 110(14):5713–5718

Miyagawa K, Ishiwata-Kimata Y, Kohno K et al (2014) Ethanol stress impairs protein folding in the endoplasmic reticulum and activates Ire1 in *Saccharomyces cerevisiae*. Biosci Biotechnol Biochem 78(8):1389–1391

Mogk A, Bukau B (2017) Role of sHsps in organizing cytosolic protein aggregation and disaggregation. Cell Stress Chaperones 22(4):493–502

Moraitis C, Curran BP (2004) Reactive oxygen species may influence the heat shock response and stress tolerance in the yeast *Saccharomyces cerevisiae*. Yeast 21(4):313–323

Morano KA, Grant CM, Moye-Rowley WS (2012) The response to heat shock and oxidative stress in *Saccharomyces cerevisiae*. Genetics 190(4):1157–1195

Mori K (2009) Signalling pathways in the unfolded protein response: development from yeast to mammals. J Biochem 146(6):743–750

Mori K, Sant A, Kohno K et al (1992) A 22 bp cis-acting element is necessary and sufficient for the induction of the yeast KAR2 (BiP) gene by unfolded proteins. EMBO J 11(7):2583–2593

Mori K, Ma W, Gething MJ et al (1993) A transmembrane protein with a cdc2+/CDC28-related kinase activity is required for signaling from the ER to the nucleus. Cell 74(4):743–756

Mori K, Kawahara T, Yoshida H et al (1996) Signalling from endoplasmic reticulum to nucleus: transcription factor with a basic-leucine zipper motif is required for the unfolded protein-response pathway. Genes Cells 1(9):803–817

Mulder HJ, Nikolaev I (2009) HacA-dependent transcriptional switch releases hacA mRNA from a translational block upon endoplasmic reticulum stress. Eukaryot Cell 8(4):665–675

Nicholls S, MacCallum DM, Kaffarnik FA et al (2011) Activation of the heat shock transcription factor Hsf1 is essential for the full virulence of the fungal pathogen *Candida albicans*. Fungal Genet Biol 48(3):297–305

Nillegoda NB, Theodoraki MA, Mandal AK et al (2010) Ubr1 and Ubr2 function in a quality control pathway for degradation of unfolded cytosolic proteins. Mol Biol Cell 21(13):2102–2116

Nishikawa SI, Fewell SW, Kato Y et al (2001) Molecular chaperones in the yeast endoplasmic reticulum maintain the solubility of proteins for retrotranslocation and degradation. J Cell Biol 153(5):1061–1070

Normington K, Kohno K, Kozutsumi Y et al (1989) *S. cerevisiae* encodes an essential protein homologous in sequence and function to mammalian BiP. Cell 57(7):1223–1236

Ogura T, Wilkinson AJ (2001) AAA+ superfamily ATPases: common structure--diverse function. Genes Cells 6(7):575–597

Okamura K, Kimata Y, Higashio H et al (2000) Dissociation of Kar2p/BiP from an ER sensory molecule, Ire1p, triggers the unfolded protein response in yeast. Biochem Biophys Res Commun 279(2):445–450

Outeiro TF, Lindquist S (2003) Yeast cells provide insight into alpha-synuclein biology and pathobiology. Science 302(5651):1772–1775

Parlati F, Dignard D, Bergeron JJ et al (1995a) The calnexin homologue cnx1+ in *Schizosaccharomyces pombe*, is an essential gene which can be complemented by its soluble ER domain. EMBO J 14(13):3064–3072

Parlati F, Dominguez M, Bergeron JJ et al (1995b) *Saccharomyces cerevisiae* CNE1 encodes an endoplasmic reticulum (ER) membrane protein with sequence similarity to calnexin and calreticulin and functions as a constituent of the ER quality control apparatus. J Biol Chem 270(1):244–253

Parsell DA, Kowal AS, Singer MA et al (1994) Protein disaggregation mediated by heat-shock protein Hsp104. Nature 372(6505):475–478

Patino MM, Liu JJ, Glover JR et al (1996) Support for the prion hypothesis for inheritance of a phenotypic trait in yeast. Science 273(5275):622–626

Peisker K, Chiabudini M, Rospert S (2010) The ribosome-bound Hsp70 homolog Ssb of *Saccharomyces cerevisiae*. Biochim Biophys Acta 1803(6):662–672

Pincus D, Chevalier MW, Aragon T et al (2010) BiP binding to the ER-stress sensor Ire1 tunes the homeostatic behavior of the unfolded protein response. PLoS Biol 8(7):e1000415

Plemper RK, Bohmler S, Bordallo J et al (1997) Mutant analysis links the translocon and BiP to retrograde protein transport for ER degradation. Nature 388(6645):891–895

Prasad R, Kawaguchi S, Ng DT (2010) A nucleus-based quality control mechanism for cytosolic proteins. Mol Biol Cell 21(13):2117–2127

Prinz WA, Grzyb L, Veenhuis M et al (2000) Mutants affecting the structure of the cortical endoplasmic reticulum in *Saccharomyces cerevisiae*. J Cell Biol 150(3):461–474

Promlek T, Ishiwata-Kimata Y, Shido M et al (2011) Membrane aberrancy and unfolded proteins activate the endoplasmic reticulum stress sensor Ire1 in different ways. Mol Biol Cell 22 (18):3520–3532

Rapoport TA, Li L, Park E (2017) Structural and mechanistic insights into protein translocation. Annu Rev Cell Dev Biol 33:369–390

Rose MD, Misra LM, Vogel JP (1989) KAR2, a karyogamy gene, is the yeast homolog of the mammalian BiP/GRP78 gene. Cell 57(7):1211–1221

Rosenbaum JC, Fredrickson EK, Oeser ML et al (2011) Disorder targets misorder in nuclear quality control degradation: a disordered ubiquitin ligase directly recognizes its misfolded substrates. Mol Cell 41(1):93–106

Ruan L, Zhou C, Jin E et al (2017) Cytosolic proteostasis through importing of misfolded proteins into mitochondria. Nature 543(7645):443–446

Rubio C, Pincus D, Korennykh A et al (2011) Homeostatic adaptation to endoplasmic reticulum stress depends on Ire1 kinase activity. J Cell Biol 193(1):171–184

Ruegsegger U, Leber JH, Walter P (2001) Block of HAC1 mRNA translation by long-range base pairing is released by cytoplasmic splicing upon induction of the unfolded protein response. Cell 107(1):103–114

Saad S, Cereghetti G, Feng Y et al (2017) Reversible protein aggregation is a protective mechanism to ensure cell cycle restart after stress. Nat Cell Biol 19(10):1202–1213

Sakono M, Zako T (2010) Amyloid oligomers: formation and toxicity of Abeta oligomers. FEBS J 277(6):1348–1358

Saloheimo M, Valkonen M, Penttila M (2003) Activation mechanisms of the HAC1-mediated unfolded protein response in filamentous fungi. Mol Microbiol 47(4):1149–1161

Sanchez Y, Lindquist SL (1990) HSP104 required for induced thermotolerance. Science 248 (4959):1112–1115

Sato K, Nakano A (2007) Mechanisms of COPII vesicle formation and protein sorting. FEBS Lett 581(11):2076–2082

Schlissel G, Krzyzanowski MK, Caudron F et al (2017) Aggregation of the Whi3 protein, not loss of heterochromatin, causes sterility in old yeast cells. Science 355(6330):1184–1187

Schoebel S, Mi W, Stein A et al (2017) Cryo-EM structure of the protein-conducting ERAD channel Hrd1 in complex with Hrd3. Nature 548(7667):352–355

Seufert W, Jentsch S (1990) Ubiquitin-conjugating enzymes UBC4 and UBC5 mediate selective degradation of short-lived and abnormal proteins. EMBO J 9(2):543–550

Shamu CE, Walter P (1996) Oligomerization and phosphorylation of the Ire1p kinase during intracellular signaling from the endoplasmic reticulum to the nucleus. EMBO J 15(12):3028–3039

Sharma D, Masison DC (2009) Hsp70 structure, function, regulation and influence on yeast prions. Protein Pept Lett 16(6):571–581

Shorter J, Lindquist S (2005) Prions as adaptive conduits of memory and inheritance. Nat Rev Genet 6(6):435–450

Sidrauski C, Walter P (1997) The transmembrane kinase Ire1p is a site-specific endonuclease that initiates mRNA splicing in the unfolded protein response. Cell 90(6):1031–1039

Sidrauski C, Cox JS, Walter P (1996) tRNA ligase is required for regulated mRNA splicing in the unfolded protein response. Cell 87(3):405–413

Silberstein S, Schlenstedt G, Silver PA et al (1998) A role for the DnaJ homologue Scj1p in protein folding in the yeast endoplasmic reticulum. J Cell Biol 143(4):921–933

Simons JF, Ferro-Novick S, Rose MD et al (1995) BiP/Kar2p serves as a molecular chaperone during carboxypeptidase Y folding in yeast. J Cell Biol 130(1):41–49

Sinclair D, Mills K, Guarente L (1998) Aging in *Saccharomyces cerevisiae*. Annu Rev Microbiol 52:533–560

Solis EJ, Pandey JP, Zheng X et al (2016) Defining the essential function of yeast Hsf1 reveals a compact transcriptional program for maintaining eukaryotic proteostasis. Mol Cell 63(1):60–71

Specht S, Miller SB, Mogk A et al (2011) Hsp42 is required for sequestration of protein aggregates into deposition sites in *Saccharomyces cerevisiae*. J Cell Biol 195(4):617–629

Stein A, Ruggiano A, Carvalho P et al (2014) Key steps in ERAD of luminal ER proteins reconstituted with purified components. Cell 158(6):1375–1388

Suzuki CK, Suda K, Wang N et al (1994) Requirement for the yeast gene LON in intramitochondrial proteolysis and maintenance of respiration. Science 264(5156):273–276

Suzuki G, Shimazu N, Tanaka M (2012) A yeast prion, Mod5, promotes acquired drug resistance and cell survival under environmental stress. Science 336(6079):355–359

Szathmary R, Bielmann R, Nita-Lazar M et al (2005) Yos9 protein is essential for degradation of misfolded glycoproteins and may function as lectin in ERAD. Mol Cell 19(6):765–775

Takatsuki A, Kohno K, Tamura G (1975) Inhibition of biosynthesis of polyisoprenol sugars in chick embryo microsomes by tunicamycin. Agric Biol Chem 39(10):2089–2091

Tamas MJ, Sharma SK, Ibstedt S et al (2014) Heavy metals and metalloids as a cause for protein misfolding and aggregation. Biomolecules 4(1):252–267

Tanaka K (2009) The proteasome: overview of structure and functions. Proc Jpn Acad Ser B Phys Biol Sci 85(1):12–36

Tanaka M, Shintani T, Gomi K (2015) Unfolded protein response is required for *Aspergillus oryzae* growth under conditions inducing secretory hydrolytic enzyme production. Fungal Genet Biol 85:1–6

Teixeira MC, Mira NP, Sa-Correia I (2011) A genome-wide perspective on the response and tolerance to food-relevant stresses in *Saccharomyces cerevisiae*. Curr Opin Biotechnol 22 (2):150–156

Thibault G, Ng DT (2012) The endoplasmic reticulum-associated degradation pathways of budding yeast. Cold Spring Harb Perspect Biol 4(12)

Travers KJ, Patil CK, Wodicka L et al (2000) Functional and genomic analyses reveal an essential coordination between the unfolded protein response and ER-associated degradation. Cell 101 (3):249–258

True HL, Lindquist SL (2000) A yeast prion provides a mechanism for genetic variation and phenotypic diversity. Nature 407(6803):477–483

True HL, Berlin I, Lindquist SL (2004) Epigenetic regulation of translation reveals hidden genetic variation to produce complex traits. Nature 431(7005):184–187

Tyedmers J, Treusch S, Dong J et al (2010a) Prion induction involves an ancient system for the sequestration of aggregated proteins and heritable changes in prion fragmentation. Proc Natl Acad Sci USA 107(19):8633–8638

Tyedmers J, Mogk A, Bukau B (2010b) Cellular strategies for controlling protein aggregation. Nat Rev Mol Cell Biol 11(11):777–788

Umebayashi K, Hirata A, Fukuda R et al (1997) Accumulation of misfolded protein aggregates leads to the formation of russell body-like dilated endoplasmic reticulum in yeast. Yeast 13 (11):1009–1020

Van Dyck L, Pearce DA, Sherman F (1994) PIM1 encodes a mitochondrial ATP-dependent protease that is required for mitochondrial function in the yeast *Saccharomyces cerevisiae*. J Biol Chem 269(1):238–242

Vashist S, Ng DT (2004) Misfolded proteins are sorted by a sequential checkpoint mechanism of ER quality control. J Cell Biol 165(1):41–52

Verghese J, Abrams J, Wang Y et al (2012) Biology of the heat shock response and protein chaperones: budding yeast (*Saccharomyces cerevisiae*) as a model system. Microbiol Mol Biol Rev 76(2):115–158

Wagner I, Arlt H, van Dyck L et al (1994) Molecular chaperones cooperate with PIM1 protease in the degradation of misfolded proteins in mitochondria. EMBO J 13(21):5135–5145

Weids AJ, Ibstedt S, Tamas MJ et al (2016) Distinct stress conditions result in aggregation of proteins with similar properties. Sci Rep 6:24554

Wickner RB (2016) Yeast and fungal prions. Cold Spring Harb Perspect Biol 8(9):a023531

Willingham S, Outeiro TF, DeVit MJ et al (2003) Yeast genes that enhance the toxicity of a mutant huntingtin fragment or alpha-synuclein. Science 302(5651):1769–1772

Xu P, Duong DM, Seyfried NT et al (2009) Quantitative proteomics reveals the function of unconventional ubiquitin chains in proteasomal degradation. Cell 137(1):133–145

Yeger-Lotem E, Riva L, Su LJ et al (2009) Bridging high-throughput genetic and transcriptional data reveals cellular responses to alpha-synuclein toxicity. Nat Genet 41(3):316–323

Zabrocki P, Bastiaens I, Delay C et al (2008) Phosphorylation, lipid raft interaction and traffic of alpha-synuclein in a yeast model for Parkinson. Biochim Biophys Acta 1783(10):1767–1780

Zhang S, Zheng H, Chen Q et al (2017) The lectin chaperone calnexin is involved in the endoplasmic reticulum stress response by regulating Ca(2+) homeostasis in *Aspergillus nidulans*. Appl Environ Microbiol 83(15):e00673–e00617

Zhao Y, Macgurn JA, Liu M et al (2013) The ART-Rsp5 ubiquitin ligase network comprises a plasma membrane quality control system that protects yeast cells from proteotoxic stress. Elife 2:e00459

Zheng X, Krakowiak J, Patel N et al (2016) Dynamic control of Hsf1 during heat shock by a chaperone switch and phosphorylation. Elife 5:e18638

Zhou C, Slaughter BD, Unruh JR et al (2014) Organelle-based aggregation and retention of damaged proteins in asymmetrically dividing cells. Cell 159(3):530–542

Importance of Stress Response Mechanisms in Filamentous Fungi for Agriculture and Industry

6

Razieh Karimi Aghcheh and Gerhard H. Braus

Abstract

Sustainable production of food crops is urgently required to meet the demands of a growing human population worldwide. Global food security has to be elevated to decline increasing rates of cancer or various other serious diseases caused by contaminated food. Rapid climate change including global warming increases detrimental effects on the food-supply chain and its safety caused by polluted environments due to utilization of fossil fuels and use of hazardous chemical substances during the past century. Soil health and crop productivity improvement strategies are of vital importance. Necessary measures include the reduction of food spoilage before and after harvesting, increased application of biological control methods to reduce pesticide application, improved bioremediation of polluted sites, and increased generation of biofuels from sustainable sources. Fungi are ubiquitous and are able to survive under very challenging conditions as part of an important ecological function in recycling carbon and nitrogen sources. They produce a wide range of enzymes for numerous primary and secondary metabolites. Fungi play prominent roles in agriculture, food production, and processing but also in crop loss. Fungal biotechnology supports pharmaceutical industries, energy production, and food industries. The other side of fungal life includes harmful mycotoxins damaging human health. Like other living organisms, fungi must cope with a variety of stresses to survive. *Fungal stress* is a major issue affecting agriculture as well as industrial microbiology. This includes stresses encountered or caused by fungi. They can act as pathogens to plants or can spoil food after harvesting during storage or transport. Fungal species can be used in bioremediation or as biocontrol agents that may enhance plant health and crop production. Especially species of the genera *Trichoderma*,

R. K. Aghcheh (✉) · G. H. Braus (✉)
Department of Molecular Microbiology and Genetics, University of Goettingen, Goettingen, Germany
e-mail: rkarimi@gwdg.de; gbraus@gwdg.de

© Springer Nature Switzerland AG 2018
M. Skoneczny (ed.), *Stress Response Mechanisms in Fungi*,
https://doi.org/10.1007/978-3-030-00683-9_6

Penicillium, or *Aspergillus* can produce biomass-degrading enzymes, antibiotics, and useful chemicals, respectively. Understanding how fungi deal with stress while infecting a plant host or during industrial processes will help to optimize the use of fungi in agricultural and industrial applications and improve the environment and combat with plant dieses caused by fungi.

6.1   Introduction

A sexual life cycle has not been described in all filamentous fungi, but usually these microorganisms possess an asexual life cycle, in which normally many asexual spores termed conidia are produced. These asexual dispersal structures can render fungi the capability to travel long distances through the air and thereby to propagate to remote environments that can be disparate from the original location. Fungi must continually adjust their physiology to inhabit highly diverse ecological niches. At the molecular level, fungi have been endowed with complex signal transduction pathways that allow them to respond appropriately to alterations in external stimuli. These stimuli can include biotic and abiotic stresses such as damages caused by living organism like bacteria, viruses, and fungi and changes in temperature, pH, nutrient availability, oxidative stress, osmotic stress, and UV radiation. This chapter is divided into fungi important for agriculture or for industry. We will focus on the molecular mechanisms and pathways, which help fungi to enhance their survival under stress conditions and gain the ability to proliferate and thrive through environmental changes.

6.2   Plant Pathogenic Fungi and Stress Mechanism

Plant-fungal interactions are multifaceted and range from mutually highly beneficial symbioses to hostile necrotrophic relationships. Vascular plants provide a unique ecological niche for diverse communities. They include cryptic symbiotic microbes, which often contribute multiple benefits, such as enhanced photosynthetic efficiency, nutrient and water use, and tolerance to abiotic and biotic stress. Among the symbiotic microbes, fungi have been associated with plants since at least 400 million years ago. Two major classes of fungal symbionts associated with plants are defined: (1) endophytic fungi, which reside entirely within plant tissues and may be associated with roots, stems, and/or leaves, and (2) mycorrhizal fungi, which reside only in roots but extend out into the rhizosphere. In response to host genotype and environmental factors, fungal endophytes can express/form a range of different life cycle/relationships with different hosts including symbiotic, mutualistic, commensalistic, and parasitic. Mutualistic fungi may confer tolerance to drought, metals, disease, extreme temperature (heat and cold), and herbivory and/or promote growth and nutrient acquisition. Commensal symbioses have no beneficial or detrimental effects on hosts, whereas parasitic symbioses negatively affect host fitness (Singh et al. 2011a, b; Ellouze et al. 2014).

Fungi, like bacteria, are among the most widespread and destructive parasites of plants. These plant pathogenic fungi may threaten plant by decreasing growth rates and/or fecundity or inducing disease symptoms that could result in lethality. The survey by Dean et al. based on 495 votes from the international community of fungal pathologists with an association with the journal *Molecular Plant Pathology* resulted in the list of Top 10 fungal plant pathogens. The Top 10 list includes, in rank order, (1) *Magnaporthe oryzae*, (2) *Botrytis cinerea*, (3) *Puccinia* spp., (4) *Fusarium graminearum*, (5) *Fusarium oxysporum*, (6) *Blumeria graminis*, (7) *Mycosphaerella graminicola*, (8) *Colletotrichum* spp., (9) *Ustilago maydis*, and (10) *Melampsora lini* (Dean et al. 2012). In the following, for a comprehensive understanding of fungal stress mechanism and the importance of studying this area of biology for agriculture, we will focus on key aspects of plant pathogenic fungal stress by including not only those Top 10 fungi of Dean's survey but also other phytopathogenic fungi studied thus far.

Mitogen-activated protein kinase (MAPK) pathways represent one of the most prominent signal transduction systems in eukaryotes including fungi consisting of MAPK extracellular signal-regulated kinase (ERK) activated by an MAPK/ERK kinase (MEK), which in turn is activated by an MEK kinase. Eukaryotic organisms use different MAPK cascades to regulate various aspects of cellular function (Gustin et al. 1998). MAPKs that specifically transmit environmental stress signals are also known as stress-activated protein kinases (SAPKs). This pathway is called the high-osmolarity glycerol (HOG) response pathway in the unicellular fungus *Saccharomyces cerevisiae* (Fig. 6.1), which serves as a reference for filamentous fungi (Hohmann 2009). Numerous studies conducted in the yeast *S. cerevisiae* have contributed to the understanding of the high-osmolarity glycerol (HOG) pathway [reviewed in Saito and Posas (2012) and discussed in Chap. 2 of this book].

HOG1 encodes a mitogen-activated protein kinase that regulates cellular turgor in yeast. However, other deleterious environmental conditions also activate the HOG pathway. Deletion of the yeast *hog1* homologs of filamentous fungi resulted in altered or defected morphology. This correlated with loss or reduced pathogenicity in a number of phytopathogenic fungi such as *Colletotrichum orbiculare*, *Bipolaris oryzae*, *Botrytis cinerea*, *Mycosphaerella graminicola*, *Verticillium dahliae*, *Fusarium oxysporum*, and *Bipolaris sorokiniana* (Kojima et al. 2004; Moriwaki et al. 2006; Segmüller et al. 2007; Mehrabi et al. 2006; Wang et al. 2016; Segorbe et al. 2017; Leng and Zhong 2015). The Hog1-type MAPKs are required for virulence of some, but not all, plant pathogens, and the reasons for these findings are not always elucidated.

Osm1 of *Magnaporthe grisea* represented the first characterized Hog1-type MAPK from a plant pathogenic fungus. *M. grisea* develops an appressorium as specialized infection cell, which ruptures mechanically the plant cuticle. During hyperosmotic stress, this fungus accumulates arabitol and to a lesser amount glycerol. Glycerol is required to generate mechanical force in the form of hydrostatic

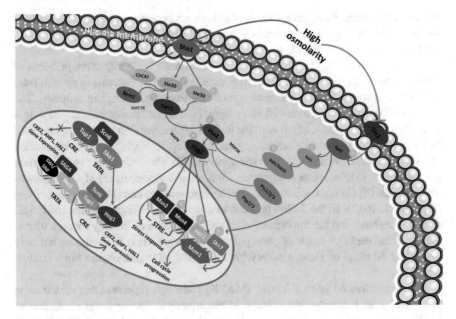

Fig. 6.1 Schematic illustration of signaling pathway leading to osmotic stress resistance in the budding yeast *Saccharomyces cerevisiae*. The upstream part of the HOG pathway comprises Sln1 and Sho1 branches that are functionally redundant but mechanistically distinct. The Sln1 branch activates the redundant Ssk2 and Ssk22 MAPKKKs, which thereafter activate Pbs2. The Sho1 branch activates the Ste11 MAPKKK, which also activates Pbs2. Once the pathway becomes activated, a substantial fraction of the Hog1 MAPK is transported into the nucleus (the central bleu part), where it regulates transcription and the cell cycle, although there are also Hog1 targets in the cytoplasm (the gray part represents the cytoplasm) (In all figures arrows indicate positive regulation, and lines with a bar at the end indicate negative regulation. Solid lines are used when there is direct experimental evidence and dotted lines when the role is still hypothetical)

turgor produced by appressoria. Deletion of *M. grisea osm1*, which corresponds to a functional homolog of the yeast Hog1, reduced dramatically the ability to accumulate arabitol in the mycelium, whereas glycerol accumulation and turgor generation in appressoria were unaffected in *osm1* null mutant strains. These mutant strains were fully pathogenic (Dioxon et al. 1999).

Alternaria species are also a highly successful group of fungal pathogens that can cause allergic symptoms in humans and diseases in a wide variety of economically important crops, including apple, broccoli, cauliflower, carrot, citrus, pear, rice, strawberry, tomato, potato, and tobacco, as well as many ornamental and weed species. The ability of the necrotrophic fungus *Alternaria alternata* to detoxify reactive oxygen species (ROS) produced by plant defense systems is crucial for pathogenesis to citrus. In *A. alternata*, the ability to coordinate signaling pathways is essential for detoxification of cellular responses induced by ROS and thus its pathogenicity. A regulatory network comprising Yap1, Hog1, Skn7, Nox, or Nps6 is assembled to underscore the intricate interplays among these signaling pathways in *A. alternata* (Chen et al. 2014; Lin et al. 2009). Resistance to oxidative stress of this fungus can be controlled via regulating siderophore-mediated iron acquisition, which is not far from expectation, because iron is required for the activities of antioxidants (e.g., catalase and SOD). This regulation requires the NADPH oxidase (NOX), the redox-activating yeast AP-1 (Yap1)-like transcription factor corresponding to mammalian activator protein 1 (AP-1), and the high-osmolarity glycerol 1 (Hog1) mitogen-activated protein kinase (MAPK). Deletion of *nox*, *yap1*, *hog1*, or *nps6* (an essential nonribosomal peptide synthetase for biosynthesis of siderophores) resulted in increased sensitivity to ROS. Importantly, expression of the *nps6* gene and biosynthesis of siderophores are regulated by Nox, Yap1, and Hog1, supporting a functional link among these regulatory pathways (Chung 2012; Chen et al. 2014). In this sophisticated signaling pathways, Skn7 physically interacts with Yap1, regulating the genes involved in ROS detoxification.

The earlier identified Yap1 (for yeast AP-1-like) regulates the oxidative stress response in *Saccharomyces cerevisiae* (Toone et al. 2001). Yap1-mediated detoxification of ROS is an essential virulence determinant in the opportunistic human pathogen *Candida albicans* (Enjalbert et al. 2007). Deleted strains in *U. maydis yap1* displayed higher sensitivity to H_2O_2 than wild-type cells, and their virulence was significantly reduced. Mutations in two Cys residues of this protein prevented nuclear accumulation of *U. maydis yap1* after H_2O_2 treatment and resulted in the same mutants as *Δyap1* with attenuated virulence ability (Molina and Kahmann 2007). In contrast, Yap1 is not required for virulence in the plant pathogen *Cochliobolus heterostrophus* (Lev et al. 2005).

Hog1 and Skn7 are two downstream regulators of the two-component histidine kinase Hsk1 in the unicellular fungus *S. cerevisiae* (Fig. 6.1). The two-component HSK-mediated signal transduction is vital for sensing and adapting to environmental changes in microorganisms (Wurgler-Murphy and Saito 1997). Upon osmotic stress, the yeast Sln1 histidine kinase transmits signals by a phosphotransfer process to the two response regulators Ssk1 and Skn7 (Li et al. 1998).

Filamentous ascomycetous fungi possess in their two-component signaling systems several histidine kinases and the additional two conserved response regulators Ssk1 and Skn7 (Fassler and West 2013). *B. cinerea* Skn7 (BcSkn7) together with the transcription factor Bap1 are two key players in the oxidative stress response (OSR) of *B. cinerea* and control typical OSR genes as *gsh1* and *grx1*. It has been shown in some fungi that genes that are regulated by Skn7 are also regulated by Ap1. Glutaredoxin 1, *grx1*, and γ-glutamyl-cysteine synthetase, *gsh1*, were among OSR genes whose transcription was clearly coregulated by both BcSkn7 and Bap1 under oxidative stress triggered by H_2O_2. The function of Bap1 is restricted to the regulation of oxidative stress, whereas BcSkn7 is in addition involved in cell wall stress, development, and virulence (Viefhues et al. 2015).

The two-component response regulators ChSsk1 and ChSkn7 of the maize pathogen *C. heterostrophus* control high-osmolarity adaptation and fungicide resistance of this fungus to the iprodione/fludioxonil fungicides (Izumitsu et al. 2007). Phosphorylation of the Hog1-type MAPK induced by high-osmolarity stress and fungicide treatments depends only on ChSsk1p. Genetic analysis of the corresponding double-deletion strain revealed that the Chssk1 and Chskn7 two-component regulators cooperate to protect against osmotic stress and fungicide application.

Ssk1 responds to diverse environmental stimuli and plays a critical role in *Alternaria alternata* pathogenesis. Loss of function of *ssk1* fulfills similar regulatory roles in signaling pathways involving a Hog1 MAP kinase during ROS resistance, osmotic resistance, fungicide sensitivity, and fungal virulence in *A. alternata*. The mutant strains lacking the gene for Ssk1 display elevated sensitivity to oxidants. They also fail to detoxify H_2O_2 effectively, induce minor necrosis on susceptible citrus leaves, and display resistance to dicarboximide and phenylpyrrole fungicides. Unlike the Skn7 response regulator, Ssk1 and Hog1 confer resistance to salt-induced osmotic stress by a yet unknown kinase sensor rather than the two-component histidine kinase HSK1 (Yu et al. 2016). Ssk1 and Hog1 play a moderate role in sugar-induced osmotic stress. It has been also shown that *ssk1* mutant strains are involved in cell differentiation and multidrug resistance in *A. alternata* (Yu et al. 2016).

S. cerevisiae also utilizes a non-HSK-related protein Sho1 to cope with osmotic stress (Maeda et al. 1994). However, deletion of a *sho1* homolog in *A. alternata* resulted in no noticeable phenotypes and had no impact on cellular tolerance to oxidative and osmotic stress, fungicide sensitivity, or fungal virulence (Yu et al. 2016).

The unique group III histidine kinases of filamentous fungi have also been linked to the regulation of high-osmolarity adaptation and the phosphorylation status of Hog1-type MAPKs (Yoshimi et al. 2005). The group III histidine kinase Dic1 in *C. heterostrophus* confers dicarboximide resistance and osmotic adaptation to this fungus. The Hog1-type MAPKs were activated at high fungicide (100 µg/ml) and osmotic stress (0.8 M KCl) levels in the histidine kinase deletion mutants of both *C. heterostrophus* and *N. crassa* indicating that group III histidine kinase positively regulates the activation of Hog1-related MAPK in filamentous fungi by phosphorylation (Yoshimi et al. 2005). Dic2 and Dic3 are involved in the Skn7 pathway; Dic2 encodes an Skn7-type response regulator, ChSkn7. Although Dic2 and Dic3 also conferred high-osmolarity adaptation and dicarboximide/phenylpyrrole fungicide sensitivity in *C. heterostrophus*, their action is independent of the HOG pathway in this fungus (Izumitsu et al. 2009).

Other components of the mitogen-activated protein kinases (MAPKs) play important functions in the regulation of fungal development, stress response, and virulence of several plant pathogenic fungi. *F. oxysporum* Fmk1 represents the MAPK orthologue to *S. cerevisiae* Fus3/Kss1 and controls vegetative hyphal fusion, stress response, and fungal virulence on plant and animal hosts. Loss of FoMpk1 also led to increased sensitivity to cell wall and heat stress, which was exacerbated by simultaneous inactivation of *fmk1* (Segorbe et al. 2017). Similarly, two identified MAPK genes (*CsSlt2* and *CsFus3*) in the genome of the fungal cereal pathogen *Bipolaris sorokiniana* were shown to be required for the pathogenicity of *B. sorokiniana*. Deletion of *Csslt2* and *Csfus3* genes resulted in hypersensitivity mutants to oxidative stress, whereas ΔCsslt2 mutants were more sensitive to cell wall-degrading enzymes. Like *Csfus3*, deletion of the MAPK kinase kinase (MAPKKK) gene (*Csste11*) also led to morphological defects, hypersensitivity to oxidative stress, and loss of pathogenicity (Leng and Zhong 2015).

The Pbs2 kinase represents another conserved component of the HOG-mediated signaling machinery which is located upstream of Hog1 and which was originally identified in yeast. The corresponding VdPbs2 protein of vascular plant pathogen *V. dahliae* is required for virulence to plants, but also for oxidative stress, fungicide response and for microsclerotia formation as fungal resting structures in the soil. VdPbs2 and VdHog1 function, therefore, in a cascade that regulates microsclerotia formation and virulence (Tian et al. 2016; Wang et al. 2016).

The process of osmotic adaptation by activation of the HOG pathway, in unicellular as well as in filamentous fungi, results in the biosynthesis and accumulation of compatible molecules such as proline, trehalose, polyols, and glycerol to counterbalance the osmotic pressure and prevent loss of water (D'Amore et al. 1999; Fillinger et al. 2001). Trehalose is an important disaccharide that can be found in bacteria, fungi, invertebrates, and plants. In fungi, trehalose formation protects against several environmental stresses and can be important for virulence.

Sorbitol-stressed cells of *U. maydis* show higher trehalase activity compared with control cells and to those stressed by NaCl and high temperature. Trehalose accumulation during sorbitol stress in *U. maydis* might be related to the adaptation of this organism during plant infection (Salmerón-Santiago et al. 2011). The proteomic analysis of the response of *U. maydis* to temperature, sorbitol, and salt stresses indicated a complex pattern which highlights the change of 18 proteins involved in carbohydrate and amino acid metabolism, protein folding, redox regulation, ion homeostasis, and stress response (Salmerón-Santiago et al. 2011). In this fungus, mutant strains deleted in the conserved gene for trehalose-6-phosphate phosphatase (*tps2*) displayed increased sensitivity to oxidative, heat, acid, ionic, and osmotic stresses as compared to the wild-type strains. Virulence of *Δtps2* mutants to maize plants was extremely reduced compared to wild-type strains, possibly due to reduced capability to deal with the hostile host environment (Cervantes-Chávez et al. 2016).

U. maydis was examined for the relation between unfolded protein response (UPR) with its pathogenicity and development. This fungus experiences endoplasmic reticulum (ER) stress during plant colonization and relies on the UPR to cope with this stress. The unfolded protein response (UPR), as another conserved eukaryotic signaling pathway, is a cellular response to ensure endoplasmic reticulum (ER) homeostasis during ER stress, which results, for example, from an increased demand for protein secretion. In *U. maydis*, characterization of the homologs of the central UPR regulatory proteins Hac1 (for homologous to Atf/Creb1) and Ire1 (inositol-requiring enzyme1) showed that exact timing of the UPR is required for virulence in this microorganism. This is as a consequence of the tightly interlinked relation of UPR with the b mating-type-dependent signaling pathway, which regulates pathogenic development. The UPR also interplays with b-controlled gene for Clampless1 (Clp1), which is essential for cell cycle release and proliferation in planta. Notably, in the interaction between Clp1 and the *U. maydis* Hac1 homolog, Cib1 promotes the stabilization of Clp1 which leads to enhanced ER stress tolerance that ultimately prevents deleterious UPR hyperactivation (Heimel et al. 2013). The exact timing of UPR is presumably connected to the elevated expression of secreted effector proteins during infection of the host plant *Zea mays*. In *S. cerevisiae*, expression of UPR target genes is induced upon binding of Hac1 to unfolded protein response elements (UPREs) in their promoters. *U. maydis* Cib1 binds to the UPREs containing promoter fragments of the two effector genes *pit2* and *tin1-1*, as bona fide UPR target genes in *U. maydis*. In this fungus, ER stress increased strongly *pit2* expression and secretion, while Cib1-dependent expression of *pit2* was abolished upon targeted deletion of the UPRE which affected also significantly virulence of *U. maydis* (Hampel et al. 2016). Furthermore, a conserved co-chaperone, Dnj1, was identified as part of the conserved cellular response to ER stress in *U. maydis*. This protein that localizes in the ER and interacts with the luminal chaperon Bip1 is required for *U. maydis* pathogenicity too (Lo Presti et al. 2015).

The genome of *U. maydis* encodes a number of small heat-shock proteins (HSPs). HSPs are a family of proteins that are produced by cells in response to exposure to stressful conditions. As they constitute mainly a family of molecular chaperones, they aid in the stabilization and correct folding of nascent polypeptides. In addition to their initially described relation to heat shock, they are expressed during other stresses including exposure to cold and UV light and during wound healing or tissue remodeling (Atalay et al. 2009). In some cases small HSPs play crucial roles in microbial pathogenesis. Quantitative real-time PCR analysis showed an increased level of differential expression of small *hsp* genes (*hsp12*, *hsp20*, and *hsp30*) in contrast to other HSPs, during infection of the host plant *Zea mays* and when the *U. maydis* was subjected to an abiotic stress such as oxidative stress, indicating a possible role of small HSPs in the pathogenicity and stress response of *U. maydis*. This observation was supported further by testing the deletion and complementation strains of three putative small HSPs (Ghosh 2014).

Summarizing, the HOG pathway is also conserved but not identical in different fungi and is intimately linked to morphogenesis and pathogenicity of most of fungi examined so far. Other upstream mitogen-activated kinases, MAPKK and MAPKKK, are too involved in this process. Furthermore, the role of unfolded protein response (UPR) as another conserved eukaryotic signaling pathway and the link between intracellular induction of heat-shock proteins in response to various stresses and pathogenicity of filamentous fungi have been already described in these organisms.

6.3 Interplay Between Secondary Metabolism/Mycotoxin Production and Stress Tolerance

In some cases the response to osmotic stress by fungi is linked to the biosynthesis of natural product secondary metabolites. Secondary metabolism describes the capacity of microbes, plants, or animals to produce molecules which are often bioactive but which are not required for growth in the laboratory. Filamentous fungi as well as actinomycetes and plants are producing a particularly wide range of secondary metabolites especially during developmental and aging phases of their life cycle and when confronted with other cells, which often corresponds to a stress situation. Intensive research on fungal genomics of the last decades including efforts like the 1000 Fungal Genomes Project (https://genome.jgi.doe.gov/programs/fungi/1000fungalgenomes.jsf) revealed that there are multiple gene clusters for potential secondary metabolites where the chemical nature of the compounds is elusive. Even the exact functions of secondary metabolites which include antibiotics, growth stimulants, flavors or pigments, and mycotoxins in their natural environment and in the context of other organisms are in most cases unknown. In soilborne fungi, secondary metabolites have been shown to possess diverse functions including antibiosis, competitive inhibition of other microbes, and abiotic stress alleviation (Daguerre et al. 2014). Mitchell et al. (2004) investigated the effects of environmental parameters including water activity (a_w: 0.85–0.987) and temperature (10–40 °C) on growth and ochratoxin A production by two strains of *Aspergillus carbonarius* isolated from wine grapes from three different European countries and Israel. This work demonstrated that for all strains the optimal conditions for growth and ochratoxin A production is different, thereby offering possible benefits for the use of the generated information on a regional basis through providing critical knowledge for the development and prediction of the risk models of contamination of grapes and grape products by this fungus under fluctuating and interacting environmental parameters (Mitchell et al. 2004). In a similar study, it was revealed that single and interacting abiotic environmental stress factors such as water and temperature interactions and type of solute used to induce water stress have impact on tolerance mechanisms, molecular ecology, and the relationship with secondary metabolite production by a group of mycotoxigenic species of economic importance species of *Aspergillus flavus*, *A. ochraceus*, *A. carbonarius*, *Penicillium nordicum*, and *P. verrucosum* (Medina et al. 2015). It was shown that these abiotic stresses are conquered by the increased synthesis of low molecular weight sugar alcohols, especially glycerol and erythritol. Furthermore, differential and temporal expression of the genes in the secondary metabolite clusters of those selected fungal species in response to a_w × temperature stress was also assessed through a microarray approach (Medina et al. 2015).

The study of the HOG pathway and its implications in development and secondary metabolism in filamentous fungi are still in its infancy. The amenable organism *Aspergillus nidulans* has been used extensively as a model fungus for the study of the molecular genetics of responses to osmotic stress as well as secondary metabolism and fungal development. Therefore, in this session we focus on describing the already known mechanism of the osmotic stress response pathway of this fungus and its role on fungal development and secondary metabolism. The HogA (SakA) pathway of *A. nidulans* (Fig. 6.2) is activated in the osmotic and oxidative response by the *A. nidulans* Ssk1 ortholog, SskA. In contrast to yeast, deletion of the gene encoding SakA of *A. nidulans* corresponding to yeast *S. cerevisiae* HogA resulted in mutant strains with only slight sensitivity to high-osmolarity stress. *A. nidulans* SakA is a repressor of sexual development and asexual spore stress resistance and survival (Furukawa et al. 2005). Other upstream components of this pathway are the *sln1* homolog, *tcsB*; the *ypd1* ortholog, *ypdA*; and also *nikA* (Duran et al. 2010; Fig. 6.2).

A. nidulans has a sophisticated NikA-dependent phosphorelay signaling system coupled with a MAPK cascade. The two downstream response regulators (*sskA* and *srrA* as *A. nidulans* orthologues of *S. cerevisiae skn7*) are central components of the

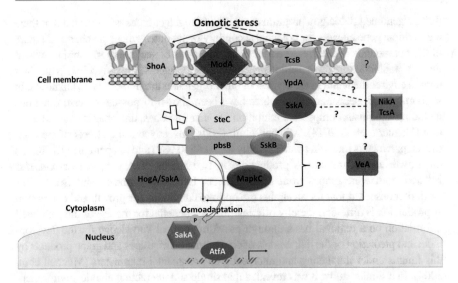

Fig. 6.2 HOG regulatory pathway of the model system *Aspergillus nidulans* and the role of osmotic stress response pathway on fungal development and secondary metabolism. The HogA (SakA) pathway of *A. nidulans* is activated in the osmotic and oxidative response by the *A. nidulans* Ssk1 ortholog, SskA. Osmoregulation in *A. nidulans* differs from that of the yeast. Upstream components of this pathway are the Sln1 homolog, TcsB; the Ypd1 ortholog, YpdA; and also NikA, a dispensable Mak2-type histidine kinase. The Sho1p signaling pathway in *A. nidulans* might not be involved in osmoregulation but instead may be carrying out other signaling functions. *A. nidulans* PbsB MAPKK activates another Hog1 ortholog, MpkC, which is not present in yeast. Proteins such as ModA, similar to Sho1, and SteC, similar to Ste11, have roles in morphogenesis and sexual development. There are SakA-dependent transcription regulators in *A. nidulans* such as RpdA (Rpd3p ortholog), AtfA (putative ortholog of *S. pombe* Atf1), and RcoA (*S. cerevisiae* Tup1 ortholog). VeA which is known to control sexual/asexual morphogenesis and secondary metabolism in filamentous fungi is also involved in modulating osmotic stress-induced conidiation, but little is known about the exact link between VeA and components of the *A. nidulans* Hog1 pathway (In all figures arrows indicate positive regulation, and lines with a bar at the end indicate negative regulation. Solid lines are used when there is direct experimental evidence and dotted lines when the role is still hypothetical)

phosphorelay system involved in oxidative and osmotic stress signal transduction and asexual sporulation in *A. nidulans*. While NikA is responsible for the growth inhibitory action of potent fungicides (iprodione and fludioxonil) in *A. nidulans*, deletion of *sskA* and *srrA* resulted in mutant strains which were not resistant as for growth against iprodione and fludioxonil (Hagiwara et al. 2007).

Deletion of *A. nidulans tscB* did not result in any osmosensitive phenotype, whereas TcsB-YpdA-SskA system of *A. nidulans* might play a similar role as the yeast Sln1-Ypd1-Ssk1 proteins (Furukawa et al. 2002, 2005). There are also structural and functional differences between *A. nidulans* Sho1 and *A. nidulans* PbsB (homolog to *S. cerevisiae* Pbs1) with the corresponding yeast proteins (Furukawa et al. 2005). Several genes for SakA-dependent regulators [RpdA (corresponding to Rpd3), AtfA (similar to *S. pombe* Atf1), and RcoA (as *S. cerevisiae* Tup1)] and the Msn2p-/Msn4p-type protein MsnA have been identified in the genome of *A. nidulans* (Han and Prade 2002). In addition to these SakA-dependent regulators with implicated roles in *A. nidulans* development and stress tolerance, studies on *A. nidulans* SteC revealed its role in morphogenesis and that this gene controls the phosphorylation of two putative MAP kinases (Wei et al. 2003). Regulatory pathways including those signaling pathways responsive to stress are relatively conserved but are not identical in the genus *Aspergillus* (de Vries et al. 2017). One example is the differences between the role of *A. nidulans* SakA and SakA from the human pathogenic *Aspergillus fumigatus*, in response to stresses. *A. fumigatus* SakA is crucial for oxidative stress response and production of secondary metabolites including gliotoxin, whereas *A. nidulans* SakA has hardly any impact on osmoregulation.

Toxic compounds named as mycotoxins are quite stable secondary metabolites of low molecular mass, which are produced by several specific fungal strains. These pharmacologically active metabolites are extremely difficult to remove and enter the food and feed chain while keeping their toxic properties. Mycotoxin contamination of the feed and food is a global problem because more than 25% of world grain production is contaminated by mycotoxins. Aflatoxins are the most health-threatening and poisonous carcinogens that are produced by certain fungi including *Aspergillus flavus* and *Aspergillus parasiticus* (Kumar et al. 2016). *A. flavus* has the ability to produce mycotoxins (aflatoxins B1, B2, G1, and G2 and sterigmatocystin), thus presenting as the most predominant mycotoxigenic species in wheat grain and several crops such as maize and peanuts. This fungus harbors two orthologs of the *A. nidulans* SskA response regulator in its genome and two orthologs of the *A. nidulans* ortholog RpdA histone deacetylase suggesting that *A. flavus* has a more sophisticated osmo-adaptation system that may include genes with redundant functions (Miskei et al. 2009). In *A. flavus*, oxidative stress plays roles in sclerotial differentiation and biosynthesis of aflatoxin B1 (Grintzalis et al. 2014). In accordance with this, recently, it was observed that different isolates of *A. flavus* show differences in oxidative stress tolerance which is correlated with their aflatoxin production capabilities (Fountain et al. 2016). Drought stress is considered as a major factor to contribute to preharvest aflatoxin contamination caused by *A. flavus* implying that the infection with aflatoxin is exacerbated by this type of abiotic stress (Guo et al. 2008). Drought stress-responding compounds such as reactive oxygen species (ROS) are associated with fungal stress-responsive signaling and secondary metabolite production and can stimulate the production of aflatoxin by *A. flavus* in vitro (Fountain et al. 2016). Transcriptional responses of field isolates of *A. flavus* with varying levels of aflatoxin production to H_2O_2-induced oxidative stress were assessed using an RNA-sequencing approach. The results obtained by transcriptomic analyses revealed that fungal development-related genes are regulated by oxidative stress. Furthermore, genes involved in the biosynthesis of aflatoxin were also differentially expressed in the toxigenic isolates with distinguished aflatoxin production capabilities in response to increasing oxidative stress. These data suggest that the secondary metabolites maybe produced as part of coordinated oxidative stress responses in *A. flavus* and emphasize on the biological importance of aflatoxin production in stress responses and in the competition with other soil-dwelling organisms (Fountain et al. 2016).

Similar to *A. flavus* production of the harmful carcinogenic aflatoxins by *Aspergillus parasiticus* has been postulated to be a mechanism to relieve oxidative stress. In *S. cerevisiae* Msn2 that encodes a C_2H_2-type zinc finger regulator, Msn2, is required for yeast cells to cope with a broad range of environmental and physiological stresses (Gasch et al. 2000). Msn2 mediates expression of a number of genes that are induced by stress conditions by binding to STRE (stress response element) motifs, CCCCT, which are located in the promoters of the regulated genes (Peterbauer et al. 2002; Dinamarco et al. 2012). The Δ*msnA* strains produced slightly higher amounts of aflatoxins and elevated amounts of kojic acid on mixed cereal medium. Deletion of this gene in *A. parasiticus* and *A. flavus* resulted in mutant strains with retarded colony growth and increased conidiation (Chang et al. 2011). Microarray assays showed that expression of genes encoding oxidative stress defense enzymes, i.e., superoxide dismutase, catalase, and cytochrome c peroxidase in *A. parasiticus* Δ*msnA* and the catalase A gene in *A. flavus* Δ*msnA*, was upregulated. Both *A. parasiticus* and *A. flavus* Δ*msnA* strains produced higher levels of reactive oxygen species (ROS) which implies that *msnA* gene may be required for the maintenance of the normal oxidative state (Chang et al. 2011).

The response to osmotic stress is linked to the biosynthesis of other mycotoxins in different fungi. Changes in the osmolarity of the medium, which can be caused by NaCl or sugars, are usually transmitted to the transcriptional level by the aid of signal cascades. The HOG signal cascade is the most important cascade for this environmental signal. The phosphorylated form of HOG as the last protein kinase in MAP kinase cascade is the form that subsequently activates the downstream transcription factors. *Penicillium* species like *P. nordicum* and *P. verrucosum* are adapted to NaCl-rich environments like salt-rich dry cured foods or even saline. Biosynthesis of the nephrotoxic mycotoxin ochratoxin A in these fungi plays an adaptive role in their habitat. In both cases, the NaCl-induced production of ochratoxin A is correlated to the phosphorylation status of the HOG MAP kinase (Cabañes et al. 2010). The involvement of this MAP kinase cascade in the regulation of trichothecene biosynthesis and for osmotic stress response has been shown in *Fusarium graminearum* as a causal agent of *Fusarium* head blight (FHB) (Ochiai et al. 2007) or alternariol biosynthesis in the case of *Alternaria alternata* (Graf et al. 2012).

The regulation of stress responses and virulence in phytopathogenic fungus *F. graminearum* was examined. Fimbrin as a cytoskeletal protein is an actin crosslinking protein important in the formation of filopodia. Fimbrin is present in several distinct structures in different cell types, including intestinal microvilli, hair cell stereocilia, and fibroblast filopodia. Fimbrin is highly conserved from yeast to humans. In yeast mutants lacking fimbrin are defective in morphogenesis and endocytosis (Bretscher 1981). FgFim of *F. graminearum* Y2021A, which is highly resistant to the fungicide JS399-19, was shown as a key protein regulating asexual and sexual development. Furthermore, deletion of *Fgfim* resulted in mutants exhibited increased sensitivity to diverse metal cations, to agents that induce osmotic stress and oxidative stress, and to agents that damage the cell membrane and cell wall. Decreased resistance to the fungicide JS399-19 and impaired virulence of FgFim loss-of-function mutants were consistent with reduced production of the toxin deoxynivalenol in host tissue (Zheng et al. 2014).

The involvement of the highly conserved *V. dahliae* Thi4 protein in stress response and pathogenicity of this fungus is another example for participation of nonstress-related proteins in regulation of aforementioned fungal processes. Biosynthesis of the vitamin thiamine is connected to oxidative stress. Thi4p, thiazole synthase which is required under vitamin B1-limiting conditions, confers increased stress tolerance to UV damage or oxidative stress and is essential for vascular disease induction in tomato caused by *Verticillium dahliae* (Hoppenau et al. 2014).

The medically important opportunistic fungal pathogen *A. fumigatus*, which is found frequently in different compost commodities, forms spores that are inhaled into our lungs and are phagocytosed in neutrophil immune cells. These spores encounter a ROS stress signal within neutrophils, which result in an apoptosis-like suicide of the conidia. Overexpression of the *bir1* gene prevents this ROS response. Bir1 corresponds to human survivin and *Saccharomyces cerevisiae* Bir1, which all act as inhibitors of apoptosis. The survivin protein functions to inhibit caspase activation, thereby leading to negative regulation of apoptosis or programmed cell death. The *A. fumigatus* Bir1 protein opposes lung neutrophils—which are triggered by programmed cell death with apoptosis-like features in *A. fumigatus* conidia—by inhibiting fungal caspase activation and DNA fragmentation in the murine lung (Shlezinger et al. 2017).

Epigenetics constitutes an important area to understand the mechanism of fungal secondary metabolism, development, and fungal stress responses. In contrast to genes for the primary metabolism, fungal genes encoding biosynthetic enzymes for secondary metabolites (SMs) often organized in clusters. Regulation of these SM gene clusters in *Aspergillus nidulans* includes gene cluster-specific transcription factors and in addition global regulators of chromatin structure. A prominent example of a global regulator of fungal secondary metabolite is LaeA, a highly conserved master regulator of secondary metabolites with a methyltransferase domain. Loss of function of the LaeA-encoding gene resulted in mutants which were unable to activate cluster-specific transcription factors as the AflR transcription factor supporting the synthesis of the aflatoxin toxin family (LaeA: loss of *aflR* expression A) (Bok and Keller 2004). LaeA-deficient strains are also crippled in formation of various antimicrobial secondary metabolites such as penicillin (PN) and terrequinone (Bok and Keller 2004). The *A. nidulans* putative mitogen-activated protein kinase encoded by *mpkB* was found to be necessary for normal expression of *laeA* and has role in sterigmatocystin gene expression and to be involved in penicillin and terrequinone A synthesis (Atoui et al. 2008). Similarly, deletion of *laeA* gene led to attenuation in biosynthesis of harmful toxins (e.g., gliotoxin, fumagilin, or helvolic acid), coupled with reduced secondary metabolism in all pathogenic fungi examined so far. Loss of LaeA abolished the virulence of the human pathogen *A. fumigatus* and plant pathogenic fungi such as *Fusarium fujikuroi* and *Cochliobolus heterostrophus* (Bok et al. 2005; Wiemann et al. 2010; Wu et al. 2012).

LaeA directly interacts in the nucleus with transcription factors in the trimeric velvet complex consisting of the velvet domain proteins VeA and VelB which are bridged through VeA to LaeA (Bayram et al. 2008). The *A. nidulans* velvet complex is required to coordinate fungal development and secondary metabolism in response to environmental triggers such as light. In response to light, the amount of the bridging protein VeA in the nucleus is reduced by the fungal cell, and therefore less of the trimeric velvet complex is formed. Increased amounts of the trimeric complex in darkness result in increased secondary metabolite biosynthesis and increased formation of sexual fruiting bodies which are overwintering structures formed in the soil (Bayram et al. 2008; Sarikaya-Bayram et al. 2010). Further studies showed that the bridging velvet domain protein VeA interacts not only with the LaeA methyltransferase but also with additional methyltransferases (e.g., VapA, VipC) as well as light regulators which affect fungal development and secondary metabolism including blue light receptors as the white-collar proteins LreA and LreB, the CryA cryptochrome as UV receptor, or the phytochrome FphA (Sarikaya-Bayram et al. 2010, 2014; Blumenstein et al. 2005; Purschwitzt et al. 2008).

The yeast Ssn6-Tup1 corepressor complex is a conserved system of transcriptional repression in eukaryotes which affects the expression of 7% of all genes under certain physiological conditions with emphasis on stress responses (Smith and Johnson 2000). In addition to the negative regulatory role of the Tup1/Ssn6 complex, it was also shown that the Tup1p/Ssn6 complex can activate gene expression under certain physiological conditions (Conlan et al. 1999). Another example is activation of transcriptional repressor Sko1, which inhibits hyperosmotic stress response genes in conjunction with Ssn6-Tup1 (Proft and Struhl 2002).

The Tup1 homolog Rco1 in the filamentous fungus *Neurospora crassa* participates in the control of vegetative growth, sexual reproduction, and asexual sporulation. *N. crassa* Rco1 forms a complex with RCM-1 (Ssn6) which is involved in photoadaptation (Olmedo et al. 2010). In *A. nidulans* it was shown that RcoA has a strong role in the development, vegetative growth, and production of the carcinogenic secondary metabolite of the aflatoxin family, sterigmatocystin (ST) (Hicks et al. 2001). The *A. nidulans rcoA* gene is required for *veA*-dependent sexual development. In this fungus, sexual development resulting from *veA* overexpression is *rcoA* dependent. This indicates that *rcoA* acts downstream of VeA in the sexual development pathway (Todd et al. 2006). The *veA* gene encodes in filamentous fungi a light-responsive factor that can integrate external stimuli, including osmotic stress to bring about physiological responses that are often manifested as alterations in secondary metabolism and/or morphogenesis.

Cochliobolus heterostrophus produces a wealth of secondary metabolites including the host-selective T-toxin. ChLae1 and ChVel1 positively regulate T-toxin biosynthesis, pathogenicity and super virulence, oxidative stress responses, sexual development, and aerial hyphal growth and negatively control melanin biosynthesis and asexual differentiation (Wu et al. 2012).

The regulation of secondary metabolites of the human pathogen *A. fumigatus* by LaeA and VeA members of velvet complex is temperature-responsive, which supports the hypothesis that fungal secondary metabolism is regulated by an intertwined network of multiple environmental factors (Lind et al. 2016). Production of toxic cyclopiazonic acid, aflatrem, or aflatoxin is linked to sclerotial formation in

A. flavus by the control of the *veA* gene-encoded regulatory protein (Duran et al. 2007). VeA exerts a profound effect on the oxidative stress response in the aflatoxin-producing fungus *A. flavus*, whereas the osmotic stress response is only slightly affected.

The expression of *A. flavus* orthologues providing the HOG signaling pathway is regulated by *veA*. Deletion of the *veA* gene reduced transcription levels of oxidative stress response genes after exposure to hydrogen peroxide, causing more sensitivity to ROS, and altered DNA-protein complex formation at the *cat1* promoter. The VeA protein promotes the expression of the transcription factor encoding *aftB* which is required for normal formation of protein-DNA complexes in the *cat1* promoter (Baidya et al. 2014).

Posttranslational modifications (PTMs) of eukaryotic histones by histone modifiers alter gene expression by changes of the chromatin structure. Especially the histone tails are subjected to a variety of PTMs, such as methylation, phosphorylation, adenosine diphosphate-ribosylation, biotinylation, acetylation, and ubiquitination. Chromatin modifications are important for the control of secondary metabolite production. The histone deacetylase (HDAC) HdaA is required for the silencing of two telomere-proximal gene clusters. Deletion of the corresponding gene resulted in transcriptional cluster activation and increased toxin and antibiotic levels as products of increased cluster expression (Lee et al. 2009; Schwab et al. 2007).

Among histone modifications, methylation plays important biological roles in eukaryotic cells. The putative H3K27 methyltransferase KMT6 of the cereal pathogen *Fusarium graminearum* corresponds to *Drosophila* enhancer of zeste (E(z)). KMT6 modifies lysine 27 at histone 3 and regulates development and the expression of secondary metabolite gene clusters (Connolly et al. 2013). Methylation of lysine 9 at histone H3 (H3K9me) is critical for regulating chromatin structure and gene transcription. Dim5 is a lysine histone methyltransferase (KHMTase) enzyme, which is responsible for the methylation of H3K9. *Fusarium verticillioides* is the most toxigenic fungus known to cause ear, stalk, and kernel diseases in maize crops throughout the world. This fungus produces fumonisins. FvDim5 regulates the trimethylation of H3K9 (H3K9me3). The *Fvdim5* deletion mutant showed significant defects in conidiation, perithecium production, and fungal virulence. Deletion of *Fvdim5* led to increased tolerance to osmotic stresses and upregulated FvHog1 phosphorylation (Gu et al. 2017). To date, the relationships between histone modifications and involvement of several transcription activators or repressors as well as methyltransferases and multiple stress responses, which have impacts on fungal fitness, remain to be clarified.

6.4 Stress Mechanism of Fungi as Biocontrol Agent and Its Importance for Agriculture

Some filamentous fungi can enhance our crop productivity, in contrast to those fungi acting as plant pathogens. Agricultural production has significantly improved during the past decades as a result of the application of new technologies. However, some modern practices affected considerably the environment. A major challenge in agriculture is to achieve higher yields in environment-friendly manner for a growing human population. Eco-friendly solutions include the application of biocontrol agents (Borriss 2011; Auld 2002; Askary 2010; Li et al. 2010; Vega et al. 2009). Biocontrol (or biological control) is defined as the use of natural or genetically modified organisms, genes, or gene products, which reduce the effects of undesirable organisms and favor organisms as crops, trees, animals, and beneficial microorganisms, which are useful to humans. Various types of fungal or bacterial species are used as biocontrol agents.

Trichoderma spp. (e.g., T. harzianum, T. atroviride, T. asperellum, and T. virens) have been known for decades as biocontrol fungi which colonize mainly the plant roots. They produce different kinds of enzymes which play a major role in biocontrol activity like degradation of cell wall, tolerance to biotic or abiotic stresses, hyphal growth, etc. Species belonging to this genus are free-living fungi common in soil and root ecosystems. Their success as biocontrol agents is in part due to mycoparasitism, a life cycle in which one fungus is parasitic on another fungus (Druzhinina et al. 2011). In the following we will discuss mechanism of stress responses of the biocontrol agents which include the HOG pathway and epigenetic control.

Mitogen-activated protein (MAP) kinases and heterotrimeric G-proteins affect biocontrol-relevant processes such as the production of hydrolytic enzymes and antifungal metabolites and the formation of infection structures. In addition to the involvement of MAPK signaling in the induction of plant systemic resistance in Trichoderma virens (Viterbo et al. 2005; Mukherjee et al. 2003), components of MAPK pathway are involved in the hyperosmotic stress response in Trichoderma harzianum. The Trichoderma genomes harbor genes that encode three MAPKs. Transcript profiling revealed a subset of genes induced in T. harzianum under hyperosmotic shock. The Th. MAPK hog1 gene controls the hyperosmotic stress in T. harzianum (Delgado-Jarana et al. 2006). Stress cross-resistance experiments showed evidences of a secondary role of ThHog1 in oxidative stress. Mutant strains deleted in hog1 showed a strongly reduced antagonistic activity against the plant pathogens Phoma betae and Colletotrichum acutatum, which points to a role of ThHog1 protein in fungus-fungus interactions (Delgado-Jarana et al. 2006). The presence of a high-osmolarity glycerol (HOG) response pathway in E. herbariorum and the significance of EhHogA in osmotic regulation, heat stress, freeze stress, and oxidative stress were studied (Jin et al. 2005). Eurotium herbariorum, teleomorph synonym of Aspergillus herbariorum, is the most common species isolated from the Dead Sea water, from the surface to a depth of 300 m in all investigated seasons. This organism needs to adapt to the extremely high salinity of the Dead Sea brines. The Dead Sea is therefore potentially an excellent model for studies of evolution under extreme environments and is an important gene pool for future agricultural genetic engineering prospects. EhHogA of the filamentous fungus E. herbariorum confers stress tolerance to lithium salt and freezing-thawing to this fungus. $\Delta hog1$ mutant of S. cerevisiae complemented by EhHogA outperformed the wild type or had higher genetic fitness under high Li^+ and freezing-thawing conditions (Jin et al. 2005).

Trichoderma species are applied for bioremediation of pesticide-polluted environments. The tolerance of these fungi to organophosphorus pesticides is of pivotal importance. In some cases, such a requirement is also key to the synergistic use of these fungi with chemical pesticides, aiming to broaden the scope of control targets to include both plant pathogens and insect pests. Proteomic analysis of *Trichoderma atroviride* mycelia stressed by organophosphate pesticide dichlorvos identified several proteins linked to stress tolerance including glutathione peroxidase-like protein (Gpx), 1,4-benzoquinone reductase, and Hex1. The *hex1*-deleted mutant showed decreased tolerance to the organophosphate, dichlorvos (Tang et al. 2010). Among the invertebrate fungal pathogens, *Beauveria bassiana* has a key role in the management of numerous arthropod agricultural, veterinary, and forestry pests. Similarly, *hog1* gene of this fungus was shown to encode a functional homolog of the yeast high-osmolarity glycerol 1 (HOG1). *Bbhog1* null mutants were more sensitive to hyperosmotic stress, high temperature, and oxidative stress than the wild-type strain demonstrating the conserved function of Hog1 MAPKs in the regulation of abiotic stress responses. The *ΔBbhog1* mutant strains exhibited greatly reduced pathogenicity, most likely due to a decrease in spore viability, a reduced ability to attach to insect cuticle, and a reduction in appressorium formation (Zhang et al. 2009). The transcript levels of two hydrophobin-encoding genes, *hyd1* and *hyd2*, were significantly decreased in a *ΔBbhog1* mutant strain suggesting that *Bbhog1* might regulate the expression of the gene associated with hydrophobicity or adherence (Zhang et al. 2009). This is in accordance with the observed role of Hfb2-6 in interactions with both biotic and abiotic environmental conditions of the biocontrol agent *Trichoderma asperellum* ACCC30536 (Huang et al. 2015). Furthermore, hydrophobins are induced by light in *Trichoderma* species, and this control requires the G-protein alpha subunit Gna1 (Schmoll et al. 2010).

Trichoderma atroviride is applied as an agent for biological control against plant pathogenic fungi and serves as model for the research on the regulation of asexual sporulation by environmental stimuli such as light and/or mechanical injury. In this fungus light impacts carbon metabolism and reaction to oxidative stress (Friedl et al. 2008). The *T. atroviride* photoreceptors Blr1 and Blr2 contribute also to adjustment of hydrophobin levels to environmental conditions (Mikus et al. 2009). Production of *Trichoderma*-specific secondary metabolites peptaibols, which are important for its antagonistic activity and represent a potential alternative to conventional antibiotics, is also associated with initiation of sporulation and influenced by light. Therefore it is not surprising that no peptaibols have been detected in *T. atroviride* mutants of *blr1* or *blr2*, hence indicating that the photoreceptors of *Trichoderma* play an important role in regulation of peptaibol production (Tisch and Schmoll 2009). In this fungus Lae1, the orthologue of *A. nidulans*, LaeA, is a key regulator of mycoparasitism and sensitivity to oxidative stress. Lae1 is essential for asexual development and involved in sporulation triggered by mechanical injury of mycelia (Karimi Aghcheh et al. 2013). The cross talk between light, conidiation, carbon metabolism and secondary metabolism, and reaction to oxidative stress will be discussed in more details in Sect. 6.5.

Prevention of ROS toxicity in plant requires a large gene network, the so-called *ROS gene network*. This is composed of a large number of genes including several families of plant proteins like superoxide dismutase (SOD), catalase (CAT), ascorbate peroxidase (APX), and glutathione peroxidase (GPX) that can reduce ROS concentration (Singh et al. 2016). The specific group of phytomolecules such as phenolics are also members of the *ROS gene network* and have the capacity to scavenge or quench deleterious ROS.

Several studies have shown that plant growth-promoting fungi (PGPF) and rhizobacteria (PGPR) play important role in activating plant resistance to various biotic or abiotic stresses. These biocontrol agents also showed the capability in activating this *ROS gene network*, which are associated with PGPR-mediated-induced resistance and providing protection against diverse pathogens (Singh et al. 2016). *Trichoderma harzianum* as a widespread mycoparasitic fungus is able to successfully colonize a wide range of substrates under different environmental conditions. However, only some species of *T. harzianum* are in use as effective biocontrol agents (Chaverri et al. 2015). Despite its wide application in agriculture, the mechanisms of biocontrol are not yet fully understood. Mycoparasitism and antibiosis are suggested, but may not be sole cause of disease reduction. The protection afforded by the biocontrol agent *T. harzianum* against the devastating plant pathogen *Rhizoctonia solani* is associated with the accumulation of the ROS gene network: maximum activity of CAT (11.0 times) was observed at 8 dapi, SOD (7.0 times) at 7 dapi, and GPX (5.4 times) and APX (8.1 times) at 7 dapi in *T. harzianum* NBRI-1055 (MCFT)-treated plants challenged with the pathogen. These results suggest that *T. harzianum*-mediated biocontrol may be related to alleviating *R. solani*-induced oxidative stress (Singh et al. 2011a, b).

Trichoderma has also maintained stress-related gene expression (e.g., *hsp70* and *hex1*) to adapt to varied environments. Overexpression of *T. harzianum hsp70* gene conferred tolerance to heat and other abiotic stresses to this fungus (Montero-Barrientos et al. 2010). Transgenic expression of this gene in *Arabidopsis* increased *Arabidopsis* resistance to heat and other abiotic stresses such as osmotic, salt, and oxidative stresses. Transgenic lines also had increased transcript levels of the Na(+)/H(+) exchanger 1 (*sos1*) and ascorbate peroxidase 1 (*apx1*) genes, involved in salt and oxidative stress responses, respectively (Montero-Barrientos et al. 2010).

The involvement of platelet-activating factor acetylhydrolase (PAF-AH) was also suggested to play a crucial role in *T. harzianum* stress response and antagonism under diverse environmental conditions (Yu et al. 2014). This protein belongs to the phospholipase A2 (Pla2) enzyme superfamily. These enzymes hydrolyze ester bonds at the sn-2 position of glycerophospholipids to release free fatty acids and lysophospholipids, which often trigger signal transduction pathways in animal or human cells and thereby playing important roles in the inflammatory response (Dan et al. 2012). Upregulation of PAF-AH expression was observed in *T. harzianum* in response to carbon starvation and strong heat shock. In addition, PAF-AH expression significantly increased under a series of maize root induction assay. Eicosanoic acid and ergosterol levels decreased in the *paf-ah* knockouts compared to wild-type cells. Stress responses mediated by PAF-AH were related to proteins Hex1, Cu/Zn superoxide dismutase, and cytochrome c. Moreover, PAF-AH exhibited antagonistic activity against *R. solani* in plate confrontation assays (Yu et al. 2014).

Several sophisticated pathways employ the conserved high-osmolarity glycerol (HOG) response pathway to provide mycoparasitic traits. Other conserved signaling pathways include heterotrimeric G-proteins, cAMP-dependent protein kinase A (cAMP-PKA), the photoreceptor complex Blr1/Blr2, HSPs, the global regulator of fungal fitness, LaeA/1, and genes for hydrophobins, antifungal metabolite formation, hydrolytic enzymes, or PAF-AH, which play roles in biocontrol capacity of biocontrol agents. The cellular choreography of dynamic interplay between these pathways, which is currently unknown, is important for a specific manipulation of these fungal biocontrol agents to decrease the infection of plants caused by other phytopathogenic fungi, bacteria, or nematodes.

6.5 Stress Responses in Industrially Important Filamentous Fungi

Filamentous fungi are required for many industrial processes to produce essential products for meeting human society needs. Prominent examples are the production of enzymes and useful pharmaceutical and industrial chemicals using *Aspergilli* strains, biosynthesis of medically important antibiotics utilizing *Penicillium* species, and production of lignocellulose-degrading enzymes for bioethanol production using *Trichoderma* spp.

The fungal genus *Aspergillus* affects humans on many levels. Some species are important pathogens of humans, animals, and crops. *Aspergilli* serve as a source of potent carcinogenic contaminants of food and as an important genetic model. Several species including *A. niger*, *A. oryzae*, and *A. terreus* have a broad range of industrial applications. Stress response proteins of several *Aspergilli* (*A. nidulans*, *A. flavus*, *A. fumigatus*, *A. niger*, *A. terreus*, *A. oryzae*, *A. clavatus*, and *N. fischeri*) were annotated and grouped according to stress types (Miskei et al. 2009). This annotation revealed existence of the orthologues of those encoding components of the *S. cerevisiae* HOG pathway in the genome of selected *Aspergilli*. Similar to other pathogenic and nonpathogenic *Aspergillus* species, *A. niger*, *A. oryzae*, and *A. terreus* also harbor elements of the SskA-HogA/SakA stress signaling pathway. Moreover, putative orthologues of both *Saccharomyces cerevisiae* Msn2/Msn4 C_2H_2 zinc finger-type and *Schizosaccharomyces pombe* Atf1 bZip-type general stress transcription factors were annotated in all analyzed *Aspergilli*, suggesting complex and robust stress defense systems in these fungi. An *Aspergillus niger* TcsB-type histidine kinase homologs could not be identified (Miskei et al. 2009). *A. niger* is used widely for the production of industrially important citric acid, glucuronic acid, and hydrolytic enzymes such as amylases, glucoamylases, lipidases, and pectinases. *A. niger* mycelia growth, production, and secretion of glucose oxidase, which is used as enzyme for the biotransformation of D-glucose to glucuronic acid, are affected by osmotic stress (Fiedurek 1998). Increased production and secretion of glucose oxidase during fermentation with *A. niger* were found as a result of increased osmotic pressure caused by NaCl (Fiedurek 1998).

A. niger has a high capacity secretory system and is widely exploited for the industrial production of native and heterologous proteins. In most cases the yields of non-fungal proteins are significantly lower than those obtained for fungal proteins. One well-studied bottleneck is misfolding of heterologous proteins in the ER during early stages of secretion, with related stress responses in the host as the unfolded protein response (UPR). In filamentous fungi and plants, a pathway termed as RESS (repression under secretion stress) decreases the protein load of ER. Secretory stress triggers transcriptional downregulation of several genes in fungi that encode for secreted proteins which results in lowering the burden on the secretory pathway and alleviating ER stress (Heimel 2015). In *A. niger*, endoplasmic reticulum stress leads to the selective downregulation of glucoamylase encoding gene expression mediated through the promoter of *glaA* in a region more than 1 kb upstream of the translational start (Al-Sheikh et al. 2004). In filamentous fungi, various approaches to employ the UPR for improved production of homologous and heterologous proteins have been investigated. HacA/Xbp1 is the conserved bZIP transcription factor in eukaryotic cells, which regulates gene expression in response to various forms of secretion stress and as part of secretory cell differentiation. In addition to the well-known HacA targets such as the ER-resident foldases and chaperones, genome-wide expression analysis upon constitutive expression of this TF (HacA^{CA}) in batch cultures of *A. niger* revealed upregulation of genes involved in protein glycosylation, phospholipid biosynthesis, intracellular protein transport, exocytosis, and protein complex assembly in the HacA^{CA} mutant strain. In contrast, genes involved in central metabolic pathways or for translation and transcription factors were downregulated (Carvalho et al. 2012). These data indicate that constitutive activation of HacA leads to a coordinated regulation of the folding and secretion capacity of the cell but with consequences on growth and fungal physiology to reduce secretion stress.

Aspergillus oryzae, which is an economically important filamentous fungus like *A. niger*, is used in the manufacture of fermented foods particularly in East Asian countries and in the production of enzymes (e.g., α-amylase and glucoamylase) for medical and food-grade use. Similar to *A. niger*, osmotic stress has also been shown to be a critical factor in enzyme production by *A. oryzae*. Activity of glucoamylase produced by *A. oryzae* showed an increase of about 20-fold as water content of the wheat bran substrate was increased (Kobayashi et al. 2007). The fungal cell wall confers cell morphology and protection against environmental insults. Fungal cells remodel their walls over time in response to environmental change, a process controlled by evolutionarily conserved stress (Hog1) and cell integrity signaling pathways. These mitogen-activated protein kinase (MAPK) pathways modulate cell wall gene expression, leading to the construction of a new, modified cell wall (Ene et al. 2015). In *A. oryzae*, disruption of the *kexB* gene disordered cell integrity signaling in this fungus (Mizutani et al. 2004). Kexin is a Ca^{+2}-dependent transmembrane serine protease that cleaves the secretory proproteins on the carboxyl side of Lys-Arg and Arg-Arg in a late Golgi compartment (Fuller et al. 1989). Fungi including *Aspergillus* species secrete large amounts of proteins which are subjected to modifications via proteolysis in Golgi; kexins are therefore speculated to be key enzymes involved in proteolytic processes in fungal Golgi apparatus. It has been also shown that this protein plays roles in fungal morphology and cell wall integrity; however, the mechanism underlying the involvement of *kexB* in morphogenesis of fungi is still unclear. Although disruption of the yeast *kex2* does not exhibit any morphological defects but double mutations in the *S. cerevisiae* MPK1 (Slt2), encoding a mitogen-activated protein (MAP) kinase, and *kex2* is lethal (Roelants et al. 2002). Deletion of *kexB* in *A. oryzae* results in morphological defects in this fungus which is concomitant with significant increase in transcriptional level of *mapkA* in Δ*kexB* strain and likewise transcripts of cell wall-related genes encoding β-1,3-glucanosyltransferase and chitin synthases, which is presumably attributable to cell integrity signaling through the increased gene expression of *mpkA*. High osmotic stress greatly downregulated the increased levels of both transcripts of *mpkA* and the phosphorylated form of MpkA in Δ*kexB* cells, simultaneously suppressing the morphological defects (Mizutani et al. 2004). This study emphasizes on the importance of KexB in the cell wall integrity pathway for the precise proteolytic processing of sensor proteins or cell wall-related enzymes under transcriptional control of MAPK pathway.

Biotechnological production of two valuable metabolites itaconic acid and lovastatin is mainly associated with the genus of *A. terreus*. Itaconic acid has wide range of applications in polymer manufacturing, whereas lovastatin is among the best-selling pharmaceuticals in the USA which is a statin drug capable of lowering cholesterol in blood and thus useful for those with hypercholesterolemia to reduce risk of cardiovascular disease. Lovastatin can be produced by solid-state fermentation (SSF) and/or submerged fermentation (SmF). Cultivated *Aspergillus terreus* in solid-state and in liquid-submerged fermentation showed differentially expression of genes (*lovE* and *lovF*) for lovastatin biosynthesis (Barrios-González et al. 2008). Higher lovastatin production in SSF was related to a more intense transcription of these two biosynthetic genes. Strong expression of glycerol dehydrogenase-encoding gene, *gldB*, that is essential for osmotolerance in *A. nidulans* and in lovastatin SSF indicated that *A. terreus* senses osmotic stress during the course of SSF, but not in SmF (Barrios-González et al. 2008). Onset of production of this fungal secondary metabolite during the idiophase growth of *A. terreus* coincides with the increase of reactive oxygen species (ROS) implying the importance of mass transfer of oxygen in the cultivation broth and within the mycelia in order to promote the oxidative state associated with lovastatin biosynthesis. Increase in ROS is concomitant with downregulation of *sod1* gene that encodes the oxidative stress defense enzyme and notably downregulation of the transcription factor *yap1* suggesting that Yap1 could provide a link between initiation of lovastatin biosynthesis and the accumulation of ROS (Miranda et al. 2013, 2014).

Aside from their importance in the natural environment like food, the ascomycetous fungi *Penicillium* are among the most biotechnologically important fungi due to their capability for production of drugs as β-lactam antibiotics. Cephalosporin and penicillin are the most frequently used β-lactam antibiotics for the treatment of bacterial infections worldwide. *Penicillium chrysogenum* is considered as one of the main industrial producer of these antibiotics. The entire developmental program and penicillin biosynthesis in *P. chrysogenum* are regulated by the heterotrimeric Gα-protein Pga1-mediated signaling pathway (García-Rico et al. 2007). The regulation of conidiation by Pga1 is mediated by repression of *brlA* and *wetA* genes rather than a control mainly by a cAMP-dependent mechanism (García-Rico et al. 2008). Pga1 also affects negatively the resistance to different stress conditions. Inactivation of the heterotrimeric G-alpha subunit Pga1 allowed growth on SDS-containing minimal medium, increase resistance of conidia to thermal and oxidative stress, and increase resistance of vegetative mycelium to thermal and osmotic stress. In contrast, constitutive activation of Pga1 causes a decrease in the resistance of conidia to thermal stress and of vegetative mycelium to thermal and osmotic stress (García-Rico et al. 2011). In agreement with these observations, proteomics analysis of the signaling pathways mediated by *P. chrysogenum Pga1* identified 30 proteins whose abundance was regulated by the Pga1-mediated signaling pathway including proteins involved in primary metabolism, stress response, development, and signal transduction (Carrasco-Navarro et al. 2016).

Members of the genus *Trichoderma* are also utilized in various industry branches—mainly in the production of antibiotics, different metabolites, and the enzymes which are used for bioethanol and biorefinery industries. The filamentous fungus *Trichoderma reesei* is a widely used industrial host organism for protein production. This fungus serves as a major producer of biomass-degrading enzymes for the use of renewable lignocellulosic material toward production of biofuels and biorefineries. In industrial cultivations, it can produce over 100 g/l of extracellular protein, mostly constituting of biomass-degrading enzymes cellulases and hemicellulases (Pakula et al. 2016).

The opportunistic *Trichoderma* species have evolved with several sophisticated molecular mechanisms in order to survive in wider habitats (Fig. 6.3). The main mechanism for survival and dispersal of *Trichoderma* is through the production of conidia. Conidiation in this fungus is induced by several factors like light, nutrients, pH, volatile compounds (VOCs), and mechanical injury. Among these light and mechanical injury have predominant influence (Hernández-Oñate and Herrera-Estrella 2015; Tisch and Schmoll 2009).

The photoreceptor complex Blr-1/Blr-2, ENVOY, VELVET, and NADPH oxidases are key participants in this process. Together with these elements, conserved signaling pathways, such as those involving heterotrimeric G-proteins, mitogen-activated protein kinases (MAPKs), and cAMP-dependent protein kinase A (cAMP-PKA), are involved in this molecular orchestration (Carreras-Villaseñor et al. 2012). In the last part of this chapter and connected to Sect. 6.4, we pay particular attention to more details about the light signaling pathway implicated in oxidative response, cellulase gene expression, peptaibol biosynthesis, and development in *Trichoderma* as industrial workhorse and as biocontrol agent. Light signaling machinery impacts diverse downstream pathways rather than only vegetative growth and reproduction, including those implicated in response to oxidative stress (Schmoll et al. 2010). Transcriptomic analyses have revealed that components of the oxidative stress pathway are responsive to light: genome-based analyses of gene

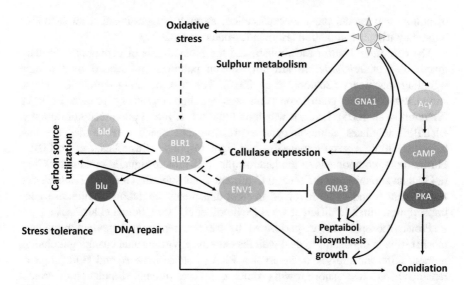

Fig. 6.3 Signaling network involved in the regulation of light responses in *Trichoderma* species. The photoreceptor Blr1/Blr2 complex, which is crucial for light-dependent carbon source utilization, upregulates a set of genes (*blu*) and downregulates another set (*bld*) and triggers asexual reproduction in *Trichoderma*. *blu* gene impacts stress tolerance and DNA repair, while this gene together with *bld* control the growth of *Trichoderma*. The Blr1/Blr2 complex controls also transcription of the *env1* gene, which in turn establishes a feedback regulatory loop with the BLR proteins, and is involved in photoadaptation. Oxidative stress caused by light influences peptaibol biosynthesis, which is also regulated by the G-protein alpha subunit Gna3. The Blr1/Blr2 complex is influenced by a second light input involving the cAMP pathway, constituted by adenylate cyclase (ACY), cAMP, and protein kinase A (PKA). Light-induced cellulase gene expression involves the function of the photoreceptors Blr1/Blr2 and ENV1, the sulfur signaling pathway, the heterotrimeric G-protein alpha subunits Gna1 and Gna3, and the cAMP pathway (In all figures arrows indicate positive regulation, and lines with a bar at the end indicate negative regulation. Solid lines are used when there is direct experimental evidence and dotted lines when the role is still hypothetical)

expression of *Trichoderma* showed that, in addition to 331 early light-regulated genes comprising about >45% *blr*-dependent, 39 out of the 178 light-induced identified genes are related to stress responses. Significantly, 17 of these stress-induced genes are related to oxidative stress. This set of genes includes key elements such as components of the MAPK (p38/Hog1) cascade (Mukherjee et al. 2013). In *N. crassa*, the corresponding gene (*os-2*) is also induced by light, and its phosphorylation shows circadian oscillations, regulating rhythmic expression of output genes (de Paula et al. 2008). These data suggest that light is perceived as a stress signal impacting a MAPK cascade, perhaps anticipating harmful effects of light, to provide protection against them. Consistent with what has been observed in response to light, recent transcriptomic analysis using high-throughput sequencing identified a significant number of injury-responsive genes encode proteins related to oxidative stress. During early stages of the response to injury, genes known to generate ROS are induced, whereas those known to scavenge ROS are repressed. In addition, the use of antioxidant agents prevents conidiation, strongly suggesting that an oxidative burst may trigger conidiation (Hernández-oñate et al. 2012).

The carbon and sulfur metabolism and the biosynthesis of peptaibol secondary metabolites of *Trichoderma* are downstream pathways influenced by the light signaling machinery (Schmoll et al. 2010). The first signaling component which stimulated cellulase gene expression was the light-regulatory protein ENVOY (Schmoll et al. 2005). Later, it was found that the *T. reesei* photoreceptor complex Blr1/Blr2 regulates cellulase gene expression and hence corroborates the link between cellulase gene expression and light response (Castellanos et al. 2010). Furthermore, components of the heterotrimeric G-protein signaling including Gna3 (a G-alpha subunit) and the second G-protein alpha subunit (Gna1) of *Trichoderma reesei* were found to be involved in light-dependent regulation of cellulase gene transcription albeit in different ways (Schmoll et al. 2009; Seibel et al. 2009).

Protein phosphorylation performed by kinases represent important regulatory mechanisms in organisms to deal with the extreme environmental changing including nutrient shifts and stresses. In eukaryotes, PKA (protein kinase A) and TOR (target of rapamycin) are two major growth-promoting kinases through stimulation of protein synthesis, inhibition of stress responses, and regulation of other cellular processes in response to nutrient availability (Zurita-Martinez and Cardenas 2005). The pathways controlled by PKA and TOR display diverse ways of cross talk sharing several common targets including Yak1 and Sch9 (Zurita-Martinez and Cardenas 2005). Null mutations in *yak1*, whose product may act downstream of the cAMP-dependent protein kinase, could suppress loss of function of the cAMP-PKA pathway (Garrett et al. 1991). This protein can be phosphorylated by the catalytic PKA subunit Tpk1 (Budovskaya et al. 2005). Yak1 functions also as a bridge between PKA and stress-responsive transcription factors Hsf1 and Msn2/4 to regulate stress response (Lee et al. 2008). Sch9 is structurally related to the catalytic subunits of PKA. PKA and Sch9 oppositely regulate given target genes including those involved in stress responses and post-diauxic shift (PDS) in yeast (Pascual-Ahuir and Proft 2007). In *A. fumigatus*, modulation of SakA^{HOG1} is regulated by SchA^{Sch9} kinase which is also essential for its virulence (Alves de Castro et al. 2016). Basal and stress-induced synthesis of the components of the highly conserved heat-shock protein Hsp90 chaperone complex requires the heat-shock transcription factor (Hsf). In *S. cerevisiae*, loss of *sch9* expressing the Hsf allele Hsf (1-583) could be capable of restoring its high-temperature growth by regulating Hsp90 chaperone complex activity (Morano and Thiele 1999). Adenylate cyclase (Acy1), as one of the crucial components of cAMP signaling pathway, and cAMP-dependent protein kinase A (PKA) are involved in vegetative growth and light-modulated cellulase regulation in *T. reesei* (Schuster et al. 2012). Disruption of Sho9 and Yak1 of *T. reesei* resulted in decreased growth of the fungus on different carbon sources and less conidiation compared to wild type. Moreover, absence of these two kinases caused abnormal sensitivity to various stresses and perturbation of cell wall integrity. Surprisingly, loss of function of *Tryak1* and *Trsch9* increased induced production of cellulases which is in contrast to the *Aspergillus nidulans* protein kinases SchA and YakA, whose functions are important for the production of cellulases, which might be due to different functions in evolutionary-related filamentous fungi (Lv et al. 2015).

As in other studied filamentous fungi with predominant impacts on agriculture, members of the velvet complex (i.e., VeA/Vel1 and LaeA/1) play crucial roles in production of value-added chemicals from abovementioned industrially important fungi and their development (Niu et al. 2015—for review see also Karimi Aghcheh and Kubicek 2015; Park and Yu 2016). Besides a large number of works about regulatory role of velvet complex in fungal development and biosynthesis of their valuable metabolites or enzymes, there are several reports where the mechanism of fungal responses to various stresses have been related to this complex. However, the exact interplay between stress tolerance in fungi and the regulation of their developmental processes and metabolism of their biotechnological treasures awaits yet further researches.

6.6   Conclusion

Filamentous fungi are exposed to a complex set of abiotic or biotic stresses in their natural habitats as well as in industrial processes where they are utilized to produce various biotechnological important products. Effective stress perception and signal transduction mechanisms are necessary for adaptation and survival of these microorganisms. Several mechanisms including mainly signal transduction pathways [e.g., mitogen-activated protein kinase (MAPK) pathways specifically the high-osmolarity glycerol (HOG) response pathway, cAMP-dependent protein kinase pathway, and those involving heterotrimeric G-proteins] have been studied extensively in numerous filamentous fungi. In addition to their roles in sensing environmental stresses, MAPKs have been too demonstrated to be involved in fungal development, pathogenicity, and/or virulence in many filamentous plant pathogenic fungi. Additional routes include fungal-specific protein complexes exemplified in the velvet complex, which coordinated epigenetic and transcriptional control elements of stress responses in fungi. The link between components of signal transduction pathways affecting fungal stress responses with velvet complex proteins is not fully understood. Furthermore, understanding of the cellular choreography of the dynamic interplay between velvet domain proteins and methyltransferases as LaeA/1 and their target genes during vegetative growth, development, and secondary metabolism as well as biosynthesis of their important enzymes is currently unknown and is important for a specific manipulation of fungi to control production of fungal toxins and biotechnologically relevant products with significance for agriculture and industry.

Acknowledgment RKA is supported by a Schrödinger Fellowship from Austria. The research in the laboratory has been funded by the Deutsche Forschungsgemeinschaft (DFG) to GHB. RKA and GHB are grateful to Christian P. Kubicek and Kai Heimel for their critical comments.

References

Al-Sheikh H, Watson AJ, Lacey GA, Punt PJ, Mackenzie DA, Jeenes DJ, Penttilä M, Alcocer MJ, Archer DB (2004) Endoplasmic reticulum stress leads to the selective transcriptional downregulation of the glucoamylase gene in *Aspergillus niger*. Mol Microbiol 53(6):1731–1742

Alves de Castro P, Dos Reis TF, Dolan SK, Oliveira Manfiolli A, Brown NA, Jones GW, Doyle S, Riaño-Pachón DM, Squina FM, Caldana C, Singh A, Del Poeta M, Hagiwara D, Silva-Rocha R, Goldman GH (2016) The *Aspergillus fumigatus* SchASCH9 kinase modulates SakAHOG1 MAP kinase activity and it is essential for virulence. Mol Microbiol 102(4):642–671. https://doi.org/10.1111/mmi.13484.137

Askary TH (2010) Nematodes as biocontrol agents. In: Lichtfouse E (ed) Sociology, organic farming, climate change and soil science. Sustainable agriculture reviews, vol 3. Springer, Dordrecht

Atalay M, Oksala N, Lappalainen J, Laaksonen DE, Sen CK, Roy S (2009) Heat shock proteins in diabetes and wound healing. Curr Protein Pept Sci 10(1):85–95

Atoui A, Bao D, Kaur N, Grayburn WS, Calvo AM (2008) *Aspergillus nidulans* natural product biosynthesis is regulated by mapkB, a putative phremone response mitogen-activated protein kinase. Appl Environ Microbiol 74(11):3596–3600. https://doi.org/10.1128/AEM.02842-07

Auld B (2002) Fungi as biocontrol agents: progress, problems and potential. Plant Pathol 51:518. https://doi.org/10.1046/j.1365-3059.2002.07351.x

Baidya S, Duran RM, Lohmar JM, Harris-Coward PY, Cary JW, Hong SY, Roze LV, Linz JE, Calvo AM (2014) VeA is associated with the response to oxidative stress in the aflatoxin producer *Aspergillus flavus*. Eukaryot Cell 13(8):1095–1103. https://doi.org/10.1128/EC.00099-14

Barrios-González J, Baños JG, Covarrubias AA, Garay-Arroyo A (2008) Lovastatin biosynthetic genes of *Aspergillus terreus* are expressed differentially in solid-state and in liquid submerged fermentation. Appl Microbiol Biotechnol 79(2):179–186. https://doi.org/10.1007/s00253-008-1409-2

Bayram Ö, Krappmann S, Ni M, Bok JW, Helmstaedt K, Valerius O, Braus-Stromeyer S, Known NJ, Keller NP, Yu JH, Braus GH (2008) VelB/VeA/LaeA complex coordinates light signal with fungal development and secondary metabolism. Science 320(5882):1504–1506

Blumenstein A, Vienken K, Tasler R, Purschwitz J, Veith D, Frankenberg-Dinkel N, Fischer R (2005) The *Aspergillus nidulans* phytochrome FphA represses sexual development in red light. Curr Biol 15(20):1833–1838

Bok JW, Keller NP (2004) LaeA, a regulator of secondary metabolism in *Aspergillus spp*. Eukaryot Cell 3(2):527–535

Bok JW, Balajee SA, Marr KA, Andes D, Nielsen KF, Frisvad JC, Keller NP (2005) LaeA, a regulator of morphogenetic fungal virulence factors. Eukaryot Cell 4(9):1574–1582

Borriss R (2011) Use of plant-associated *Bacillus* strains as biofertilizers and biocontrol agents in agriculture. In: Maheshwari D (ed) Bacteria in agrobiology: plant growth responses. Springer, Berlin

Bretscher A (1981) Fimbrin is a cytoskeletal protein that crosslinks F-actin *in vitro*. Proc Natl Acad Sci USA 78(11):6849–6853

Budovskaya YV, Stephan JS, Deminoff S, Herman PK (2005) An evolutionary proteomics approach identifies substrates of the cAMP-dependent protein kinase. Proc Natl Acad Sci USA 102:13933–13938

Cabañes FJ, Bragulat MR, Castellá C (2010) Ochratoxin A producing species in the genus *Penicillium*. Toxins (Basel) 2(5):1111–1120. https://doi.org/10.3390/toxins2051111

Carrasco-Navarro U, Vera-Esterlla R, Barkla BJ, Zúñiga-León E, Reyes-Vivas H, Fernández FJ, Fierro F (2016) Proteomic analysis of the signaling pathway mediated by the heterotrimeric Gα protein Pga1 of *Penicillium chrysogenum*. Microb Cell Factories 15(1):173

Carreras-Villaseñor N, Sánchez-Arreguín JA, Herrera-Esterlla AH (2012) *Trichoderma*: sensing the environment for survival and disposal. Microbiology 158(Pt 1):3–16. https://doi.org/10.1099/mic.0.052688-0

Carvalho ND, Jørgensen TR, Arentschorst M, Nitsche MM, Van den Hondel CA, Archer DB, Ram AF (2012) Genome-wide expression analysis upon constitutive activation of the HacA bZIP transcription factor in *Aspergillus niger* reveals a coordinated cellular response to counteract ER stress. BMC Genomics 13:350. https://doi.org/10.1186/1471-2164-13-350

Castellanos F, Schmoll M, Martinez P, Tisch D, Kubicek CP, Herrera-Estrella AH, Esquivel-Naranjo EU (2010) Crucial factors of the light perception machinery and their impact on growth and cellulase gene transcription in *Trichoderma reesei*. Fungal Genet Biol 47:468–476

Cervantes-Chávez JA, Valdés-Santiago L, Bakkeren G, Hurtado-Santiago E, León-Ramírez CG, Esquivel-Naranjo EU, Landeros-Jaime F, Rodríguez-Aza Y, Ruiz-Herrera J (2016) Trehalose is required for stress resistance and virulence of the Basidiomycota pathogen *Ustilago maydis*. Microbiology 162(6):1009–1022. https://doi.org/10.1099/mic.0.000287

Chang PK, Scharfenstein LL, Luo M, Mahoney N, Molyneuux RJ, Yu J, Brown RL, Campbell BC (2011) Loss of *msnA*, a putative stress regulatory gene, in *Aspergillus parasiticus* and *Aspergillus flavus* increased production of conidia, aflatoxins and kojic acid. Toxins (Basel) 3(1):82–104. https://doi.org/10.3390/toxins3010082

Chaverri P, Branco-Rocha F, Jaklitsch W, Gazis R, Degenkolb T, Samuels GJ (2015) Systematics of the *Trichoderma harzianum* species complex and the re-identification of commercial biocontrol strains. Mycologia 107(3):558–590. https://doi.org/10.3852/14-147

Chen LH, Yang SL, Chung KR (2014) Resistance to oxidative stress via regulating siderophore-mediated iron acquisition by the citrus fungal pathogen *Alternaria alternata*. Microbiology 160 (Pt 5):970–979. https://doi.org/10.1099/mic.0.076182-0

Chung RK (2012) Stress response and pathogenicity of the necrotrophic fungal pathogen *Alternaria alternate*. Scientifica (Cairo) 635431. https://doi.org/10.6064/2012/635431

Conlan RS, Gounalaki N, Hatzis O, Tzamarias D (1999) The Tup1-Cyc8 protein complex can shift from a transcriptional co-repressor to a transcriptional activator. J Biol Chem 274(1):205–210

Connolly LR, Smith KM, Freitag M (2013) The *Fusarium graminearum* histone H3K27 methyltransferase KMT6 regulates development and expression of secondary metabolite gene clusters. PLoS Genet 9(10):e1003916. https://doi.org/10.1371/journal.pgen.1003916

D'Amore T, Crumplen R, Stewart GG (1999) The involvement of trehalose in yeast stress tolerance. J Ind Mirobiol 3(3):191–195

Daguerre Y, Siegel K, Edel-Hermann V, Steinberg C (2014) Fungal proteins and genes associated with biocontrol mechanisms of soil-borne pathogens: a review. Fungal Biol Rev 28(4):97–125

Dan P, Rosenblat G, Yedgar S (2012) Phospholipase A2 activities in skin physiology and pathology. Eur J Pharmacol 691:1–8

de Paula RM, Lamb TM, Bennett L, Bell-Pedersen DA (2008) Connection between MAPK pathways and circadian clocks. Cell Cycle 7:2630–2634

de Vries RP, Riley R, Wiebenga A, Aguilar-Osorio G, Amillis S, Uchima CA, Anderluh G, Asadollahi M, Askin M, Barry K, Battaglia E, Bayram Ö, Benocci T, Braus-Stromeyer SA, Caldana C, Cánovas D, Cerqueira GC, Chen F, Chen W, Choi C, Clum A, Dos Santos RA, Damásio AR, Diallinas G, Emri T, Fekete E, Flipphi M, Freyberg S, Gallo A, Gournas C, Habgood R, Hainaut M, Harispe ML, Henrissat B, Hildén KS, Hope R, Hossain A, Karabika E, Karaffa L, Karányi Z, Kraševec N, Kuo A, Kusch H, LaButti K, Lagendijk EL, Lapidus A, Levasseur A, Lindquist E, Lipzen A, Logrieco AF, MacCabe A, Mäkelä MR, Malavazi I, Melin P, Meyer V, Mielnichuk N, Miskei M, Molnár ÁP, Mulé G, Ngan CY, Orejas M, Orosz E, Ouedraogo JP, Overkamp KM, Park HS, Perrone G, Piumi F, Punt PJ, Ram AF, Ramón A, Rauscher S, Record E, Riaño-Pachón DM, Robert V, Röhrig J, Ruller R, Salamov A, Salih NS, Samson RA, Sándor E, Sanguinetti M, Schütze T, Sepčić K, Shelest E, Sherlock G, Sophianopoulou V, Squina FM, Sun H, Susca A, Todd RB, Tsang A, Unkles SE, van de Wiele N, van Rossen-Uffink D, Oliveira JV, Vesth TC, Visser J, Yu JH, Zhou M, Andersen MR,

Archer DB, Baker SE, Benoit I, Brakhage AA, Braus GH, Fischer R, Frisvad JC, Goldman GH, Houbraken J, Oakley B, Pócsi I, Scazzocchio C, Seiboth B, vanKuyk PA, Wortman J, Dyer PS, Grigoriev IV (2017) Comparative genomics reveals high biological diversity and specific adaptations in the industrially and medically important fungal genus *Aspergillus*. Genome Biol 18(1):28. https://doi.org/10.1186/s13059-017-1151-0

Dean R, Van Kan JA, Pretorius ZA, Hammond-Kosack KE, Di Pietro A, Spanu PD, Rudd JJ, Dickman M, Kahmann R, Ellis J, Foster GD (2012) The Top 10 fungal pathogens in molecular plant pathology. Mol Plant Pathol 13(4):414–430. https://doi.org/10.1111/j.1364-3703.2011. 00783.x

Delgado-Jarana J, Sousa S, González F, Rey M, Llobella A (2006) ThHog1 controls the hyperosmotic stress response in *Trichoderma harzianum*. Microbiology 152(Pt6):1687–1700

Dinamarco TM, Almeida RS, de Castro PA, Brown NA, dos Reis TF, Ramalho LN, Savoldi M, Goldman MH, Goldman GH (2012) Molecular characterization of the putative transcription factor SebA involved in virulence in *Aspergillus fumigatus*. Eukaryot Cell 11(4):518–531. https://doi.org/10.1128/EC.00016-12

Dioxon KP, Xu JR, Smironoff N, Talbot NJ (1999) Independent signaling pathways regulate cellular turgor during hyperosmotic stress and appressorium-mediated plant infection by *Magnaporthe grisea*. Plant Cell 11(10):2045–2058

Druzhinina IS, Seidl-Seiboth V, Herrera-Estrella A, Horwitz BA, Kenerley CM, Monte E, Mukherjee PK, Zeilinger S, Grigoriev IV, Kubicek CP (2011) *Trichoderma*: the genomics of opportunistic success. Nat Rev Microbiol 9(10):749–759. https://doi.org/10.1038/nrmicro2637

Duran RM, Cary JW, Calvo AM (2007) Production of cyclopiazonic acid, aflatrem, and aflatoxin by *Aspergillus flavus* is regulated by *veA*, a gene necessary for sclerotial formation. Appl Microbiol Biotechnol 73(5):1158–1168

Duran R, Cary JW, Calvo AM (2010) Role of the osmotic stress regulatory pathway in morphogenesis and secondary metabolism in filamentous fungi. Toxins (Basel) 2(4):367–381. https://doi.org/10. 3390/toxins2040367

Ellouze W, Esmaeili Taheri A, Bainard LD, Yang C, Bazghaleh N, Navarro-Borrell A, Hanson K, Hamel C (2014) Soil fungal resources in annual cropping systems and their potential for management. Biomed Res Int 2014:531824. https://doi.org/10.1155/2014/531824

Ene IV, Walker LA, Schiavone M, Lee KK, Martin-Yken H, Dague E, Gow NAR, Munro CA, Brown AJP (2015) Cell wall remodeling enzymes modulate fungal cell wall elasticity and osmotic stress resistance. MBio 6(4):e00986-15. https://doi.org/10.1128/mBio.00986-15

Enjalbert B, MacCallum DM, Odds FC, Brown AJ (2007) Niche-specific activation of the oxidative stress response by the pathogenic fungus *Candida albicans*. Infect Immun 75(5):2143–2151

Fassler JS, West AH (2013) Histidine phosphotransfer proteins in fungal two-component signal transduction pathways. Eukaryot Cell 12(8):1052–1060. https://doi.org/10.1128/EC.00083-13

Fiedurek J (1998) Effect of osmotic stress on glucose oxidase production and secretion by *Aspergillus niger*. J Basic Microbiol 38:107–112

Fillinger S, Chaveroche MK, van Dijck P, de Vries R, Ruijter G, Thevelein J, d'Enfert C (2001) Trehalose is required for the acquisition of tolerance to a variety of stresses in the filamentous fungus *Aspergillus nidulans*. Microbiology 147:1851–1862

Fountain JC, Bajaj P, Nayak SN, Yang L, Pandey MK, Kumar V, Jayale AS, Chitikineni A, Lee RD, Kemerait RC, Varshney RK, Guo B (2016) Responses of *Aspergillus flavus* to oxidative stress are related to fungal development regulator, antioxidant enzyme, and secondary metabolite biosynthesis gene expression. Front Microbiol 7:2048. https://doi.org/10.3389/fmicb.2016.02048

Friedl MA, Schmoll M, Kubicek CP, Druzhinina IS (2008) Photostimulation of *Hypocrea atroviridis* growth occurs due to a cross-talk of carbon metabolism, blue light receptors and response to oxidative stress. Microbiology 154:1229–1241

Fuller RS, Brake A, Thorner J (1989) Yeast prohormone processing enzyme (KEX2 gene product) is a Ca^{+2}-dependent serine protease. Proc Natl Acad Sci USA 86:1434–1438

Furukawa K, Katsuno Y, Urao T, Yabe T, Yamada-Okabe T, Yamada-Okabe H, Yamagata Y, Abe K, Nakajima T (2002) Isolation and functional analysis of a gene, *tcsB*, encoding a transmembrane hybrid-type histidine kinase from *Aspergillus nidulans*. Appl Environ Microbiol 68:5304–5310

Furukawa K, Hoshi Y, Maeda T, Nakajama T, Abe K (2005) *Aspergillus nidulans* HOG pathway is activated only by two-component signaling pathway in response to osmotic stress. Mol Microbiol 56(5):1246–1261

García-Rico RO, Martin JF, Fierro F (2007) The *pga1* gene of *Penicillium chrysogenum* NRRL 1951 encodes a heterotrimeric G protein alpha subunit that controls growth and development. Res Microbiol 158(5):437–446

García-Rico RO, Fierro F, Martin JF (2008) Heterotrimeric Galpha protein Pga1 of *Penicillium chrysogenum* controls conidiation mainly by a cAMP-independent mechanism. Biochem Cell Biol 86(1):57–69. https://doi.org/10.1139/o07-148

García-Rico RO, Martin JF, Fierro F (2011) Heterotrimeric Gα protein Pga1 from *Penicillium chrysogenum* triggers germination in response to carbon sources and affects negatively resistance to different stress conditions. Fungal Genet Biol 48(6):641–649. https://doi.org/10.1016/j.fgb.2010.11.013

Garrett S, Menold MM, Broach JR (1991) The *Saccharomyces cerevisiae* YAK1 gene encodes a protein kinase that is induced by arrest early in the cell cycle. Mol Cell Biol 11:4045–4052

Gasch AP, Spellman PT, Kao CM, Carmel-Harel O, Eisen MB, Storz G, Botstein D, Brown PO (2000) Genomic expression programs in the response of yeast cells to environmental changes. Mol Biol Cell 11(12):4241–4257

Ghosh A (2014) Small heat shock proteins (HSP12, HSP20 and HSP30) play a role in *Ustilago maydis* pathogenesis. FEMS Microbiol Lett 361(1):17–24

Graf E, Schmidt-Heydt M, Geisen R (2012) HOG MAP kinase regulation of alternariol biosynthesis in *Alternaria alternata* is important for substrate colonization. Int J Food Microbiol 157 (3):353–359. https://doi.org/10.1016/j.ijfoodmicro.2012.06.004

Grintzalis K, Vernardis SI, Klapa MI, Georgiou CD (2014) Role of oxidative stress in sclerotial differentiation and aflatoxin B1 biosynthesis in *Aspergillus flavus*. Appl Environ Microbiol 80 (18):5561–5571. https://doi.org/10.1128/AEM.01282-14

Gu Q, Ji T, Sun X, Huang H, Zhang H, Lu X, Wu L, Huo R, Wu H, Gao X (2017) Histone H3 lysine 9 methyltransferase FvDim5 regulates fungal development, pathogenicity and osmotic stress responses in *Fusarium verticillioides*. FEMS Microbiol Lett 364(19). https://doi.org/10.1093/femsle/fnx184

Guo B, Chen ZY, Lee RD, Scully BT (2008) Drought stress and preharvest aflatoxin contamination in agricultural commodity: genetics, genomics and proteomics. J Integr Plant Biol 50 (10):1281–1291. https://doi.org/10.1111/j.1744-7909.2008.00739.x

Gustin MC, Albertyn J, Alexander M, Davenport K (1998) MAP kinase pathway in the yeast *Saccharomyces cerevisiae*. Microbiol Mol Biol Rev 62(4):1264–1300

Hagiwara D, Matsubayashi Y, Marui J, Furukawa K, Yamashino T, Kanamaru K, Kato M, Abe K, Kobayashi T, Mizuno T (2007) Characterization of the NikA histidine kinase implicated in the phosphorelay signal transduction *of Aspergillus nidulans*, with special reference to fungicide responses. Biosci Biotechnol Biochem 71(3):844–847

Hampel M, Jakobi M, Schmity L, Meyer U, Finkernagel F, Doehlemann G, Heimel K (2016) Unfolded protein response (UPR) regulator Cib1 controls expression of genes encoding secreted virulence factors in *Ustilago maydis*. PLoS One 11(4):e0153861. https://doi.org/10.1371/journal.pone.0153861

Han KH, Prade RA (2002) Osmotic stress-coupled maintenance of polar growth in *Aspergillus nidulans*. Mol Microbiol 43(5):1065–1078

Heimel K (2015) Unfolded protein response in filamentous fungi – implications in biotechnology. Appl Microbiol Biotechnol 99(1):121–132. https://doi.org/10.1007/s00253-014-6192-7

Heimel K, Freitag J, Hampel M, Ast J, Bölker M, Kämper J (2013) Cross talk between the unfolded protein response and pathways that regulate pathogenic development in *Ustilago maydis*. Plant Cell 25(10):4262–4277. https://doi.org/10.1105/tpc.113.115899

Hernández-Oñate MA, Herrera-Estrella A (2015) Damage response involves mechanisms conserved across plants, animals and fungi. Curr Genet 61:359–372. https://doi.org/10.1007/s00294-014-0467-5

Hernández-Oñate MA, Esquivel-Naranjo EU, Mendoza-Mendoza A, Steward A, Herrera-Esterlla AH (2012) An injury-response mechanism conserved across kingdoms determines entry of the fungus *Trichoderma atroviride* into development. Proc Natl Acad Sci USA 109(37):14918–14923. https://doi.org/10.1073/pnas.1209396109

Hicks J, Lockington RA, Strauss J, Dieringer D, Kubicek CP, Kelly J, Keller NP (2001) RcoA has pleiotropic effects on *Aspergillus nidulans* cellular development. Mol Microbiol 39 (6):1482–1493

Hohmann S (2009) Control of high osmolarity signaling in the yeast *Saccharomyces cerevisiae*. FEBS Lett 583(24):4025–4029. https://doi.org/10.1016/j.febslet.2009.10.069

Hoppenau CE, Tran VT, Kusch H, Aßhauer KP, Landesfeind M, Meinicke P, Popova B, Braus-Stromeyer SA, Braus GH (2014) *Verticillium dahliae* VdTHI4, involved in thiazole biosynthesis, stress response and DNA repair functions, is required for vascular disease induction in tomato. Environ Exp Bot 108:14–22

Huang Y, Mijiti G, Wang Z, Yu W, Fan H, Zhang R, Liu Z (2015) Functional analysis of the class II hydrophobin gene HFB2-6 from the biocontrol agent *Trichoderma asperellum* ACCC30536. Microbiol Res 171:8–20. https://doi.org/10.1016/j.micres.2014.12.004

Izumitsu K, Yoshimi A, Tanaka C (2007) Two-component response regulators Ssk1p and Skn7p additively regulate high-osmolarity adaptation and fungicide sensitivity in *Cochliobolus heterostrophus*. Eukaryot Cell 6(2):171–181

Izumitsu K, Yoshimi A, Hamada S, Morita A, Saitoh Y, Tanaka C (2009) Dic2 and Dic3 loci confer osmotic adaptation and fungicidal sensitivity independent of the HOG pathway in *Cochliobolus heterostrophus*. Mycol Res 113(Pt 10):1208–1215. https://doi.org/10.1016/j.mycres.2009.08.005

Jin Y, Weining S, Nevo E (2005) A MAPK gene from Dead Sea confers stress tolerance to lithium salt and freezing-thawing: prospects for saline agriculture. Proc Natl Acad Sci USA 102 (52):18992–18997

Karimi Aghcheh R, Kubicek CP (2015) Epigenetics as an emerging tool for improvement of fungal strains used in biotechnology. Appl Microbiol Biotechnol 99(15):6167–6181. https://doi.org/10.1007/s00253-015-6763-2

Karimi Aghcheh R, Druzhinina IS, Kubicek CP (2013) The putative protein methyltransferase LAE1 of *Trichoderma atroviride* is a key regulator of asexual development and mycoparasitism. PLoS One 8(6):e67144. https://doi.org/10.1371/journal.pone.0067144

Kobayashi A, Sano M, Oda K, Hisada H, Hata Y, Ohashi S (2007) The glucoamylase-encoding gene (*glaB*) is expressed in solid-state culture with a low water content. Biosci Biotechnol Biochem 71:1797–1799

Kojima K, Takano Y, Yoshimi A, Tanaka C, Kikuchi T, Okuno T (2004) Fungicide activity through activation of a fungal signaling pathway. Mol Microbiol 53(6):1785–1796

Kumar P, Mahato DK, Kamle M, Mohanta TK, Kang SG (2016) Aflatoxins: a global concern for food safety, human health and their management. Front Microbiol 7:2170. https://doi.org/10.3389/fmicb.2016.02170

Lee P, Cho BR, Joo HS, Hahn JS (2008) Yeast Yak1 kinase, a bridge between PKA and stress-responsive transcription factors, Hsf1 and Msn2/Msn4. Mol Microbiol 70:882–895

Lee I, Oh JH, Dagenais TR, Andes D, Keller NP (2009) HdaA, a class 2 histone deacetylase of *Aspergillus fumigatus*, affects germination and secondary metabolite production. Fungal Genet Biol 46(10):782–790

Leng Y, Zhong S (2015) The role of mitogen-activated protein (MAP) kinase signaling in the fungal development, stress response and virulence of the fungal cereal pathogen *Bipolaris sorokiniana*. PLoS One 10(5):e0128291. https://doi.org/10.1371/journal.pone.0128291

Lev S, Hadar R, Amedeo P, Baker SE, Yoder OC, Horwitz BA (2005) Activation of an AP1-like transcription factor of the maize pathogen *Cochliobolus heterostrophus* in response to oxidative stress and plant signals. Eukaryot Cell 4(2):443–454

Li S, Ault A, Malone CL, Raitt D, Dean S, Johnston LH, Deschenes RJ, Fassler JS (1998) The yeast histidine protein kinase, Sln1p, mediates phosphotransfer to two response regulators, Ssk1p and Skn7p. EMBO J 17(23):6952–6962. https://doi.org/10.1093/emboj/17.23.6952

Li ZZ, Alves SB, Roberts DW, Fan MZ, Delalibera I, Tang J, Lopes RB, Faria M, Rangel DEN (2010) Biological control of insects in Brazil and China: history, current programs and reasons for their successes using entomopathogenic fungi. Biocontrol Sci Tech 20:117–136. https://doi.org/10.1080/09583150903431665

Lin CH, Yang SL, Chung KR (2009) The YAP1 homolog-mediated oxidative stress tolerance is crucial for pathogenicity of the fungus *Alternaria alternata*. Mol Plant-Microbe Interact 22 (8):942–952. https://doi.org/10.1094/MPMI-22-8-0942

Lind AL, Smith TD, Saterlee T, Calvo AM, Rokas A (2016) Regulation of secondary metabolism by the velvet complex is temperature-responsive in *Aspergillus*. G3 (Bethesda) 6 (12):4023–4033. https://doi.org/10.1534/g3.116.033084

Lo Presti L, López Díaz C, Turrà D, Di Pietro A, Hampel M, Heimel K, Kahmann R (2015) A conserved co-chaperone is required for virulence in fungal plant pathogens. New Phytol 209 (3):1135–1148. https://doi.org/10.1111/nph.13703

Lv X, Zhang W, Chen G, Liu W (2015) *Trichoderma reesei* Sch9 and Yak1 regulate vegetative growth, conidiation, and stress response and induced cellulase production. J Microbiol 53 (4):236–242. https://doi.org/10.1007/s12275-015-4639-x

Maeda T, Wurgler-Murphy SM, Saito H (1994) A two-component system that regulates an osmosensing MAP kinase cascade in yeast. Nature 369(6477):242–245

Medina A, Schmidt-Heydt M, Rodríguez A, Parra R, Geisen R, Magan N (2015) Impacts of environmental stress on growth, secondary metabolite biosynthetic gene clusters and metabolite production of xerotolerant/xerophilic fungi. Curr Genet 61:325–334. https://doi.org/10.1007/s00294-014-0455-9

Mehrabi R, Zwiers LH, De Waard MA, Kema GH (2006) MgHog1 regulates dimorphism and pathogenicity in the fungal wheat pathogen *Mycosphaerella graminicola*. Mol Plant-Microbe Interact 19(11):1262–1269

Mikus M, Hatvani L, Neuhof T, Komon-Zelazowska M, Dieckmann R, Schwecke T, Druzhinina IS, von Dohren H, Kubicek CP (2009) Differential regulation and posttranslational processing of the class II hydrophobin genes from the biocontrol fungus *Hypocrea atroviridis*. Appl Environ Microbiol 75:3222–3229

Miranda RU, Gómez-Quiro LE, Mejía A, Barrios-González J (2013) Oxidative state in idiophase links reactive oxygen species (ROS) and lovastatin biosynthesis: differences and similarities in submerged- and solid-state fermentations. Fungal Biol 117:85–93

Miranda RU, Gómez-Quiroz L e, Mendoza M, Pérez-Sánchez A, Fierro F, Barrios-González J (2014) Reactive oxygen species regulate lovastatin biosynthesis in *Aspergillus terreus* during submerged and solid-state fermentations. Fungal Biol 118:979–989

Miskei M, Karányi Z, Pócsi I (2009) Annotation of stress-response proteins in the *Aspergilli*. Fungal Genet Biol 46(Suppl 1):S105–S120. https://doi.org/10.1016/j.fgb.2008.07.013

Mitchell D, Parra R, Aldred D, Magan N (2004) Water and temperature relations of growth and ochratoxin A production by *Aspergillus carbonarius* strains from grapes in Europe and Israel. J Appl Microbiol 97:439–445. https://doi.org/10.1111/j.1365-2672.2004.02321.x

Mizutani O, Nojima A, Yamamoto M, Furukawa K, Fujioka T, Yamagata Y, Abe K, Nakajima T (2004) Disordered cell integrity signaling caused by disruption of the *kexB* gene in *Aspergillus oryzae*. Eukaryot Cell 3(4):1036–1048

Molina M, Kahmann R (2007) An *Unsilago maydis* gene involved in H_2O_2 detoxification is required for virulence. Plant Cell 19(7):2293–2309

Montero-Barrientos M, Hermosa R, Cardoza RE, Gutiérrez S, Nicolás C, Monte E (2010) Transgenic expression of the *Trichoderma harzianum hsp70* gene increases Arabidopsis resistance to heat and other abiotic stresses. J Plant Physiol 167(8):659–665. https://doi.org/10.1016/j.jplph.2009.11.012

Morano K, Thiele DJ (1999) The Sch9 protein kinase regulates Hsp90 chaperone complex signal transduction activity *in vivo*. EMBO J 18(21):5953–5962

Moriwaki A, Kubo E, Arase S, Kihara J (2006) Disruption of SRM1, a mitogen-activated protein kinase, affects sensitivity to osmotic and ultraviolet stressors in the phytopathogenic fungus *Bipolaris oryzae*. FEMS Microbiol Lett 257(2):253–261

Mukherjee PK, Latha J, Hadar R, Horwitz BA (2003) TmkA, a mitogen-activated protein kinase of *Trichoderma virens*, is involved in biocontrol properties and repression of conidiation in the dark. Eukaryot Cell 2(3):446–455

Mukherjee PK, Horwitz BA, Singh US, Mukherjee M, Schmoll M (2013) *Trichoderma*: biology and applications. https://doi.org/10.1079/9781780642475.0000

Niu J, Arentshorst M, Nair PD, Dai Z, Baker SE, Frisvad JC, Nielsen KF, Punt PJ, Ram AF (2015) Identification of a classical mutant in the industrial host *Aspergillus niger* by systems genetics: LaeA is required for citric acid production and regulates the formation of some secondary metabolites. G3 (Bethesda) 6(1):193–204. https://doi.org/10.1534/g3.115.024067

Ochiai N, Tokai T, Nishiuchi T, Takahashi-Ando N, Fujimura M, Kimura M (2007) Involvement of the osmosensor histidine kinase and osmotic stress-activated protein kinases in the regulation of secondary metabolism in *Fusarium graminearum*. Biochem Biophys Res Commun 363 (3):639–644

Olmedo M, Navarro-Sampedro L, Ruger-Herreros C, Kim SR, Jeong BK, Lee BU, Corrochano LM (2010) A role in the regulation of transcription by light for RCO-1 and RCM-1, the *Neurospora* homologs of the yeast Tup1-Ssn6 repressor. Fungal Genet Biol 47:939–952. https://doi.org/10.1016/j.fgb.2010.08.001

Pakula TM, Nygren H, Barth D, Heinonen M, Castillo S, Penttilä M, Arvas M (2016) Genome wide analysis of protein production load in *Trichoderma reesei*. Biotechnol Biofuels 9:132. https://doi.org/10.1186/s13068-016-0547-5

Park HS, Yu JH (2016) Velvet regulators in *Aspergillus* spp. Korean J Microbiol Biotechnol 44 (4):409. https://doi.org/10.4014/mbl.1607.07007

Pascual-Ahuir A, Proft M (2007) The Sch9 kinase is a chromatin-associated transcriptional activator of osmostress-responsive genes. EMBO J 26:3098–3108

Peterbauer CK, Litscher D, Kubicek CP (2002) The *Trichoderma atroviride seb1* (stress response element binding) gene encodes an AGGGG-binding protein which is involved in the response to high osmolarity stress. Mol Gen Genomics 268(2):223–231

Proft M, Struhl K (2002) Hog1 kinase converts the Sko1-Cyc8-Tup1 repressor complex into an activator that recruits SAGA and SWI/SNF in response to osmotic stress. Mol Cell 9:1307–1317

Purschwitz J, Muller S, Kastner C, Schoser M, Haas H, Espeso EA, Atoui A, Calvo AM, Fischer R (2008) Functional and physical interaction of blue- and red-light sensors in *Aspergillus nidulans*. Curr Biol 18(4):255–259

Roelants FM, Torrance PD, Bezman N, Thorner J (2002) Pkh1 and Pkh2 differentially phosphorylate and activate Ypk1 and Ykr2 and define protein kinase modules required for maintenance of cell wall integrity. Mol Biol Cell 13:3005–3028

Saito H, Posas F (2012) Response to hyperosmotic stress. Genetics 192(2):289–318. https://doi.org/10.1534/genetics.112.140863

Salmerón-Santiago KG, Pardo JP, Flores-Herrera O, Mendoza-Hernández G, Miranda-Arango M, Guerra-Sánchez G (2011) Response to osmotic stress and temperature of the fungus *Ustilago maydis*. Arch Microbiol 193(10):701–709. https://doi.org/10.1007/s00203-011-0706-9

Sarikaya Bayram Ö, Bayram Ö, Valerius O, Park HS, Irniger S, Gerke J, Ni M, Han KH, Yu JH, Braus GH (2010) LaeA control of *velvet* family regulatory proteins for light-dependent development and fungal cell-type specificity. PLoS Genet 6(12):e1001226

Sarikaya-Bayram Ö, Bayram Ö, Feussner L, Kim JH, Kim HS, Kaever A, Feussner I, Chae KS, Han DM, Han KH, Braus GH (2014) The membrane-bound VapA-VipC-VapB methyltransferase complex guides signal transduction for epigenetic and transcriptional control of fungal development. Dev Cell 29(4):406–420

Schmoll M, Franchi L, Kubicek CP (2005) Envoy, a PAS/LOV domain protein of *Hypocrea jecorina* (Anamorph *Trichoderma reesei*), modulates cellulase gene transcription in response to light. Eukaryot Cell 4:1998–2007

Schmoll M, Schuster A, do Nascimento Silva R, Kubicek CP (2009) The G-alpha protein GNA3 of *Hypocrea jecorina* (anamorph *Trichoderma reesei*) regulates cellulase gene expression in the presence of light. Eukaryot Cell 8:410–420

Schmoll M, Esquivel-Naranjo EU, Herrera-Estrella A (2010) *Trichoderma* in the light of day – physiology and development. Fungal Genet Biol 47(11–2):909–916. https://doi.org/10.1016/j.fgb. 2010.04.010

Schuster A, Tisch D, Seidl-Seiboth V, Kubicek CP, Schmoll M (2012) Roles of protein kinase A and adenylate cyclase in light-modulated cellulase regulation in *Trichoderma reesei*. Appl Environ Microbiol 78(7):2168–2178. https://doi.org/10.1128/AEM.06959-11

Segmüller N, Ellendorf U, Tudzynski B, Tudzynski P (2007) BcSAK1, a stress-activated mitogen-activated protein kinase, is involved in vegetative differentiation and pathogenicity in *Botrytis cinerea*. Eukaryot Cell 6(2):211–221

Segorbe D, Di Pietro A, Pérez-Nadales E, Turrá D (2017) Three *Fusarium oxysporum* mitogen-activated protein kinases (MAPKs) have distinct and complementary roles in stress adaptation and cross-kingdom pathogenicity. Mol Plant Pathol 18(7):912–924. https://doi.org/10.1111/mpp.12446

Seibel C, Gremel G, Silva RD, Schuster A, Kubicek CP, Schmoll M (2009) Light dependent roles of the G-protein subunit GNA1 of *Hypocrea jecorina* (anamorph *Trichoderma reesei*). BMC Biol 7:58

Shlezinger N, Iremer H, Dhingra S, Beattie SR, Cramer RA, Braus GH, Sharon A, Hohl TM (2017) Sterilizing immunity in the lung relies on targeting fungal apoptosis-like programmed cell death. Science 37(6355):1037–1041. https://doi.org/10.1126/science.aan0365

Shwab EK, Bok JW, Tribus M, Galher J, Graessle S, Keller NP (2007) Histone deacetylase activity regulates chemical diversity in *Aspergillus*. Eukaryot Cell 6(9):1656–1664

Singh BN, Singh A, Singh SP, Singh HB (2011a) *Trichoderma harzianum*-mediated reprogramming of oxidative stress response in root apoplast sunflower enhances defense against *Rhizoctonia solani*. Eur J Plant Pathol 131:121–134. https://doi.org/10.1007/s10658-011-9792-4

Singh LP, Gill SS, Tuteja N (2011b) Unraveling the role of fungal symbionts in plant abiotic stress tolerance. Plant Signal Behav 6(2):175–191

Singh R, Singh S, Parihar P, Mishra RK, Tripathi DK, Singh VP, Chauhan DK, Prasad SM (2016) Reactive oxygen species (ROS): beneficial companions of plant's developmental processes. Front Plant Sci 7:1299. https://doi.org/10.3389/fpls.2016.01299

Smith RL, Johnson AD (2000) Turning genes off by Ssn6-Tup1: a conserved system of transcriptional repression in eukaryotes. Trends Biochem Sci 25:325–330

Tang J, Liu L, Huang X, Li Y, Chen Y, Chen J (2010) Proteomic analysis of *Trichoderma atroviride* stressed by organophosphate pesticide dichlovors. Can J Microbiol 56(2):121–127. https://doi.org/10.1139/W09-110

Tian L, Wang Y, Yu J, Xiong D, Zhao H, Tian C (2016) The mitogen-activated protein kinase Vdpbs2 of *Verticillium dahliae* regulates microsclerotia formation, stress response, and plant infection. Front Microbiol 7:1532

Tisch D, Schmoll M (2009) Light regulation of metabolic pathways in fungi. Appl Microbiol Biotechnol 85:1259–1277

Todd RB, Hynes MJ, Andrianopoulos A (2006) The *Aspergillus nidulans rcoA* gene is required for *veA*-dependent sexual development. Genetics 174(3):1685–1688

Toone WM, Morgan BA, Jones N (2001) Redox control of AP-1-like factors in yeast and beyond. Oncogene 20(19):2336–2346

Vega FE, Goettel MS, Blackwell M, Chandler D, Jackson MA, Keller S, Koike M, Maniania NK, Monzon A, Ownley BH, Pell JK, Rangel DEN, Roy HE (2009) Fungal entomopathogens: new insights on their ecology. Fungal Ecol 2:149–159

Viefhues A, Schlathoelter I, Simon A, Viaud M, Tudzynski P (2015) Unraveling the function of the response regulator BcSkn7 in the stress signaling network of *Botrytis cinerea*. Eukaryot Cell 14 (7):636–651. https://doi.org/10.1128/EC.00043-15

Viterbo A, Harel M, Horwitz BA, Chet I, Mukherjee PK (2005) *Trichoderma* mitogen-activated protein kinase in involved in induction of plant systemic resistance. Appl Environ Microbiol 71 (10):6241–6245

Wang Y, Tian L, Xiong D, Klosterman SJ, Xiao S, Tian C (2016) The mitogen-activated protein kinase gene, VdHopg1, regulates osmotic stress response, microsclerotia formation and virulence in *Verticillium dahliae*. Fungal Genet Biol 88:13–23. https://doi.org/10.1016/j.fgb.2016.01.011

Wei H, Requena N, Fischer R (2003) The MAPKK kinase SteC regulates conidiophore morphology and is essential for heterokaryon formation and sexual development in the homothallic fungus *Aspergillus nidulans*. Mol Microbiol 47(6):1577–1588

Wiemann P, Brown DW, Kleigrewe K, Bok JW, Keller NP, Humpf HU, Tudzynski B (2010) FfVel1 and FfLae1, components of a *velvet*-like complex in *Fusarium fujikuroi* affect differentiation, secondary metabolism and virulence. Mol Microbiol 77(4):972–994

Wu D, Oide S, Choi MY, Turgeon BG (2012) ChLae1 and ChVel1 regulate T-toxin production, virulence, oxidative stress response, and development of the maize pathogen *Cochliobolus heterostrophus*. PLoS Pathog 8(2):e1002542

Wurgler-Murphy SM, Saito H (1997) Two-component signal transducers and MAPK cascades. Trends Biochem Sci 22(5):172–176

Yoshimi A, Kojima K, Takano Y, Tanaka C (2005) Group III histidine kinase is a positive regulator of Hog1-type mitogen-activated protein kinase in filamentous fungi. Eukaryot Cell 4(11):1820–1828. https://doi.org/10.1128/EC.4.11.1820-1828.2005

Yu C, Fan L, Wu Q, Fu K, Gao S, Wang M, Gao J, Li Y, Chen J (2014) Biological role of *Trichoderma harzianum*-derived platelet-activating factor acetylhydrolase (PAF-AH) on stress response and antagonism. PLoS One 9(6):e100367. https://doi.org/10.1371/journal.pone.0100367

Yu PL, Chen LH, Chung KR (2016) How the pathogenic fungus *Alternaria alternata* copes with stress via the response regulators SSK1 and SHO1. PLoS One 11(2):e0149153. https://doi.org/10.1371/journal.pone.0149153

Zhang Y, Zhao J, Fang W, Zhang J, Luo Z, Zhang M, Fan Y, Pei Y (2009) Mitogen-activated protein kinase *hog1* in the entomopathogenic fungus *Beauveria bassiana* regulates environmental stress responses and virulence to insects. Appl Environ Mirobiol 75(11):3787–3795. https://doi.org/10.1128/AEM.01913-08

Zheng Z, Gao T, Zhang Y, Hou Y, Wang J, Zhou M (2014) FgFim, a key protein regulating resistance to the fungicide JS399-19, asexual and sexual development, stress responses and virulence in *Fusarium graminearum*. Mol Plant Pathol 15(5):488–499. https://doi.org/10.1111/mpp.12108

Zurita-Martinez SA, Cardenas ME (2005) Tor and cyclic AMP-protein kinase A: two parallel pathways regulating expression of genes required for cell growth. Eukaryot Cell 4(1):63–71

In the Crossroad Between Drug Resistance and Virulence in Fungal Pathogens

7

Mafalda Cavalheiro and Miguel Cacho Teixeira

Abstract

Fungal species from the *Cryptococcus*, *Candida*, and *Aspergillus* genera are recognized as important human pathogens, causing infections which are often deadly and extremely difficult to treat due to their ability to exhibit and develop antifungal resistance and virulence. Since current therapeutics are not capable of defeating these infections efficiently, there's still need to find better therapeutic strategies. Interesting targets include the mechanisms underlying antifungal resistance, virulence, or, ideally, both, as these are the key traits that allow these pathogenic fungi to persist in the human host.

In this chapter, the mechanisms that are in the crossroad between antifungal resistance and virulence are reviewed. These mechanisms include components of the cell wall and plasma membrane, such as the β-subunit of membrane lipid flippase in *Cryptococcus neoformans*, and major transcriptional regulators, such as PDR1 in *Candida glabrata*, responsible for the regulation of antifungal resistance but also for some virulence features. Also, specific intracellular pathways have been shown to have a role in both features, such as the calcium–calcineurin pathway or the unfolded protein response (UPR) pathway, which have multiple implications in resistance and virulence. Highlighting the importance of the discovery of new drugs that target pathways in the crossroad between drug resistance and virulence, some recently proposed therapeutic approaches are also analyzed.

M. Cavalheiro · M. C. Teixeira (✉)
Department of Bioengineering, Instituto Superior Técnico, University of Lisbon, Lisbon, Portugal

iBB – Institute for Bioengineering and Biosciences, Biological Sciences Research Group, Lisbon, Portugal
e-mail: mafalda.cavalheiro@tecnico.ulisboa.pt; mnpct@tecnico.ulisboa.pt

© Springer Nature Switzerland AG 2018
M. Skoneczny (ed.), *Stress Response Mechanisms in Fungi*,
https://doi.org/10.1007/978-3-030-00683-9_7

7.1   Introduction

Fungal infections are a worldwide problem that has a significant impact on human health. The major causative agents behind these infections include species of the *Cryptococcus, Candida, Aspergillus*, and *Pneumocystis* genera. These pathogenic species are able to cause superficial infections, such as skin and nail infections that affect 25% of the world population or vulvovaginal candidiasis that targets 50–75% of women in childbearing. Once the mucosal barrier is overcome, fungi can lead to invasive infections that annually result in 1.5 million deaths. Immunocompromised patients are the ones mostly affected, with an increase of incidence of invasive infections due to HIV/AIDS and immunosuppressive therapeutics (Brown et al. 2012). It is clear that existing diagnostic tools or treatment choices are insufficient.

Infections caused by *Cryptococcus* species are caused by two fungal pathogens: *Cryptococcus neoformans* and *Cryptococcus gattii. C. neoformans* comprises three varieties: serotype A *C. neoformans* var. *grubii*, serotype D *C. neoformans* var. *neoformans*, and serotype AD (Mitchell and Perfect 1995), the first being the most predominant in infected patients (Chen et al. 2014b). *C. neoformans* has the ability to cause fungal pneumonia and meningitis in immunocompromised patients (Srikanta et al. 2014). Human exposure to *C. neoformans* may occur from inhalation of aerosolized basidiospores, due to the presence of birds or avian excreta, leading to alveolar deposition (Lugarini et al. 2008; Schmalzle et al. 2016). Usually, the infection is primarily pulmonary, persisting for long periods of time and then spreading to the central nervous system (Hole and Wormley 2016). Cryptococcal meningitis occurs with high frequency globally (approximately 1 million cases yearly), leading to an estimated premature death of 624,700 people every year (Park et al. 2009). The incidence of this disease in adults is highly associated with HIV/AIDS (Williamson et al. 2016). *Cryptococcus gattii* has arisen as an important fungal pathogen, especially after an outbreak registered in the Pacific Northwest region in 1999 (Byrnes et al. 2010). *C. gattii* can be found in soil, trees, and tree hollows and acts as an opportunistic pathogen but also as a primary pathogen, with a predilection for causing lung infections. When compared to *C. neoformans*, the number of infections is relatively rare although many cases are suspected to be misdiagnosed (Bielska and May 2015). *C. gattii* has four molecular types: VGI, VGII, VGIII, and VGIV. The first two are commonly found in infections upon non-immunocompromised individuals, while the last are more frequently found in immunocompromised patients (Byrnes et al. 2011).

In turn, *Candida* species are known to belong to the human microflora, with 25–40% of the healthy population being colonized by *Candida albicans* (Polvi et al. 2015), without manifestation of disease. Nevertheless, upon failure of the host immune system, these species have the ability to become pathogenic, leading to superficial infections, like oropharyngeal and vulvovaginal candidiasis, or systemic infections (Polke et al. 2015). In the United States, these species are the fourth most frequent cause of nosocomial bloodstream infections, which are associated with a high mortality rate (~50%) (Wenzel and Gennings 2005; Geffers and Gastmeier 2011). In 2007, it was estimated that $2–4 billion were spent per year in the United States, associated with healthcare costs to treat hematogenously disseminated candidiasis (Perlroth et al. 2007).

Candida albicans is the most common *Candida* species found in immunocompromised patients with 50% of incidence in invasive infections (Perlroth et al. 2007), leading to 400,000 deaths per year (Polvi et al. 2015). *Candida glabrata* is the second cause of *Candida* infections being responsible for 15–25% of the cases (Perlroth et al. 2007; Tscherner et al. 2011). *Candida parapsilosis* and *Candida tropicalis* follow behind *C. glabrata* (Wisplinghoff et al. 2004), although the distributions of incidence are found to change according to geographic epidemiology (Perlroth et al. 2007). *C. parapsilosis* has an increased potential to become virulent and can easily be found in the hands of healthy hosts (Kuhn et al. 2004; Bonassoli et al. 2005). On the other hand, *C. tropicalis* is usually associated with infections in the urinary tract (Rho et al. 2004). Generally, the first steps of infection start in the human mucosa where different *Candida* virulence factors arise to allow epithelium colonization, followed by dissemination of these species in the host. These virulence features vary within *Candida* species but generally involve the capacity to adapt to different host niches, the development of dimorphic growth, and the capacity to undergo phenotypic switching, among other mechanisms (Polke et al. 2015).

Aspergillus species are responsible for a wide range of infections like invasive aspergillosis (IA), allergic bronchopulmonary aspergillosis (ABPA), and chronic pulmonary aspergillosis (CPA). IA occurs in individuals with compromised immunity, while ABPA and CPA are usually developed in patients suffering from chronic pulmonary diseases (Denning et al. 2013). Among the 30 species of *Aspergillus* known to cause human infections, the most frequent is *Aspergillus fumigatus* (Lelièvre et al. 2013), an airborne pathogen, inhaled by humans everyday (Latgé 1999). *Aspergillus fumigatus* natural habitat appears to be the soil, although it easily reaches the lungs of immunocompromised patients, in the form of airborne conidia. In the lungs, penetration of the epithelium or internalization of the conidia by epithelial cells has been reported, constituting the primary site of infection (Latgé 1999). This pathogen is a causative agent of IA, which is associated to a mortality rate of 90% in the case of late or misdiagnosed infections. Even in the case of timely diagnosis, among the 200,000 cases of IA registered per year, approximately 100,000 result in the death of the patient (Chauhan et al. 2006; Brown et al. 2012; Brown and Goldman 2016).

Increasing concern over fungal pathogenesis has emerged due to the absence of efficient therapies to deal with these pathogens. The increased incidence and mortality rates associated with invasive fungal infections highlight the need to better understand the virulence and antifungal drug resistance mechanisms displayed by fungal pathogens. This chapter reviews current knowledge on the mechanisms that have been found to lay on the crossroad between antifungal drug resistance and virulence (Table 7.1), pointed out as the most promising targets for the development of new drugs and therapeutic options. With this scope in mind, current trends in antifungal drug design are also analyzed.

Table 7.1 Key players of *Aspergillus fumigatus, Cryptococcus neoformans, Cryptococcus gattii, Candida albicans, Candida glabrata,* and *Candida tropicalis* signaling pathways involved in antifungal resistance and/or virulence

Signaling pathways involved in antifungal resistance and virulence		Key players of each species					
		Aspergillus fumigatus	*Cryptococcus neoformans*	*Cryptococcus gattii*	*Candida albicans*	*Candida glabrata*	*Candida tropicalis*
	The calcium–calcineurin pathway	Calcineurin and CrzA	Calcineurin and Cch1	Calcineurin	Calcineurin, Cch1 and Mid1	Calcineurin, Crz1, Rcn1 and Rcn2	Calcineurin and Crz1
	Hsp90 chaperone	Hsp90	Hsp90	Hsp90	Hsp90	–	–
	HOG pathway	PtcB	Ssk1, Skn7, Toc1, Toc2. Ena1 and Nha1	–	Hog1, Ssk1	–	–
	Osmotic and pH stress responses	PacC	–	–	Gup1	–	–
	Cell wall integrity signaling pathway	Ecm33, RodA, RodB, MpkA, SrbA and SrbB	Mpk1	–	Mkc1, Cek1, Cst20, Hst7 and Cph1	–	–
	UPR	HacA and IreA	Ire1 and Hxl1	–	Hac1	Ire1	–

7.2 The Mechanisms Behind Coupled Drug Resistance and Virulence

7.2.1 Previously Known Drug Resistance Determinants That Have a Role in Virulence

Drug resistance can be developed through different mechanisms that have been well characterized for most fungal pathogens. *Cryptococcus neoformans* azole resistance is known to be developed through mutations in the *ERG11* gene, encoding the target of azole drugs, or by its overexpression due to chromosomic duplications. Additional mechanisms of azole resistance in *C. neoformans* include the overexpression of the drug efflux transporter Afr1 and the activation of stress response pathways and oxygen-sensing pathways (Xie et al. 2014). In turn, *Candida* species display similar mechanisms of antifungal resistance that include the upregulation or modification of the drug target, Erg11, the induction of stress response pathways, but mostly the increased expression of drug transporters from the ATP-binding cassette superfamily (ABC) (Sanglard et al. 1995, 1997, 1999, 2001) and major facilitator superfamily (MFS) (Costa et al. 2013a, b, 2014, 2016; Pais et al. 2016a, b; Sanglard et al. 1995) regulated by transcription factors, such as Tac1 and Mrr1 in *C. albicans* and Pdr1 in *C. glabrata* (Perlin et al. 2017). Although not so deeply studied, similar azole resistance mechanisms have been described in *Aspergillus fumigatus*, including mutations in the *cyp51A* gene, encoding the target of azole drugs, and its overexpression, as well as the overexpression of AtfR, Mdr3, and Mdr4 drug efflux transporters (Xie et al. 2014).

Interestingly, increasing evidence points out for a new role of the referred multidrug transporters as virulence determinants. In *Candida albicans*, overexpression of the Cdr1 transporter was observed upon interaction with blood, neutrophiles, or rabbit kidney tissues (Fradin et al. 2005; Walker et al. 2009). On the other hand, the absence of a multidrug resistance (MDR) MFS transporter, Mdr1, results in *C. albicans* attenuated virulence in immunocompetent or immunocompromised mice (Becker et al. 1995). The mechanism by which these transporters affect *C. albicans* virulence has not yet been described. Other interesting transporters in this context are the QDR transporters, which are predicted Drug:H^+ antiporters. Although having no apparent effect in drug resistance, *C. albicans* $\Delta qdr3$ deletion mutant displays significant defects in biofilm formation, while $\Delta qdr2$ and $\Delta qdr3$ deletion mutants exhibit decreased virulence in immunocompetent BALB/c mice. These QDR transporters were suggested to participate in *C. albicans* virulence through their action over lipid homeostasis (Shah et al. 2014). The absence of three other genes, *NAG3*, *NAG4*, and *NAG6*, known to confer resistance to cycloheximide has been linked to attenuated virulence of *C. albicans*. Although the function of these genes is still unclear, it is known that these proteins help in the uptake of extracellular hexoses, but no virulence mechanisms have been described (Yamada-Okabe and Yamada-Okabe 2002).

Moreover, the MFS-MDR transporters Tpo1_1, Tpo1_2, and Dtr1 of *C. glabrata* have also been shown to contribute to virulence in this species. Although known to enhance resistance to azoles, amphotericin B, and flucytosine (Pais et al. 2016a), in the case of CgTpo1_1 and CgTpo1_2, or against weak acids, in the case of Dtr1, these transporters were shown to be necessary for full virulence against the *Galleria mellonella* model of infection. In terms of the underlying mechanism, Tpo1_1 was found to confer resistance toward antimicrobial peptides, such as the human histatin-5, while Tpo1_2 was found to be necessary for the expression of adhesins and biofilm formation, as well as in the survival to phagocytosis (Santos et al. 2017). Dtr1 was recently found to also enhance the ability to proliferate inside *G. mellonella* hemocytes by conferring resistance to oxidative and acidic stresses experienced in phagolysosomes (Romão et al. 2017).

In *Cryptococcus neoformans*, the ABC transporter Afr1 was also found to be important for virulence. *AFR1* overexpression enhances *C. neoformans* virulence in a murine model of infection, increasing its capacity to survive inside macrophages (Sanguinetti et al. 2006). Orsi and colleagues (2009) have shown that *AFR1* overexpression leads to increased resistance to the anticryptococcal activity of microglia, by reducing the acidification and maturation of phagosomes, while not influencing the phagocytosis process itself (Orsi et al. 2009).

In the case of *Aspergillus fumigatus*, ABC transporters with high homology with *Saccharomyces cerevisiae* Pdr5, AbcA, and AbcB, necessary for azole resistance, were shown to have a role in cell proliferation in infected *G. mellonella* larvae. AbcA was shown to enhance azole resistance during infection, while AbcB was discovered to be required for normal virulence of *A. fumigatus* (Paul et al. 2013). *AbcB* gene is regulated by the AtrR transcription factor, which is known for its importance in azole resistance, controlling *cyp51A*. This regulator was also found to be necessary for *A. fumigatus* virulence in a murine infection model of aspergillosis (Hagiwara et al. 2017). Although clearly having a participation in *A. fumigatus* virulence, further studies are required to identify the mechanism behind the contribution of AbcB and AtrR to virulence.

Similarly, *C. glabrata* transcription factor Pdr1, a multidrug resistance regulator, whose activity leads to increased expression of drug efflux transporters like Cdr1, Cdr2, and Qdr2 (Vermitsky and Edlind 2004; Vermitsky et al. 2006; Ferrari et al. 2009; Costa et al. 2013b), was shown to be a virulence factor. Ferrari et al. (2009) have shown the importance of gain-of-function (GOF) mutations in the *PDR1* gene that occur upon clinical exposure to azole drugs, leading to increased fluconazole resistance in mouse models and failure of treatment with azoles. Interestingly, the presence of GOF mutations also enhances *C. glabrata* virulence, with an increase of tissue colonization in the kidney, spleen, and liver of infected mice (Ferrari et al. 2009). According to this observation, Pdr1 must be responsible for the regulation of genes involved in *C. glabrata* virulence. In order to find such genes, Ferrari et al. (2011a) searched for the common genes differentially expressed upon seven differ-ent GOF mutations in *PDR1* gene. From the total of genes differentially expressed upon each mutation, only two were common to all transcriptomic profiles obtained: *CDR1* and *PUP1* genes. Significantly, *C. glabrata* multidrug transporter Cdr1 was also found to be a virulence factor, similarly to what had been observed for its homologue in *C. albicans*. The mitochondrial protein Pup1 was, additionally, found to contribute to *C. glabrata* virulence, although its precise role remains to be established (Ferrari et al. 2011b). Looking further on to find the processes that Pdr1 regulates upon GOF mutations in *C. glabrata* virulence, Vale-Silva et al. (2013) have studied the interaction of strains harboring different *PDR1* GOF mutations with bone marrow-derived macrophages (BMDMs), human acute monocytic leukemia cell line (THP-1)-derived macrophages, and diverse epithelial cell lines. Their in vitro interaction has shown that Pdr1 GOF mutations lead to a higher capacity of *C. glabrata* strains to evade phagocytosis by macrophages. Interestingly, adherence to the epithelial cell lines tested (CHO-Lec2, HeLa, and Caco-2) was found to increase in all strains with *PDR1* GOF mutations, when compared to the parental strain (Vale-Silva et al. 2013), possibly through the Pdr1-dependent upregulation of the *EPA1* gene, encoding an important adhesin (Vale-Silva et al. 2016).

The *C. neoformans* APSES transcription factor Mbs1 has also been identified as a determinant of antifungal drug resistance, namely, against flucytosine and amphotericin B, but also as player in virulence. Mbs1 is activated in *C. neoformans* in the presence of flucytosine, and regulates the biosynthesis of ergosterol, being necessary for amphotericin B resistance, but not for azole resistance, since it leads to decreased levels of ergosterol on the cell membrane. Moreover, Mbs1 is responsible for the regulation of membrane stability and oxidative, osmotic, and genotoxic stress response. This transcription factor is also involved in the activation of virulence factors. The absence of *MBS1* gene leads to a decreased production of melanin and capsule at 37 °C, when compared to the wild-type. This role in virulence was confirmed in a murine model of systemic cryptococcosis, where the $\Delta mbs1$ deletion mutant exhibited attenuated virulence (Song et al. 2012).

Alterations in the ergosterol biosynthesis pathway are known to allow the development of azole and polyene resistance. One of these mechanisms is the loss of function of Erg3, a sterol $\Delta^{5,6}$-desaturase in *Candida* species. Upon exposure to azole drugs, *ERG11* inhibition leads to the accumulation of 14α-methylergosta-8,24-dien-3β,6α-diol, a toxic compound that is not produced when Erg3 has suffered alterations that change the products of the ergosterol pathway, leading to the accumulation of 14α-methylfecosterol, which is not harmful to the cell. Two *Candida albicans* isolates, CAAL2 and CAAL76, recovered from two patients after fluconazole prophylaxis, exhibited alterations in *ERG3* gene sequence (an amino acid substitution L193R and a deletion of 13 bp of the gene $\Delta366$–378, respectively). Fluconazole, voriconazole, and amphotericin B resistance was observed in both isolates, but more interestingly, the loss of function of Erg3 sterol $\Delta^{5,6}$-desaturase led to a decrease in virulence, with reduced hyphal growth, in a murine model of invasive candidiasis (Miyazaki et al. 2006; Morio et al. 2012), proving that changes in the ergosterol pathway influence *C. albicans* virulence.

Another drug resistance mechanism in fungal pathogens is the loss of mitochondrial function, which leads to several alterations in the fungal cell (Shingu-Vazquez and Traven 2011). Uncoupled oxidative phosphorylation in *C. albicans* petite mutants leads to resistance against fluconazole and voriconazole, due to the upregulation of the Mdr1 azole drug transporter (Cheng et al. 2007), while

C. glabrata cells with dysfunctional mitochondria (Brun et al. 2003) exhibit upregulation of *CgCDR1*, *CgCDR2*, and/or *CgSNQ2* genes and thus increased resistance to azole drugs (Brun et al. 2004; Ferrari et al. 2011a). Significantly, mitochondrial dysfunction leads to changes in the virulence of pathogenic fungi. In *C. albicans*, mitochondrial inactivation due to loss of Goa1 mitochondrial protein results in attenuated virulence in a murine model of candidiasis and increased killing by neutrophils. Goa1 seems to be necessary for morphogenesis and chlamydospore formation (Bambach et al. 2009). Likewise, *C. glabrata* petite mutants have shown decreased virulence in immunocompromised mice, comparatively to the parental strain (Brun et al. 2005). However, another study has shown the exact opposite where a petite mutant has led to increased mortality and tissue burden in a systemic and vaginal murine model, even though this strain exhibited growth deficiency in vitro. A microarray analysis of the transcriptomic profile of in vitro grown petite cells, compared to that of the parental strain, has shown altered expression of genes involved in oxidoreductive metabolism, stress response, and cell wall remodeling processes, accompanying the petite phenotype. The observed cell wall remodeling is believed to be behind the enhanced virulence exhibited by this petite mutant strain (Ferrari et al. 2011a).

Interestingly, mitochondria loss of function also influences virulence in *Cryptococcus* and *Aspergillus* species. In *C. neoformans*, the mitochondrial superoxide dismutase was found to be necessary for virulence in intranasal and intravenously infected mice, contributing for the response to oxidative and osmotic stresses, as well as for growth at 37 °C (Narasipura et al. 2005). Also for *C. gattii*, virulence requires the full function of mitochondria. A transcriptomic profiling of 24 *C. gattii* strains recovered from within J774 macrophages revealed the upregulation of several nuclear-encoded proteins, which have a function in the mitochondria. Microscopic visualization of these strains after growth within macrophages showed a tubular morphology of the mitochondria, believed to result from mitochondrial fusion known to protect against mtDNA mutations (Ma et al. 2009). In turn, *A. fumigatus* expresses an alternative oxidase, AOX, a component of the respiratory chain, whose role in virulence appears to be involved in the protection of *A. fumigatus* cells against the oxidative stress present inside macrophages (Magnani et al. 2008). These studies highlight the different roles mitochondria might have in fungal pathogenesis and the necessity to better understand these mechanisms for a correct evaluation of isolates and their clinical relevance, as well as for the study of new therapeutic mechanisms that might have mitochondria as the primary target.

A very particular *C. neoformans* mechanism for innate resistance is based on the β-subunit of a membrane lipid flippase. Huang et al. (2016) found that the absence of this subunit results in alteration on the cellular membrane allowing the penetration of caspofungin in the cell, leading to caspofungin susceptibility. This alteration was found to enhance susceptibility not only to this echinocandin but also to fluconazole. Besides having an effect in antifungal resistance, the lipid flippase β-subunit, encoded by the *CDC50* gene, is necessary for virulence in a murine model of cryptococcosis, having a role in growth at 37 °C, and resistance to macrophage killing. The authors of this study have highlighted the fact that this subunit is overexpressed during the interaction with macrophages, regulating the phosphatidylserine content, a surface molecule believed to work as a signal for phagocytosis by macrophages, a phenomenon that might explain its importance in virulence (Huang et al. 2016).

A work focused on *Candida glabrata* isolates has highlighted that changes imposed by drug exposure leading to antifungal resistance also change the capacity of these isolates to evade the innate immune system. *C. glabrata* isolates were found to activate more macrophage responses when sub inhibitory concentrations of micafungin and caspofungin were used (Beyda et al. 2015). This effect has also been observed in *C. albicans* and *A. fumigatus* upon treatment with caspofungin, which is responsible for the unmasking of β-1,3-glucans of the cell wall of the pathogen, allowing a more efficient innate immune response of the host (Hohl et al. 2008; Wheeler et al. 2008). These studies demonstrate that acquisition of antifungal resistance is able to indirectly influence virulence in pathogenic fungi.

7.2.2 Previously Known Virulence Factors That Contribute to Drug Resistance

Once a fungal pathogen initiates the first steps of infection in the human host, a series of alterations allow an adaptive cellular response to the environment perceived by the pathogen. The pathogen usually activates different virulence factors that allow its progression and dissemination throughout the host. Generally, different phenotypes are induced allowing its survival and host immune system evasion. On the other hand, the host itself is faced with mucosal damage, toxins, and induction of inflammatory processes due to the presence of these unwelcome microorganisms (Calderone and Fonzi 2001; Alspaugh 2014; Netea et al. 2015; Ghazaei 2017). Some of the alterations initially known to increase virulence in *Cryptococcus* and *Candida* species have been described to also have a role in antifungal resistance (see Fig. 7.1).

Cryptococcus neoformans virulence factors have been described extensively, with special attention to capsule development, formation of extracellular vesicles, and changes in cell wall components, such as chitin and chitosan, melanin production, growth of titan cells, DNA repair, and others (Alspaugh 2014). Among all these, melanin production is considered a major advantage of *Cryptococcus* spp., enabling its survival in environmental and host niches. Besides conferring resistance to UV light, the presence of melanin in *C. neoformans* enables resistance to heavy metals and toxic compounds (Nosanchuk and Casadevall 2006), such as silver nitrate (García-Rivera and Casadevall 2001). Once inside the host, melanization of *C. neoformans* has been shown to have the power to modulate the inflammatory responses, with its presence leading to alterations in the cytokine content of the lungs, varying the concentration of IL-4, IL-12, and MCP-1. Moreover, melanized encapsulated cells of *C. neoformans* are capable to evade more efficiently in vivo phagocytosis then non-melanized encapsulated cells (Mednick et al. 2005). The presence of melanin in the cell wall of *C. neoformans* has been shown to confer

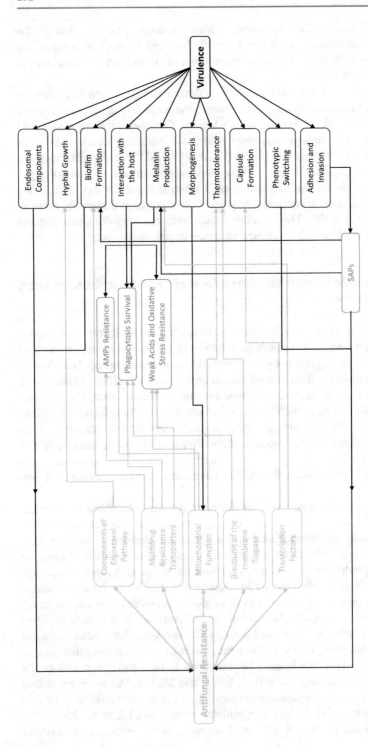

Fig. 7.1 Previously known antifungal resistance mechanisms found to be involved in virulence (yellow lines) and previously known virulence mechanisms found to be involved in antifungal resistance (black lines), according to the evidences found for *Aspergillus fumigatus*, *Cryptococcus gattii*, *Candida albicans*, *Candida glabrata*, and *Candida tropicalis*

resistance to oxygen- and nitrogen-derived radicals. Significantly, melanin has also shown to compromise the full activity of antifungal drugs. Amphotericin B and caspofungin preincubated with melanin were not so efficient against *C. neoformans*. In fact, in time-kill assays with CFU determination, non-melanized cells do not survive to amphotericin B and caspofungin while melanized cells do resist. This slight enhancement of resistance is believed to be related to the interaction of melanin with the amphotericin B or caspofungin molecules, reducing the free drug concentration. Nevertheless, the MIC values obtained for these two drugs and fluconazole, itraconazole, voriconazole, or flucytosine did not change between melanized or non-melanized cells (Duin et al. 2002, 2004; Ikeda et al. 2003). This difference in the action of melanin toward different drugs seems to be related to the molecular weight of the drug. Melanin was found to form five concentric layers on the cell wall, which have very small pores, leading to a less porous cell (Eisenman et al. 2005; Jacobson and Ikeda 2005). Due to the high molecular weight of amphotericin B and caspofungin, when compared to azoles and flucytosine, these big molecules might have their diffusion through the cell wall compromised, an hypothesis that could explain their reduced activity in melanized cells (Nosanchuk and Casadevall 2006).

Also present on fungal pathogen's cell walls are secreted aspartyl proteinases (SAPs) considered to be a virulence factor in *Candida* species, helping infection and colonization of the host. These hydrolytic enzymes are involved in the adhesion and invasion of the host mucosa by being able to degrade several human proteins, such as hemoglobin, keratin, and collagen (Ray and Payne 1990). Surprisingly, azole- and amphotericin B-resistant *C. albicans* strains have shown to exhibit increased extra-cellular production and activity of SAPs (Wu et al. 2000; Costa et al. 2010; Kumar and Shukla 2010). The same increase in activity of SAPs is also observed in biofilm cells upon exposure to sub-inhibitory concentrations of fluconazole (Mores et al. 2011). Although a clear relationship between SAPs production and antifungal resistance exists in *C. albicans* strains, more research is needed to understand how these proteinases contribute to antifungal drug resistance.

Moreover, controlling the endocytosis points in endosomal compartments, the Age3 ARF (ADP-ribosylation factor)-GAP (GTPase-activating protein) protein has also proven to be involved in the regulation of other processes in *C. albicans* cell wall. Loss of this functional ortholog of the *GCS1* gene in *S. cerevisiae* was found to lead to significant defects in filamentous growth due to the altered polarization pattern of actin cytoskeleton and absence of lipid raft polarization, linking this protein to its role in virulence. Interestingly, these alterations in the cell wall seem to have a side effect due to the increased susceptibility of the deletion mutant Δ*age3* to miconazole and itraconazole, even though Cdr1 and Cdr2 transporters are well localized on the membrane, in the same amount as in the parental strain (Lettner et al. 2010).

Another important virulence factor deployed by *Candida* species that also includes the cell wall is phenotypic switching. Although phenotypic switching is employed by these species to evade the innate immune system or to better adapt to certain niches in the host, in *Candida tropicalis* phenotypic switching has been recognized as an important virulence factor that contributes to changes in hemolytic and proteolytic activities, biofilm formation, and itraconazole resistance. In fact, Moralez and colleagues have described that phenotypic variants from the same parental strain exhibited differences in itraconazole resistance, being the rough variant the most resistant (Moralez et al. 2013).

7.2.3 Stress Resistance Pathways Contributing to Both Virulence and Drug Resistance in Pathogenic Fungi

There are several signaling pathways that are known to have multiple roles in protecting major cellular functions that secure the survival of pathogenic fungi. This subchapter highlights several *Candida*, *Cryptococcus*, and *Aspergillus* spp. intrinsic pathways, focusing on specific components that have a general role in these species survival, pathogenicity, virulence, and resistance to antifungal drugs. Given their wide range of action inside the cell, these pathways constitute possible targets for the development of new therapeutic drugs.

7.2.3.1 The Calcium–Calcineurin Pathway

Calcium is a very important secondary messenger needed for the survival and adaptation of several fungi. The intracellular calcium homeostasis is regulated by calcineurin, through different signaling and regulatory cellular processes. Calcineurin is a Ca^{2+}-calmodulin-activated protein phosphatase 2B, composed by two subunits: a catalytic subunit (CnA) and a regulatory subunit (CnB). The existence of an autoinhibitory domain blocks the catalytic subunit, being removed upon a conformational change provoked by the association of Ca^{2+}-calmodulin. Calmodulin is a Ca^{2+} sensor protein responsible for detecting an increase in the intracellular levels of Ca^{2+} that results from an external stimulus. Upon this increase, calmodulin associates with Ca^{2+}, activating the calcineurin phosphatase (Liu et al. 2015; Juvvadi et al. 2017). This signal is known to activate the Crz1 transcription factor in *S. cerevisiae* and *C. albicans*, indirectly inducing the expression of several target genes (Stathopoulos-gerontides et al. 1999; Yoshimoto et al. 2002; Karababa et al. 2006).

The calcium–calcineurin pathway has been shown to be extremely important in the development of different virulence features and also in antifungal resistance of *Candida*, *Cryptococcus*, and *Aspergillus* (Liu et al. 2015; Juvvadi et al. 2017). Studies have highlighted the importance of calcineurin itself in the virulence features of *Cryptococcus neoformans* (Odom et al. 1997a; Fox et al. 2001). Upon the absence of calcineurin, *C. neoformans* is no longer able to grow at 37 °C or perform hyphal elongation, becoming sensitive to CO_2 and alkaline pH (Odom et al. 1997a; Cruz et al. 2001). Growth at 37 °C is also affected in *C. gattii* calcineurin mutants that exhibit lower capacity to tolerate this temperature when compared to the wild-type. Nevertheless, certain molecular types of *C. gattii* seem more dependent than others on calcineurin control of thermotolerance. In addition, *C. gattii* calcineurin was found to have a minor role in capsule production. Both *C. neoformans* and *C. gattii* cell membrane requires calcineurin for complete integrity and ER stress response, although only *C. gattii* calcineurin was found to be required for flucona-zole tolerance (Chen et al. 2013). *C. gattii* virulence is affected by the absence of calcineurin activity in a murine inhalation model, for all three molecular types tested (VGIIa, VGIIb, and VGI), exhibiting reduced fungal burden in the brain and lung tissues of infected animal. In turn, *C. neoformans* shows reduced virulence in rabbits immunosuppressed with steroids and infected intrathecally, representing an animal model of cryptococcal meningitis (Odom et al. 1997a).

Calcineurin's role in *Aspergillus fumigatus* has been described for its importance in hyphal growth, cell wall integrity, conidium production, virulence, and antifungal resistance (Steinbach et al. 2006; Ferreira et al. 2007; Cramer et al. 2008; Fortwendel et al. 2010). The absence of the catalytic subunit, CnA, was shown to impair filamentous growth in vitro (Steinbach et al. 2006; Ferreira et al. 2007) and in a murine inhalational model of invasive pulmonary aspergillosis, displaying no hyphal formation in the lungs of infected animals (Steinbach et al. 2006). The role of calcineurin in the control of hyphal growth seems to be related to the formation of the hyphal septum. The CnA subunit was found to be co-localized with actin in the contractile ring during the early phase of septum formation, as well as CnB that was found to have septal localization dependent of CnA (Juvvadi et al. 2011). Alterations in the morphology of conidium (Steinbach et al. 2006; Ferreira et al. 2007) and the incapacity to grow in fetal bovine serum (FBS) (Ferreira et al. 2007) were also observed in the absence of CnA subunit, resulting in a significant decrease of *A. fumigatus* virulence in intranasal, intravenous, and inhalational murine models, as well as in the *G. mellonella* infection model (Steinbach et al. 2006). This clear action in *A. fumigatus* virulence is accompanied by a role in antifungal resistance. At high concentrations, echinocandins have attenuated activity toward *A. fumigatus*. This so-called caspofungin paradoxical effect is lost in the absence of the catalytic subunit of calcineurin or in the absence of the transcription factor CrzA. Calcineurin seems to be responsible for the activation of CrzA transcription factor, since the loss of this regulator results in a similar outcome to the absence of the catalytic subunit of calcineurin, with alterations in hyphal growth, cell wall structure, germination of conidium, as well as attenuated virulence (Cramer et al. 2008; Fortwendel et al. 2010). Interestingly, the absence of either of these components leads to reduction of the β-1,3-glucan levels in the cell wall, indicating a role of the calcium–calcineurin pathway in the control of this component, probably helping in the resistance to echinocandins, which lead to the inhibition of β-1,3-glucan synthesis (Cramer et al. 2008).

Regarding *Candida* species, the role of calcineurin is very extensive, varying from species to species. Its importance in virulence has been shown in systemic, ocular, and urinary tract murine models of infection (Blankenship et al. 2003; Sanglard et al. 2003; Miyazaki et al. 2010; Chen et al. 2011, 2012, 2014a; Yu et al. 2015). For instance, *C. albicans* calcineurin is essential for virulence in a systemic murine model of infection but not for vaginal or pulmonary models (Bader et al. 2006), while *C. glabrata* was the only one found to require calcineurin in the urinary tract model of infection so far (Yu et al. 2015). These differences might be based on the specific roles of calcineurin in each *Candida* species.

One important virulence feature of *C. albicans*, *C. glabrata*, *C. dubliniensis*, and *C. tropicalis* is the capacity to survive in serum, which mimics the adaptation that these species have to endure in systemic infections (Blankenship et al. 2003; Sanglard et al. 2003; Bader et al. 2006; Chen et al. 2012). Blankenship and Heitman (2005) have demonstrated that the endogenous level of calcium in the serum is the major stress agent causing death of *C. albicans* calcineurin mutants (Blankenship and Heitman 2005). This study gives evidence of the importance of calcineurin for *C. albicans* survival in serum. Nevertheless, for *C. glabrata*, just like in *C. neoformans*, calcineurin seems to have a more important role in the growth at host temperatures (Chen et al. 2012), a phenotype which is not observed for other *Candida* species (Blankenship et al. 2003; Yu et al. 2015). Calcineurin seems to be necessary also for hyphal development, another trait of virulence, in *C. dubliniensis* and *C. tropicalis* (Chen et al. 2011, 2014a). However, the role of calcineurin in hyphal growth is still a debating issue since other contradictory evidences were found. While Sanglard et al. (2003) have shown that in spider medium, the majority of *C. albicans* strains were not able to perform filamentous growth upon loss of calcineurin activity (Sanglard et al. 2003), in other reported cases, calcineurin did not seem to affect hyphal formation (Blankenship et al. 2003; Bader et al. 2006). Moreover, cell wall integrity, ER stress response, and pH homeostasis necessary for the full virulence capacity of *C. glabrata* were found to also depend on calcineurin. The absence of calcineurin affected all these processes leading to decreased tissue colonization in a systemic murine model of infection (Chen et al. 2012).

Besides virulence, calcineurin has also been shown to be required for azole and echinocandin resistance in *C. albicans*, *C. glabrata*, *C. tropicalis*, *C. dubliniensis*, and *C. lusitaniae* (Chen et al. 2012, 2014a; Yu et al. 2015). The genes required to enhance drug resistance by calcineurin was found to be different for azoles and echinocandins: in the case of azole resistance a pathway independent of the transcription factor Crz1 is required, while for echinocandin resistance, calcineurin acts through the Crz1 regulator (Miyazaki et al. 2010; Chen et al. 2012). In *C. tropicalis*, the development of virulence seems to be intrinsically related to the activation of the Crz1 transcription factor (Chen et al. 2014a).

Focusing further on other players in calcineurin pathway in *C. glabrata*, special attention has been given to two endogenous regulators of calcineurin, Rcn1 and Rcn2, known to modulate the activity of calcineurin signaling. Rcn1 was found to be responsible for the activation of calcineurin pathway, leading to the activation of the Crz1 transcription factor (Miyazaki et al. 2011; Chen et al. 2012), while the expression of *RCN2* gene seems to be regulated by calcineurin through Crz1 (Miyazaki et al. 2011), inhibiting the calcineurin pathway in a negative feedback process (Chen et al. 2012). Interestingly, Rcn1 was found to be necessary to micafungin, fluconazole, and tunicamycin resistance in *C. glabrata*, although resistance to tunicamycin was not dependent on the action of Crz1 (Miyazaki et al. 2011).

Moreover, the calcium influx system, which has two major components (Cch1, a plasma membrane Ca^{2+} channel, and Mid1, a regulatory subunit of Cch1), was found to be necessary for *C. albicans* virulence. The deletion mutants of these two components exhibit defects in developing a normal oxidative stress response, hyphae preservation, and invasive growth. In fact, these deletion mutants showed decreased virulence of *C. albicans* upon infection of a mouse model (Yu et al. 2012). Also in *C. neoformans*, the Cch1 protein is involved in virulence, being essential for growth in a low calcium availability in environment and mammalian body temperature (Liu et al. 2006). All these studies give evidence of the major role calcium–calcineurin pathway has in virulence and antifungal resistance of *Cryptococcus*, *Aspergillus*, and *Candida* species, as well as the importance of specific components of this pathway.

7.2.3.2 The Hsp90 Chaperone

Another important survival mechanism is regulated by the Hsp90 chaperone. This chaperone is involved in the heat-shock response, being essential for the folding, transport, maturation, and degradation of diverse key regulators of signaling pathways. The work of Cowen and Lindquist (2005) has highlighted the role of Hsp90 in acquisition of azole resistance in *S. cerevisiae* and *C. albicans*. Hsp90 allows the generation of new mutations that help in rapid phenotypic changes under stress conditions (Cowen and Lindquist 2005). By enabling the formation of new mutations, Hsp90 allows the development of alterations within the cell that help coping with the new challenges presented by azole antifungal drugs. In *C. neoformans* Hsp90 also contributes for antifungal resistance and virulence. Hsp90 has been shown to be necessary for *C. neoformans* thermotolerance, a very important virulence factor for the survival at host temperatures (Cordeiro et al. 2016; Chatterjee and Tatu 2017). This chaperone is localized in the cell wall of *C. neoformans* through the ER-Golgi classical secretory pathway (Chatterjee and Tatu 2017), inducing capsule formation and maintenance (Cordeiro et al. 2016; Chatterjee and Tatu 2017). Probably due to its location on the cell wall, Hsp90 is important for echinocandin resistance at 37 °C, although having no role in resistance at 25 °C (Chatterjee and Tatu 2017). Upon exposure to an inhibitor of Hsp90, radicicol, *C. neoformans* and *C. gattii* have been shown to become more sensitive to fluconazole, itraconazole, voriconazole, and amphotericin B. In fact, the interaction between azoles and radicicol was described as synergistic, in planktonic cells. In biofilms, the interaction of radicicol with amphotericin B and azoles helped inhibiting biofilm formation and mature biofilms in both species, although more significantly in *C. gattii* (Cordeiro et al. 2016). In *Aspergillus fumigatus*, Hsp90 is essential for cell viability. When Hsp90 is genetically repressed, decreased hyphal formation and deficiencies in conidia are observed, accompanied by loss of the paradoxical effect of caspofungin and increased susceptibility to this echinocandin. Upon exposure to cell wall stress, Hsp90 becomes localized at the cell wall instead of the cytosol, its localization in unstressed conditions (Lamoth et al. 2012).

Further analysis of how Hsp90 potentiates the development of drug resistance in *C. albicans* has led to the identification of calcineurin as a direct target of this chaperone. Both inhibition of Hsp90 and calcineurin lead to decreased resistance in *C. albicans* clinical isolates previously identified as fluconazole resistant (Cowen and Lindquist 2005). The connection between the two pathways gives evidence of the complexity of antifungal resistance development in *C. albicans*.

7.2.3.3 The High Osmolarity Glycerol (HOG) Pathway

Another important cellular pathway with impact in pathogenesis is the high osmolarity glycerol (HOG) mitogen-activated protein kinase (MAPK) signaling pathway. As indicated by its name, the HOG pathway regulates the adaptation to increased osmolarity in medium, so that the cell does not loose water, suffering a shrinking process (Hohmann 2009). In *C. neoformans*, the HOG pathway is regulated by a two-component system composed by Tco1 and Tco2. This system is responsible for the control of response regulators that activate Hog1 under particular stress conditions. *C. neoformans* expresses two homologues of *S. cerevisiae* response regulators, Ssk1 and Skn7, named accordingly. In *C. neoformans*, only Ssk1 functions in a Hog1-dependent manner, although both have a key role in maintaining antifungal resistance and virulence in this fungal pathogen. Ssk1 is an upstream regulator necessary for resistance to osmotic and oxidative stress, UV irradiation, the fludioxonil antifungal drug, but also for *C. neoformans* survival at human host temperatures, melanin, and capsule production. On the other hand, Skn7 is required for Na$^+$ stress response and fludioxonil resistance, contributing also for melanin production. By controlling these response regulators, Toc1 and Toc2 were found to be necessary for fludioxonil resistance and Toc2 for oxidative stress response. Indeed, Toc1 is required for *C. neoformans* virulence in a murine model, contributing for melanin production (Bahn et al. 2006).

In turn, Hog1 is known to be responsible for the activation of two cation transporters, Ena1 and Nha1, upon cation and osmotic shock or upon cation shock, respectively (Jung et al. 2012). *ENA1* gene deletion attenuates *C. neoformans* virulence in a mice model and confers susceptibility to alkaline pH (Idnurm et al. 2009), while increasing the susceptibility to amphotericin B (Jung et al. 2012). In the *Δena1Δnha1* mutant strain, the increase of susceptibility to this antifungal drug is more evident and clear for fluconazole and ketoconazole (Jung et al. 2012). The *C. neoformans* double deletion mutant was also found to have increased melanin and capsule production, although having attenuated virulence in a murine model of systemic cryptococcosis. Nevertheless, the *Δnha1* mutant did not show differences when comparing to the wild-type virulence (Jung et al. 2012).

In *A. fumigatus*, the HOG pathway is known to have several cellular functions, influencing virulence and antifungal resistance. A specific phosphatase in this pathway, PtcB, was shown to be involved in cell wall integrity and resistance to damaging agents, control of the levels of chitin and β-1,3-glucan, and biofilm formation, by influencing adhesion and hydrophobicity. The absence of the phosphatase in *A. fumigatus* results in an avirulent strain in a murine model of invasive pulmonary aspergillosis (Winkelströter et al. 2015).

The HOG pathway of *Candida* species has also been shown to have a role in antifungal resistance and virulence. *C. albicans* Hog1 is involved in the resistance to stress provoked by nikkomycin Z and Congo red, while being necessary for hyphal formation and full virulence in a murine model of systemic infection (Alonso-Monge et al. 1999). Moreover, a recent study revealed that Hog1 deletion leads to defects in adhesion to gastrointestinal tract mucosa and sensitivity to bile salts (Prieto et al. 2014). One of the response activators of the HOG pathway, Ssk1, is also involved in the virulence of *C. albicans*. The absence of this response regulator results in complete loss of hyphal formation, and only in nitrogen-limited conditions did the *Δssk1* deletion mutant exhibit a hyperinvasive phenotype (Calera et al. 2000). In a murine model of hematogenously disseminated candidiasis, the *Δssk1* deletion mutant was found to be avirulent, confirming the role of Ssk1 in *C. albicans* virulence (Calera et al. 2000).

7.2.3.4 The Osmotic and pH Stress Responses

The osmotic stress response has also been shown to be related to virulence in fungal pathogens. A specific putative O-acyltransferase, Gup1, known for its association with a deficiency in glycerol uptake upon osmotic stress response, is necessary for the full development of certain virulence traits, such as hyphal formation, adhesion to polystyrene, agar invasion, and biofilm formation. Interestingly, Gup1's absence was further found to increase antifungal resistance in *C. albicans* (Ferreira et al. 2010). Nevertheless, this role in antifungal resistance and virulence in *C. albicans* does not seem to be related to glycerol uptake.

The control of expression of alkaline-responsive genes is regulated by the PacC transcription factor in *Aspergillus fumigatus*. This regulator is crucial for growth at pH 8, as would be expected, but also influences several mechanisms of virulence and antifungal resistance. The deletion of *PACC* gene leads to failure of epithelial invasion, with defects in contact-mediated epithelial entry and decreased expression of genes encoding proteases. Further analysis of the deletion mutant showed an attenuated virulence in leukopenic mice, probably due to the defects previously described. The *ΔpacC* mutant displays susceptibility to caspofungin in vitro and during in vivo treatment in mice infected, with reduction of fungal burden (Bertuzzi et al. 2014). PacC is a homologue of *C. albicans* Rim101, which also has a role in virulence, being necessary for the full capacity of *C. albicans* to damage endothelial cells (Davis et al. 2000).

7.2.3.5 The Cell Wall Integrity Signaling Pathway

Another important part of the cell that requires strong regulation is the cell wall, which is extremely important in the protection of the fungus against environmental stresses that might arise in host niches. Therefore, the cell wall integrity signaling pathway and the several components of the cell wall play an important role in the protection and interaction with the immune system of the host. One of these components, a protein belonging to the ECM33/SPS2 family of glycosylphosphatidylinositol-anchored proteins, Ecm33, of *A. fumigatus*, has been shown to be involved in antifungal resistance and virulence. Curiously, the deletion of *ECM33* leads to higher resistance to Congo red and caspofungin and hypervirulence in an immunocompromised murine model for disseminated aspergillosis (Romano et al. 2006). The absence of this protein results in changes in the cell wall that seem to allow the development of higher resistance and virulence capacity of *A. fumigatus*.

Another component on the cell wall, RodA, plays a key role in *A. fumigatus* interaction with the host immune system. RodA's absence has proven to lead to an augmentation of cytokine production, neutrophil infiltration, and a faster elimination of fungal burden in C57BL/6 corneas. Carion et al. (2013) found that RodA masks dectin-1 and dectin-2 recognition, preventing a more aggravated immune response (Carion et al. 2013). Interestingly, another hydrophobin, RodB, was shown to be significantly downregulated upon exposure to amphotericin B (Gautam et al. 2008), indicating an involvement of hydrophobins in the cell wall-dependent virulence and polyene resistance.

Moreover, the *C. neoformans* MAP kinase Mpk1 is also an important player in the persistence of this pathogen in the host. Mpk1 was found to be essential for in vitro growth at 37 °C and for complete *C. neoformans* virulence in a mouse model of cryptococcosis. At the same time, Mpk1 was shown to be phosphorylated upon exposure of several drugs that affect the fungus cell wall, such as the chitin synthase inhibitor nikkomycin Z, caspofungin, and the calcineurin inhibitor FK506, while the absence of Mpk1 increases *C. neoformans* susceptibility to nikkomycin Z and caspofungin. Upon the inhibition of calcineurin by FK506, phosphorylation of the Mpk1 is observed, provoking the induced expression of the *FKS1* gene, encoding a component of the β-1,3-glucan synthase complex, the target by caspofungin. According to these results, Mpk1 has an essential function in *C. neoformans* by providing protection against several stress agents of the cell wall, while participating in this pathogen's virulence (Kraus et al. 2003).

In *Candida albicans* MAPK pathway participates in the activation of Mkc1 MAP kinase, responsible for cell integrity. This kinase has shown to be necessary for full virulence in a murine model of systemic infection (Diez-Orejas et al. 1997), probably due to its role in invasive growth and biofilm formation (Kumamoto 2005). Moreover, another MAP kinase, involved in vegetative growth and cell wall formation, Cek1, was found to have a role in the organization of the cell wall. In the absence of this MAP kinase, several alterations of the cell wall lead to an easier recognition by murine macrophages, comparatively to the wild-type. The increased exposure of β-1,3-glucans, α-1,2 and β-1,2-mannosides and a lose structure of the cell wall are believed to contribute to the increased binding of macrophages to the Δ*cek1* deletion mutant (Román et al. 2016). Furthermore, absence of Cek1 in *C. albicans* strains used to infect a mouse mastitis model failed to provoke lesions, while 65% of the mice infected with *C. albicans* wild-type cells exhibit severe lesions (Guhad et al. 1998). Cek1 involvement in hyphal development, growth in serum, and virulence in a murine model of systemic candidiasis has also been described. In fact, other participants of the MAPK pathway in *C. albicans*, Cst20, Hst7, and Cph1 also influence hyphal growth in these pathogenic fungi (Csank et al. 1998). Therefore, the correct functioning of the MAPK pathway is thus required for *C. albicans* virulence and capacity to evade the host immune system.

Also contributing to the cell wall correct function is lipid metabolism. This process is regulated by sterol regulatory element-binding proteins (SREBP) in different microorganisms. In *A. fumigatus*, SrbA is a SREBP that regulates several genes related to the ergosterol biosynthesis and which becomes induced upon conditions of hypoxia and iron depletion. The absence of *SRBA* gene affects 87 genes, affecting ergosterol biosynthesis and azole resistance, maintenance of cell polarity, hyphal morphogenesis, and iron acquisition (Willger et al. 2008; Blatzer et al. 2011). SrbA is necessary for fluconazole and voriconazole resistance, although the same was not observed for caspofungin and amphotericin B. *A. fumigatus* virulence was also affected in murine models of invasive pulmonary aspergillosis, although SrbA function in virulence is not associated with resistance to oxidative stress inside macrophages (Willger et al. 2008). In fact, it is believed that SrbA importance in *A. fumigatus* virulence is related to its role in adaptation to hypoxia and iron starvation conditions, besides hyphal formation (Willger et al. 2008; Blatzer et al. 2011). Moreover, SrbB has also been shown to co-regulate virulence genes together with SrbA (Chung et al. 2014).

7.2.3.6 The Unfolded Protein Response (UPR)

Another pathogenesis-related biological process is the endoplasmic reticulum (ER) stress response. This response is induced upon the accumulation of misfolded proteins in the ER due to hostile environmental conditions. Upon ER stress, the unfolded protein response (UPR), from which Hac1 is an important transcription factor in *C. albicans*, is initiated. Besides this important function, Hac1 has a role in *C. albicans* virulence, since the absence of Hac1 leads to decreased filamentous growth, with increase in pseudohyphae and budding cells. Adhesion is also compromised upon *HAC1* gene deletion, these cells exhibiting less sticky flocks than the wild-type. Hac1 transcription factor is responsible for the upregulation of several adhesins and protein mannosyltransferases upon ER stress (Wimalasena et al. 2008).

In *A. fumigatus*, the Hac1 ortholog is HacA, which has the same role in the activation of UPR pathway genes. But more interesting is the role of HacA in the modulation of *A. fumigatus* resistance toward amphotericin B, caspofungin, itraconazole, and fluconazole, as well as virulence in a murine model of invasive aspergillosis (Richie et al. 2009). Responsible for the activation of HacA transcription factor is the ER transmembrane sensor, IreA, which upon accumulation of misfolded proteins activates not only HacA but other independent pathways. HacA translation is activated by the cleavage of an intron in the HacA mRNA. This IreA sensor was found to be essential for virulence, its absence leading to a completely avirulent phenotype in the same type of model. IreA is also necessary for *A. fumigatus* thermotolerance, growth under hypoxia, changes in the composition of cell wall and membrane, and azole resistance (Feng et al. 2011).

C. glabrata Ire1 is a kinase/endoribonuclease that interacts with unfolded proteins in the ER, leading to the activation of stress-responsive genes (Miyazaki et al. 2013; Miyazaki and Kohno 2014). Contrary to *C. albicans*, Ire1 does not activate the Hac1 transcription factor. In fact, Ire1 together with calcineurin and Slt2 from the MAPK pathway are needed to activate the UPR in *C. glabrata*. Ire1 was also found be necessary for full virulence in *C. glabrata*. Its absence leads to lower mortality of cyclophosphamide-immunosuppressed mice and lower fungal burden in the kidney and spleen in a mouse model of disseminated candidiasis (Miyazaki et al. 2013).

In the case of *C. neoformans*, the transcription factor Hxl1, with no homology to Hac1, is activated in the same way by an Ire1 protein kinase, which is responsible for the splicing of *HXL1* mRNA upon ER stresses. Both Ire1 and Hxl1 are necessary for normal growth at 37 °C, as well as for full virulence of *C. neoformans* in a murine model of systemic cryptococcosis. Nevertheless, Ire1 seems to participate in *C. neoformans* virulence in a dependent and independent way of Hxl1. Only *Δire1* deletion mutant has shown defects in capsule formation and increased susceptibility to diamide. Still, for azoles, for phenylpyrroles, and slightly for amphotericin B, *C. neoformans* resistance depends on the UPR pathway, through Ire1 activation of Hxl1 (Cheon et al. 2011). Ire1 was also found to be necessary for caspofungin resistance (Blankenship et al. 2010).

7.2.3.7 Chromosomal Rearrangements

Other important mechanisms that have been found to be responsible for the augmentation of both resistance and virulence in *Candida* spp. are chromosomal rearrangement and aneuploidy, which happen frequently in *C. glabrata*. In fact, in a given patient, recovered sequential isolates exhibited two or three different karyotypes (Shin et al. 2007). Small chromosome formation has been observed in *C. glabrata* isolates collected from patients hospitalized at Danish hospitals. The formation of such structures is believed to happen through translocation mechanisms, generating duplicated genes. These small chromosomes include genes required for azole resistance, such as ABC transporter-encoding genes, like *CgCDR2* (Ahmad et al. 2013), as well as genes known to have a role in virulence, like aspartyl protease-encoding genes (Kaur et al. 2007; Poláková et al. 2009). In *C. albicans*, aneuploidy was also observed to cause alterations in azole resistance, where the increase or decrease in resistance was related to the gain or loss of an isochromosome composed by two left arms of chromosome 5. Genes found in this isochromosome were discovered to encode enzymes of the ergosterol pathway, efflux pumps, and the Tac1 transcription factor (Selmecki et al. 2006). The generation of these chromosomal changes is thus favorable for the pathogen, which gains new strategies to elude and evade the host immune system and resist antifungal drugs, proliferating in the host.

7.3 Finding New Targets, Drugs, and Therapeutic Strategies to Fight Fungal Infections

Given the emergence of antifungal resistance and increased number of fungal infections, there's a current need to search for alternatives that might direct scientific work to effective antifungal therapy. The last part of this chapter tries to collect the more recent work being developed in the pursuit of new targets and new antifungals that might be more adequate, based on the mechanisms behind both antifungal resistance and virulence.

Recent work has been focused on the calcium–calcineurin pathway for the discovery of new targets. As described above, this pathway is activated upon an increase in the intracellular concentration of Ca^{2+}, upon an external signal. Nevertheless, when this increase is extreme, the cell suffers a different process: Ca^{2+}-mediated cell death. Knowing this, inhibitors of this pathway may be considered as new drugs since their use would most probably lead to a continuous accumulation of intracellular Ca^{2+}, provoking cell death (Liu et al. 2015).

Some of the antifungal compounds based on the disturbance of Ca^{2+} homeostasis act on several specific sites of the calcium–calcineurin pathway. For instance, fluphenazine, calmidazolium, and W-7 analog are inhibitors of calmodulin. Their use in *S. cerevisiae* strains significantly increased their sensitivity toward azoles (Edlind et al. 2002). On the other hand, there are also specific calcineurin phosphatase inhibitors like cyclosporine (CsA) and FK506. These compounds are dependent of cytosolic proteins, cyclophilin (CsP), and FK506-binding protein (FKBP), respectively, to which they bind to form a complex that inhibits calcineurin activity (Li and Handschumachers 1993). Since these inhibitors also have an immunosuppressive effect in humans, non-immunosuppressive analogs of these inhibitors have been developed, preserving an antifungal effect toward *C. neoformans* (Odom et al. 1997b; Cruz et al. 2000). The importance of calcineurin inhibitors is also highlighted in an in vitro interaction with caspofungin, where several strains and isolates from *Aspergillus* species became susceptible to the synergistic effect caused by the combination of the inhibitors with the drug (Kontoyiannis et al. 2003; Steinbach et al. 2004). The use of cyclosporine together with fluconazole has shown synergetic results not only in vitro but also in a murine keratitis model, against *Candida albicans* (Marchetti et al. 2000; Onyewu et al. 2006). Also in *C. albicans* biofilms, in vitro or in a rat catheter model, susceptibility was observed upon the combination of fluconazole with either CsA or FK506 calcineurin inhibitors (Uppuluri et al. 2008). Moreover, FK506 has also shown a synergistic effect in vitro when combined with fluconazole, itraconazole and voriconazole, although this effect appears to be *C. albicans* strain dependent (Sun et al. 2008). Likewise, CsA synergistic effect in *C. parapsilosis* cells in vitro when combined with fluconazole, voriconazole, amphotericin B, and caspofungin was demonstrated (Cordeiro et al. 2014). Besides the use of analogs of these inhibitors in therapy, studies like the one performed by Tonthat et al. (2016) might help targeting specific sequences in *C. albicans* and *A. fumigatus* FKBP, distinct from the human FKBP, decreasing possible toxic side effects associated with the use of new drugs (Tonthat et al. 2016).

In turn, verapamil is a calcium channel blocker that has been demonstrated to have an inhibitory effect against *C. albicans* biofilms. Verapamil alone or in combination with fluconazole or tunicamycin has shown to exhibit antifungal activity, indicating that this inhibitor has potential to be used in therapy (Yu et al. 2013). The antiarrhythmic drug amiodarone is another compound that has been shown to have antifungal properties in *Saccharomyces*, *Candida*, *Cryptococcus*, and *Aspergillus*, by its effect in Ca^{2+} homeostasis (Courchesne 2002). Studies using the yeast model *S. cerevisiae* have shown the role of amiodarone in an extreme augmentation of intracellular Ca^{2+}, followed by cell death (Gupta et al. 2003; Muend and Rao 2008). Hypersensitivity to amiodarone was observed in *S. cerevisiae* mutant strains lacking Ca^{2+} transporters (Pmr1, Pmc1, Cch1/Mid1, Yvc1, and Vcx1) and other transporters (Pmr1, Pdr5, and vacuolar H^+-ATPase), as well as proteins involved in ergosterol biosynthesis, intracellular trafficking, and signaling (Gupta et al. 2003). Provoking the same imbalance of cytosolic Ca^{2+} is the marine-derived polyketide endoperoxide, designated plakortide F acid. This marine-derived compound was shown to have an inhibitory activity toward *Candida albicans*, *Cryptococcus neoformans*, and *Aspergillus fumigatus*. Like with amiodarone, mutants lacking Ca^{2+} transporters are very sensitive to plakortide F acid (Xu et al. 2011). Although all these inhibitory compounds of the calcium–calcineurin pathway are very promising, having calcineurin or the pathway as targets, they are not likely to be approved for clinical use yet. More research is still needed to find a more efficient inhibitory drug that would not be immunosuppressive for the host nor interact with the human calcineurin.

Given the importance of Hsp90 in antifungal resistance and virulence in fungal pathogens, additional research has been focused on the action of Hsp90 inhibitors in therapy. Radicicol is an Hsp90 inhibitor found to enhance the activity of fluconazole against *Cryptococcus* species, infecting the *Caenorhabditis elegans* model, or in biofilms (Cordeiro et al. 2016), as described previously. Other Hsp90 inhibitors, structurally similar to the natural product, geldanamycin, have shown to improve fluconazole action in *G. mellonella* model of infection with *C. albicans*, while the interaction with caspofungin helped fighting *A. fumigatus* in the same model of infection (Cowen et al. 2009).

Moreover, in an attempt to explore oxidative stress to fight *A. fumigatus* infections, combined utilization of redox-potent natural compounds like 2,3-dihydroxybenzaldehyde, thymol, or salicylaldehyde together with amphotericin B or itraconazole has revealed increased antifungal action in vitro (Kim et al. 2012). Once again, the combination of antifungal drugs and inhibitors of important pathways involved in antifungal resistance and virulence seems to be a promising strategy for antifungal therapy.

Inhibitors targeting cell wall biosynthesis have also been tested against *Candida* and *Aspergillus* species and *C. neoformans*. E1210 is a fungal glycosylphosphatidylinositol (GPI) biosynthesis inhibitor shown to have increased efficacy when used in combination with azoles or amphotericin B, in murine mouse models, against *Candida albicans*, *Candida tropicalis*, *Aspergillus fumigatus*, and *Aspergillus flavus* (Hata et al. 2011). Also affecting the cell wall is 1-[4-butylbenzyl] isoquinoline (BIQ), discovered to block GPI-anchored mannoprotein localization on the *S. cerevisiae* cell wall. When used against *C. albicans*, decreased adhesion to rat intestine epithelial cell monolayer was observed. In *S. cerevisiae*, the primary target of BIQ is the Gwt1 protein, necessary for the correct localization of GPI-anchored mannoproteins (Tsukahara et al. 2003). Other compounds focused on the destabilization of the cell wall are inhibitors of β-1,3-glucan synthesis. Enfumafungin, ascosterocide, arundifungin, and ergokonin A have been identified as such inhibitors, leading to a drug candidate designated MK-3118, whose target is the β-1,3-glucan synthase. MK-3118 has shown powerful in vitro and in vivo activity against *Candida* and *Aspergillus* species (Roemer and Krysan 2014). Other inhibitors of β-1,3-glucan synthase belong to the group of piperazinyl-pyridazinones that exhibit antifungal activity against *Candida* and *Aspergillus* species in vitro. One of these compounds, SCH C, was found to be active against *C. neoformans*. In vivo activity of SCH B was tested in a model of systemic *C. glabrata* infection in immunocompromised mice, exhibiting antifungal activity (Walker et al. 2011). Focused on the inhibition of β-1,6-glucan synthesis, D75-4590, a pyridobenzimidazole derivative, was found to block Kre6, one of β-1,6-glucan synthases. This inhibitor has shown effective action against *Candida* species (Kitamura et al. 2009b). Another pyridobenzimidazole derivative, D21-6076, has shown to have the same function, being able to inhibit *C. albicans* invasion in an in vitro model of vaginal candidiasis (Kitamura et al. 2009a). Combination therapy with antifungal drugs is also used to target the fungal cell wall. One example is the use of nikkomycin Z, which inhibits chitin synthesis, against *C. albicans* strains, that exhibited in vitro synergistic effect with voriconazole or caspofungin (Sandovsky-Losica et al. 2008; Verwer et al. 2012).

On the other hand, mitochondria have also been used as a target with good antifungal results. Inhibitors of mitochondrial respiration have shown to be potential new drugs against fungal pathogens. For instance, the amino acid-derived 1,2-benzisothiazolinone (BZT) have shown fungicidal activity against *Candida* species, *C. neoformans*, and *A. fumigatus*. In fact, *Candida* strains resistant to azoles, polyenes, and micafungin were found to be susceptible to BZT compounds (Alex et al. 2012). T-2307 (4-{3-[1-(3-{4-[amino(imino)-methyl]phenoxy}propyl) piperidin-4-yl]propoxy}benzamidine) is another drug candidate revealed to have fungicidal activity against *C. albicans*, *A. fumigatus*, and *C. neoformans* in mouse models of disseminated infection. Although its mechanism of action is not yet clear, it is believed to affect mitochondrial function (Mitsuyama et al. 2008).

There are also new drug candidates with more specific targets, like parnafungin, a natural product, that inhibits poly(A) polymerase. This natural product has shown antifungal activity against *Candida* and *Aspergillus* species, being very effective in fighting *C. albicans*, in a murine model of candidiasis (Jiang et al. 2008). Another example is the new inhibitor of Pdr1-dependent gene activation, iKIX1, in *C. glabrata*. This inhibitor blocks azole-stimulated gene expression and the Pdr1-dependent drug efflux pathway, leading to a re-sensitization of *C. glabrata* azole-resistant strains to azole drugs, in vitro and in *G. mellonella* and murine models (Nishikawa et al. 2016).

Studies focused on other possible targets are also underway. That is the case of the work developed by Suliman et al. (2007) where methionine synthase is proposed and has a potential mark for future therapeutics. Methionine synthase catalyzes the last step of methionine biosynthesis in fungi and humans. Nevertheless, in humans, methionine synthase is cobalamin-dependent, opposite to fungi's cobalamin-independent methionine synthase, having both enzymes differences in structure and mechanical action. This coupled to the fact that methionine synthase in

C. albicans is encoded by an essential gene, helps visualizing this enzyme as a possible new target (Suliman et al. 2007). Research on siderophores has also been undertaken due to their significance in iron uptake in the human host, together with the fact that the host will not suffer toxic effects with a potential new drug, since no siderophores are produced by human cells (Balhara et al. 2016).

7.4 Conclusion and Perspectives: Finding a Sustainable Solution

Antifungal resistance and virulence development are behind the incredible capacity exhibited by fungal pathogens to establish infections in the human host. Both traits are intrinsically related, having several pathways and components working together to allow the survival of *Cryptococcus*, *Aspergillus*, and *Candida* species (see Fig. 7.2). This chapter collects not only evidences of previously known virulence mechanisms that are involved in antifungal resistance and vice versa but also of stress resistance pathways that have a role in both virulence and antifungal resistance, presenting a complex cooperation between mechanisms. A good example of the intercommunication between the different pathways is the involvement of the

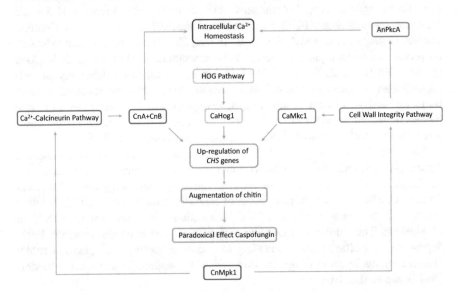

Fig. 7.2 Schematic representation of an example of the intercommunication between stress resistance pathways involved in antifungal resistance and virulence in *Aspergillus*, *Cryptococcus*, and *Candida* species. Calcium–calcineurin pathway (blue), HOG pathway (yellow), and cell wall integrity pathway (orange) participate in the upregulation of CHS genes involved in the synthesis of chitin and subsequent resistance to high concentrations of caspofungin (green) in *Candida albicans*. Both PkcA and calcineurin regulate the intracellular calcium homeostasis (purple) in *Aspergillus nidulans*. The MAP kinase Mpk1 (dark blue) participates in the calcium–calcineurin pathway and the cell wall integrity pathway in *Cryptococcus neoformans*

cell wall integrity signaling pathway, HOG pathway, and calcium–calcineurin pathway in the caspofungin paradoxical effect in *Candida albicans*. The three pathways participate in the response to caspofungin exposure, by the activation of the synthesis of chitin, whose elevated presence in the cell wall drives resistance acquisition (Wiederhold et al. 2005; Walker et al. 2008). The same effect of cooperation is believed to happen in *A. nidulans*, where the protein kinase C (PkcA) is able to overcome the absence of calcineurin, upon its overexpression, indicating that these proteins might have a common end function. Moreover, both PkcA and calcineurin have a regulating effect over MpkA activation (Colabardini et al. 2014). In turn, the *C. neoformans* Mpk1, already described above, is involved in the cell wall integrity signaling pathway and calcium–calcineurin pathway (Kraus et al. 2003), being a good example of the interaction between pathways and their role in antifungal resistance and virulence in pathogenic fungi.

The participation of these major stress resistance pathways in virulence and antifungal resistance highlight new possible targets for the development of new therapeutic strategies. Although some promising new drugs and drug targets have been described, there is still no ideal solution. Besides finding the best target and developing new drugs or therapy, economic and social issues should be taken into account when developing a new pharmacologic product due to the high rates of HIV patients, the preferential targets of fungal infections, in underdeveloped world regions. This problematic aggravates the development of new therapeutic drugs, increasing the number of restricted parameters to be considered like low cost and oral bioavailability (Krysan 2015).

Fighting fungal infections in the future will involve the adequate study of all stress resistance pathways, virulence, and antifungal resistance mechanisms to find the best target or targets, accompanied by the development of new drugs with simultaneous antifungal activity, pharmacological safety, and economic accessibility.

References

Ahmad KM, Ishchuk OP, Hellborg L, Jørgensen G, Skvarc M, Stenderup J, Jørck-Ramberg D, Polakova S, Piškur J (2013) Small chromosomes among Danish *Candida glabrata* isolates originated through different mechanisms. Antonie Van Leeuwenhoek 104(1):111–122. https://doi.org/10.1007/s10482-013-9931-3

Alex D, Gay-Andrieu F, May J, Thampi L, Dou D, Mooney A, Groutas W, Calderone R (2012) Amino acid-derived 1,2-benzisothiazolinone derivatives as novel small-molecule antifungal inhibitors: identification of potential genetic targets. Antimicrob Agents Chemother 56(9):4630–4639. https://doi.org/10.1128/AAC.00477-12

Alonso-Monge R, Navarro-Garcia F, Molero G, Gustin M, Pla J, Sánchez M, Nombela C (1999) Role of the mitogen-activated protein kinase Hog1p in morphogenesis and virulence of *Candida albicans*. J Bacteriol 181(10):3058–3068

Alspaugh JA (2014) Virulence mechanisms and *Cryptococcus neoformans* pathogenesis. Fungal Genet Biol 78:55–58. https://doi.org/10.1016/j.fgb.2014.09.004

Bader T, Schröppel K, Bentink S, Agabian N, Köhler G, Morschhäuser J (2006) Role of calcineurin in stress resistance, morphogenesis, and virulence of a *Candida albicans* wild-type strain. Infect Immun 74(7):4366–4369. https://doi.org/10.1128/IAI.00142-06

Bahn Y-S, Kojima K, Cox GM, Heitman J (2006) A unique fungal two-component system regulates stress responses, drug sensitivity, sexual development, and virulence of *Cryptococcus neoformans*. Mol Biol Cell 17:3122–3135. https://doi.org/10.1091/mbc.E06

Balhara M, Chaudhary R, Ruhil S, Singh B, Dahiya N, Parmar VS, Jaiwal PK, Chhillar AK (2016) Siderophores; iron scavengers: the novel & promising targets for pathogen specific antifungal therapy. Expert Opin Ther Targets 20(12):1477–1489. https://doi.org/10.1080/14728222.2016.1254196

Bambach A, Fernandes MP, Ghosh A, Kruppa M, Alex D, Li D, Fonzi WA, Chauhan N, Sun N, Agrellos OA, Vercesi AE, Rolfes RJ, Calderone R (2009) Goa1p of *Candida albicans* localizes to the mitochondria during stress and is required for mitochondrial function and virulence. Eukaryot Cell 8(11):1706–1720. https://doi.org/10.1128/EC.00066-09

Becker JM, Henry LK, Jiang W, Koltin Y (1995) Reduced virulence of *Candida albicans* mutants affected in multidrug resistance. Infect Immun 63(11):4515–4518

Bertuzzi M, Schrettl M, Alcazar-Fuoli L, Cairns TC, Muñoz A, Walker LA, Herbst S, Safari M, Cheverton AM, Chen D, Liu H, Saijo S, Fedorova ND, Armstrong-James D, Munro CA, Read ND, Filler SG, Espeso EA, Nierman WC, Haas H, Bignell EM (2014) The pH-responsive PacC transcription factor of *Aspergillus fumigatus* governs epithelial entry and tissue invasion during pulmonary aspergillosis. PLoS Pathog 10(10):1–20. https://doi.org/10.1371/journal.ppat.1004413

Beyda ND, Liao G, Endres BT, Lewis RE, Garey KW (2015) Innate inflammatory response and immunopharmacologic activity of micafungin, caspofungin, and voriconazole against wild-type and FKS mutant *Candida glabrata* isolates. Antimicrob Agents Chemother 59(9):5405–5412. https://doi.org/10.1128/AAC.00624-15

Bielska E, May RC (2015) What makes *Cryptococcus gattii* a pathogen? FEMS Yeast Res 16 (1):1–12. https://doi.org/10.1093/femsyr/fov106

Blankenship JR, Heitman J (2005) Calcineurin is required for *Candida albicans* to survive calcium stress in serum. Infect Immun 73(9):5767–5774. https://doi.org/10.1128/IAI.73.9.5767-5774.2005

Blankenship JR, Wormley FL, Boyce MK, Schell WA, Filler SG, Perfect JR, Heitman J (2003) Calcineurin is essential for *Candida albicans* survival in serum and virulence. Eukaryot Cell 2 (3):422–430. https://doi.org/10.1128/EC.2.3.422

Blankenship JR, Fanning S, Hamaker JJ, Mitchell AP (2010) An extensive circuitry for cell wall regulation in *Candida albicans*. PLoS Pathog 6(2):1–12. https://doi.org/10.1371/journal.ppat.1000752

Blatzer M, Barker BM, Willger SD, Beckmann N, Blosser SJ, Cornish EJ, Mazurie A, Grahl N, Haas H, Cramer RA (2011) SREBP coordinates iron and ergosterol homeostasis to mediate triazole drug and hypoxia responses in the human fungal pathogen *Aspergillus fumigatus*. PLoS Genet 7(12):1–18. https://doi.org/10.1371/journal.pgen.1002374

Bonassoli LA, Bertoli M, Svidzinski TIE (2005) High frequency of *Candida parapsilosis* on the hands of healthy hosts. J Hosp Infect 59(2):159–162. https://doi.org/10.1016/j.jhin.2004.06.033

Brown NA, Goldman GH (2016) The contribution of *Aspergillus fumigatus* stress responses to virulence and antifungal resistance. J Microbiol 54(3):243–253. https://doi.org/10.1007/s12275-016-5510-4

Brown GD, Denning DW, Gow NAR, Levitz SM, Netea MG, White TC (2012) Hidden killers: human fungal infections. Sci Transl Med 4(165):1–9. https://doi.org/10.1126/scitranslmed.3004404

Brun S, Aubry C, Lima O, Filmon R, Berge T, Chabasse D, Bouchara J (2003) Relationships between respiration and susceptibility to azole antifungals in *Candida glabrata*. Antimicrob Agents Chemother 47(3):847–853. https://doi.org/10.1128/AAC.47.3.847

Brun S, Bergès T, Poupard P, Vauzelle-Moreau C, Renier G, Chabasse D, Bouchara J-P (2004) Mechanisms of azole resistance in petite mutants of *Candida glabrata*. Antimicrob Agents Chemother 48(5):1788–1796. https://doi.org/10.1128/AAC.48.5.1788

Brun S, Dalle F, Saulnier P, Renier G, Bonnin A, Chabasse D, Bouchara JP (2005) Biological consequences of petite mutations in *Candida glabrata*. J Antimicrob Chemother 56 (2):307–314. https://doi.org/10.1093/jac/dki200

Byrnes EJ, Li W, Lewit Y, Ma H, Voelz K, Ren P, Carter DA, Chaturvedi V, Bildfell RJ, May RC, Heitman J (2010) Emergence and pathogenicity of highly virulent *Cryptococcus gattii* genotypes in the northwest United States. PLoS Pathog 6(4):1–16. https://doi.org/10.1371/journal.ppat.1000850

Byrnes EJ, Li W, Ren P, Lewit Y, Voelz K, Fraser JA, Dietrich FS, May RC, Chatuverdi S, Chatuverdi V, Heitman J (2011) A diverse population of *Cryptococcus gattii* molecular type VGIII in Southern Californian HIV/AIDS patients. PLoS Pathog 7(9):1–18. https://doi.org/10.1371/journal.ppat.1002205

Calderone R, Fonzi W (2001) Virulence factors of *Candida albicans*. Trends Microbiol 9 (7):327–335. https://doi.org/10.1016/S0966-842X(01)02094-7

Calera JA, Zhao XJ, Calderone R (2000) Defective hyphal development and avirulence caused by a deletion of the SSK1 response regulator gene in *Candida albicans*. Infect Immun 68 (2):518–525. https://doi.org/10.1128/IAI.68.2.518-525.2000

Carion S de J, Leal SM Jr, Ghannoum MA, Pearlman E (2013) The RodA hydrophobin on *Aspergillus fumigatus* spores masks Dectin-1 and Dectin-2 dependent responses and enhances fungal survival in vivo. J Immunol 191(5):2581–2588. https://doi.org/10.1007/s12020-009-9266-z.A

Chatterjee S, Tatu U (2017) Heat shock protein 90 localizes to the surface and augments virulence factors of *Cryptococcus neoformans*. PLoS Negl Trop Dis 11(8):1–20. https://doi.org/10.1371/journal.pntd.0005836

Chauhan N, Latge J-P, Calderone R (2006) Signalling and oxidant adaptation in *Candida albicans* and *Aspergillus fumigatus*. Nat Rev Microbiol 4(6):435–444. https://doi.org/10.1038/nrmicro1426

Chen YL, Brand A, Morrison EL, Silao FGS, Bigol UG, Malbas FF, Nett JE, Andes DR, Solis NV, Filler SG, Averette A, Heitman J (2011) Calcineurin controls drug tolerance, hyphal growth, and virulence in *Candida dubliniensis*. Eukaryot Cell 10(6):803–819. https://doi.org/10.1128/EC.00310-10

Chen Y-L, Konieczka JH, Springer DJ, Bowen SE, Zhang J, Silao FGS, Bungay AAC, Bigol UG, Nicolas MG, Abraham SN, Thompson DA, Regev A, Heitman J (2012) Convergent evolution of calcineurin pathway roles in thermotolerance and virulence in *Candida glabrata*. G3 Genes|Genomes|Genetics 2(6):675–691. https://doi.org/10.1534/g3.112.002279

Chen Y-L, Lehman VN, Lewit Y, Averette AF, Heitman J (2013) Calcineurin governs thermotolerance and virulence of *Cryptococcus gattii*. G3 Genes|Genomes|Genetics 3:527–539. https://doi.org/10.1534/g3.112.004242

Chen YL, Yu SJ, Huang HY, Chang YL, Lehman VN, Silao FGS, Bigol UG, Bungay AAC, Averette A, Heitman J (2014a) Calcineurin controls hyphal growth, virulence, and drug tolerance of *Candida tropicalis*. Eukaryot Cell 13(7):844–854. https://doi.org/10.1128/EC.00302-13

Chen Y, Toffaletti DL, Tenor JL, Litvintseva AP, Fang C, Mitchell TG, McDonald TR, Nielsen K, Boulware DR, Bicanic T, Perfect JR (2014b) The *Cryptococcus neoformans* transcriptome at the site of human. MBio 5(1):1–10. https://doi.org/10.1128/mBio.01087-13.Editor

Cheng S, Clancy CJ, Nguyen KT, Clapp W, Nguyen MH (2007) A *Candida albicans* petite mutant strain with uncoupled oxidative phosphorylation overexpresses MDR1 and has diminished susceptibility to fluconazole and voriconazole. Antimicrob Agents Chemother 51 (5):1855–1858. https://doi.org/10.1128/AAC.00182-07

Cheon SA, Jung KW, Chen YL, Heitman J, Bahn YS, Kang HA (2011) Unique evolution of the UPR pathway with a novel bZIP transcription factor, HxL1, for controlling pathogenicity of *Cryptococcus neoformans*. PLoS Pathog 7(8):1–16. https://doi.org/10.1371/journal.ppat.1002177

Chung D, Barker BM, Carey CC, Merriman B, Werner ER, Lechner BE, Dhingra S, Cheng C, Xu W, Blosser SJ, Morohashi K, Mazurie A, Mitchell TK, Haas H, Mitchell AP, Cramer RA (2014) ChIP-seq and in vivo transcriptome analyses of the *Aspergillus fumigatus* SREBP SrbA reveals a new regulator of the fungal hypoxia response and virulence. PLoS Pathog 10 (11):1–23. https://doi.org/10.1371/journal.ppat.1004487

Colabardini AC, Ries LNA, Brown NA, Savoldi M, Dinamarco TM, Von Zeska MR, Goldman MHS, Goldman GH (2014) Protein kinase C overexpression suppresses calcineurin-associated defects in *Aspergillus nidulans* and is involved in mitochondrial function. PLoS One 9(8):1–17. https://doi.org/10.1371/journal.pone.0104792

Cordeiro R de A, Macedo R de B, Teixeira CEC, Marques FJ de F, Bandeira T de JPG, Moreira JLB, Brilhante RSN, Rocha MFG, Sidrim JJC (2014) The calcineurin inhibitor cyclosporin A exhibits synergism with antifungals against *Candida parapsilosis* species complex. J Med Microbiol 63:936–944. https://doi.org/10.1099/jmm.0.073478-0

Cordeiro R de A, Evangelista AJ de J, Serpa R, de Farias Marques FJ, de Melo CVS, de Oliveira JS, da Silva Franco J, de Alencar LP, de Jesus Pinheiro Gomes Bandeira T, Brilhante RSN, Sidrim JJC, Rocha MFG (2016) Inhibition of heat-shock protein 90 enhances the susceptibility to antifungals and reduces the virulence of *Cryptococcus neoformans*/*Cryptococcus gattii* species complex. Microbiology 162(2):309–317. https://doi.org/10.1099/mic.0.000222

Costa CR, Jesuíno RSA, de Aquino Lemos J, de Fátima Lisboa Fernandes O, e Souza LKH, Passos XS, do Rosário Rodrigues Silva M (2010) Effects of antifungal agents in sap activity of *Candida albicans* isolates. Mycopathologia 169(2):91–98. https://doi.org/10.1007/s11046-009-9232-6

Costa C, Henriques A et al (2013a) The dual role of *Candida glabrata* drug:H+ antiporter CgAqr1 (ORF CAGL0J09944g) in antifungal drug and acetic acid resistance. Front Microbiol 4 (170):1–13

Costa C, Pires C et al (2013b) *Candida glabrata* drug:H+ antiporter CgQdr2 confers imidazole drug resistance, being activated by transcription factor CgPdr1. Antimicrob Agents Chemother 57 (7):3159–3167

Costa C et al (2014) *Candida glabrata* drug:H+ antiporter CgTpo3 (ORF CAGL0I10384g): role in azole drug resistance and polyamine homeostasis. J Antimicrob Chemother 69(7):1767–1776

Costa C et al (2016) Clotrimazole drug resistance in *Candida glabrata* clinical isolates correlates with increased expression of the drug:H+ antiporters CgAqr1, CgTpo1_1, CgTpo3 and CgQdr2. Front Microbiol 7:526

Courchesne WE (2002) Characterization of a novel, broad-based fungicidal activity for the antiarrhythmic drug amiodarone. J Pharmacol Exp Ther 300(1):195–199. https://doi.org/10.1124/jpet.300.1.195

Cowen LE, Lindquist S (2005) Hsp90 potentiates the rapid evolution of new traits: drug resistance in diverse fungi. Science 309(5744):2185–2189

Cowen LE, Singh SD, Kohler JR, Collins C, Zaas AK, Schell WA, Aziz H, Mylonakis E, Perfect JR, Whitesell L, Lindquist S (2009) Harnessing Hsp90 function as a powerful, broadly effective therapeutic strategy for fungal infectious disease. Proc Natl Acad Sci 106(8):2818–2823. https://doi.org/10.1073/pnas.0813394106

Cramer RA, Perfect BZ, Pinchai N, Park S, Perlin DS, Asfaw YG, Heitman J, Perfect JR, Steinbach WJ (2008) Calcineurin target CrzA regulates conidial germination, hyphal growth, and pathogenesis of *Aspergillus fumigatus*. Eukaryot Cell 7(7):1085–1097. https://doi.org/10.1128/EC.00086-08

Cruz MC, Poeta MDEL, Wang P, Wenger R, Zenke G, Quesniaux VFJ, Movva NR, Perfect JR, Cardenas ME, Heitman J (2000) Immunosuppressive and nonimmunosuppressive cyclosporine analogs are toxic to the opportunistic fungal pathogen *Cryptococcus neoformans* via cyclophilin-dependent inhibition of calcineurin. Antimicrob Agents Chemother 44(1):143–149

Cruz MC, Fox DS, Heitman J (2001) Calcineurin is required for hyphal elongation during mating and haploid fruiting in *Cryptococcus neoformans*. EMBO J 20(5):1020–1032. https://doi.org/10.1093/emboj/20.5.1020

Csank C, Schröppel K, Leberer E, Harcus D, Mohamed O, Meloche S, Thomas DY, Whiteway M (1998) Roles of the *Candida albicans* mitogen-activated protein kinase homolog, Cek1p , in hyphal development and systemic candidiasis. Infect Immun 66(6):2713–2721

Davis D, Edwars JJ, Mitchell AP, Ibrahim AS (2000) *Candida albicans* RIM101 pH response pathway is required for host-pathogen interactions. Infect Immun 68(10):5953–5959. https://doi.org/10.1128/IAI.68.10.5953-5959.2000.Updated

Denning DW, Pleuvry A, Cole DC (2013) Global burden of allergic bronchopulmonary aspergillosis with asthma and its complication chronic pulmonary aspergillosis in adults. Med Mycol 51 (4):361–370. https://doi.org/10.3109/13693786.2012.738312

Diez-Orejas R, Molero G, Navarro-Garcia F, Pla J, Nombela C, Sanchez-Perez M (1997) Reduced virulence of *Candida albicans* MKC1 mutants: a role for mitogen-activated protein kinase in pathogenesis. Infect Immun 65(2):833–837

Edlind T, Smith L, Henry K, Katiyar S, Nickels J (2002) Antifungal activity in *Saccharomyces cerevisiae* is modulated by calcium signalling. Mol Microbiol 46(1):257–268. https://doi.org/10.1046/j.1365-2958.2002.03165.x

Eisenman HC, Nosanchuk JD, Webber JBW, Emerson RJ, Camesano TA, Casadevall A (2005) Microstructure of cell wall-associated melanin in the human pathogenic fungus *Cryptococcus neoformans*. Biochemistry 44(10):3683–3693. https://doi.org/10.1021/bi047731m

Feng X, Krishnan K, Richie DL, Aimanianda V, Hartl L, Grahl N, Powers-Fletcher MV, Zhang M, Fuller KK, Nierman WC, Lu LJ, Latgé JP, Woollett L, Newman SL, Cramer RA, Rhodes JC, Askew DS (2011) HacA-independent functions of the ER stress sensor irea synergize with the canonical UPR to influence virulence traits in *Aspergillus fumigatus*. PLoS Pathog 7(10):1–18. https://doi.org/10.1371/journal.ppat.1002330

Ferrari S, Ischer F, Calabrese D, Posteraro B, Sanguinetti M, Fadda G, Rohde B, Bauser C, Bader O, Sanglard D (2009) Gain of function mutations in CgPDR1 of *Candida glabrata* not only mediate antifungal resistance but also enhance virulence. PLoS Pathog 5(1):1–17. https://doi.org/10.1371/journal.ppat.1000268

Ferrari S, Sanguinetti M, De Bernardis F, Torelli R, Posteraro B, Vandeputte P, Sanglard D (2011a) Loss of mitochondrial functions associated with azole resistance in *Candida glabrata* results in enhanced virulence in mice. Antimicrob Agents Chemother 55(5):1852–1860. https://doi.org/10.1128/AAC.01271-10

Ferrari S, Sanguinetti M, Torelli R, Posteraro B, Sanglard D (2011b) Contribution of CgPDR1-regulated genes in enhanced virulence of azole-resistant *Candida glabrata*. PLoS One 6(3): e17589. https://doi.org/10.1371/journal.pone.0017589

Ferreira ME de S, Heinekamp T, Härtl A, Brakhage AA, Semighini CP, Harris SD, Savoldi M, de Gouvêa PF, Goldman MH de S, Goldman GH (2007) Functional characterization of the *Aspergillus fumigatus* calcineurin. Fungal Genet Biol 44(3):219–230. https://doi.org/10.1016/j.fgb.2006.08.004

Ferreira C, Silva S, Faria-Oliveira F, Pinho E, Henriques M, Lucas C (2010) *Candida albicans* virulence and drug-resistance requires the O-acyltransferase Gup1p. BMC Microbiol 10:238. https://doi.org/10.1186/1471-2180-10-238

Fortwendel JR, Juvvadi PR, Perfect BZ, Rogg LE, Perfect JR, Steinbach WJ (2010) Transcriptional regulation of chitin synthases by calcineurin controls paradoxical growth of *Aspergillus fumigatus* in response to caspofungin. Antimicrob Agents Chemother 54(4):1555–1563. https://doi.org/10.1128/AAC.00854-09

Fox DS, Cruz MC, Sia RAL, Ke H, Cox GM, Cardenas ME, Heitman J (2001) Calcineurin regulatory subunit is essential for virulence and mediates interactions with FKBP12-FK506 in *Cryptococcus neoformans*. Mol Microbiol 39(4):835–849. https://doi.org/10.1046/j.1365-2958.2001.02295.x

Fradin C, De Groot P, MacCallum D, Schaller M, Klis F, Odds FC, Hube B (2005) Granulocytes govern the transcriptional response, morphology and proliferation of *Candida albicans* in human blood. Mol Microbiol 56(2):397–415. https://doi.org/10.1111/j.1365-2958.2005.04557.x

García-Rivera J, Casadevall A (2001) Melanization of *Cryptococcus neoformans* reduces its susceptibility to the antimicrobial effects of silver nitrate. Med Mycol 39:353–357. https://doi.org/10.1080/mmy.39.4.353.357

Gautam P, Shankar J, Madan T, Sirdeshmukh R, Sundaram CS, Gade WN, Basir SF, Sarma PU (2008) Proteomic and transcriptomic analysis of *Aspergillus fumigatus* on exposure to amphotericin B. Antimicrob Agents Chemother 52(12):4220–4227. https://doi.org/10.1128/AAC.01431-07

Geffers C, Gastmeier P (2011) Nosocomial infections and multidrug-resistant organisms in Germany: epidemiological data from KISS (The Hospital Infection Surveillance System). Medicine 108(18):320. https://doi.org/10.3238/arztebl.2011.0320a

Ghazaei C (2017) Molecular insights into pathogenesis and infection with *Aspergillus fumigatus*. Malays J Med Sci 24(1):10–20. https://doi.org/10.21315/mjms2017.24.1.2

Guhad FA, Jensen HE, Aalbaek B, Csank C, Mohamed O, Harcus D, Thomas DY, Whiteway M, Hau J (1998) Mitogen-activated protein kinase-defective *Candida albicans* is avirulent in a novel model of localized murine candidiasis. FEMS Microbiol Lett 166(1):135–139. https://doi.org/10.1016/S0378-1097(98)00323-1

Gupta SS, Ton VK, Beaudry V, Rulli S, Cunningham K, Rao R (2003) Antifungal activity of amiodarone is mediated by disruption of calcium homeostasis. J Biol Chem 278 (31):28831–28839. https://doi.org/10.1074/jbc.M303300200

Hagiwara D, Miura D, Shimizu K, Paul S, Ohba A, Gonoi T, Watanabe A, Kamei K, Shintani T, Moye-Rowley WS, Kawamoto S, Gomi K (2017) A novel Zn_2-Cys_6 transcription factor AtrR plays a key role in an azole resistance mechanism of *Aspergillus fumigatus* by co-regulating *cyp51A* and *cdr1B* expressions. PLoS Pathog. https://doi.org/10.1371/journal.ppat.1006096

Hata K, Horii T, Miyazaki M, Watanabe NA, Okubo M, Sonoda J, Nakamoto K, Tanaka K, Shirotori S, Murai N, Inoue S, Matsukura M, Abe S, Yoshimatsu K, Asada M (2011) Efficacy of oral E1210, a new broad-spectrum antifungal with a novel mechanism of action, in murine models of candidiasis, aspergillosis, and fusariosis. Antimicrob Agents Chemother 55 (10):4543–4551. https://doi.org/10.1128/AAC.00366-11

Hohl TM, Feldmesser M, Perlin DS, Pamer EG (2008) Caspofungin modulates inflammatory responses to *Aspergillus fumigatus* through stage-specific effects on fungal β-glucan exposure. J Infect Dis 198(2):176–185. https://doi.org/10.1086/589304.Caspofungin

Hohmann S (2009) Control of high osmolarity signalling in the yeast *Saccharomyces cerevisiae*. FEBS Lett 583(24):4025–4029. https://doi.org/10.1016/j.febslet.2009.10.069

Hole C, Wormley FL (2016) Innate host defenses against *Cryptococcus neoformans*. J Microbiol 54 (3):202–211. https://doi.org/10.1007/s12275-016-5625-7

Huang W, Liao G, Baker GM, Wang Y, Lau R, Paderu P, Perlin DS, Xue C (2016) Lipid flippase subunit Cdc50 mediates drug resistance and virulence in *Cryptococcus neoformans*. MBio 7 (3):1–13. https://doi.org/10.1128/mBio.00478-16

Idnurm A, Walton FJ, Floyd A, Reedy JL, Heitman J (2009) Identification of ENA1 as a virulence gene of the human pathogenic fungus *Cryptococcus neoformans* through signature-tagged insertional mutagenesis. Eukaryot Cell 8(3):315–326. https://doi.org/10.1128/EC.00375-08

Ikeda R, Sugita T, Jacobson ES, Shinoda T (2003) Effects of melanin upon susceptibility of Cryptococcus to antifungals. Microbiol Immunol 47(4):271–277. https://doi.org/10.1111/j.1348-0421.2003.tb03395.x

Jacobson ES, Ikeda R (2005) Effect of melanization upon porosity of the cryptococcal cell wall. Med Mycol 43:327–333. https://doi.org/10.1080/13693780412331271081

Jiang B, Xu D, Allocco J, Parish C, Davison J, Veillette K, Sillaots S, Hu W, Rodriguez-Suarez R, Trosok S, Zhang L, Li Y, Rahkhoodaee F, Ransom T, Martel N, Wang H, Gauvin D, Wiltsie J, Wisniewski D, Salowe S, Kahn JN, Hsu MJ, Giacobbe R, Abruzzo G, Flattery A, Gill C, Youngman P, Wilson K, Bills G, Platas G, Pelaez F, Diez MT, Kauffman S, Becker J, Harris G, Liberator P, Roemer T (2008) PAP inhibitor with in vivo efficacy identified by *Candida albicans* genetic profiling of natural products. Chem Biol 15(4):363–374. https://doi.org/10.1016/j.chembiol.2008.02.016

Jung K-W, Strain AK, Nielsen K, Jung K-H, Bahn Y-S (2012) Two cation transporters Ena1 and Nha1 cooperatively modulate ion homeostasis, antifungal drug resistance, and virulence of *Cryptococcus neoformans* via the HOG pathway. Fungal Genet Biol 49(4):332–345. https://doi.org/10.1111/j.1747-0285.2012.01428.x.Identification

Juvvadi PR, Fortwendel JR, Rogg LE, Burns KA, Randell SH, Steinbach WJ (2011) Localization and activity of the calcineurin catalytic and regulatory subunit complex at the septum is essential for hyphal elongation and proper septation in *Aspergillus fumigatus*. Mol Microbiol 82 (5):1235–1259. https://doi.org/10.1007/s11103-011-9767-z.Plastid

Juvvadi PR, Lee SC, Heitman J, Steinbach WJ (2017) Calcineurin in fungal virulence and drug resistance: prospects for harnessing targeted inhibition of calcineurin for an antifungal therapeutic approach. Virulence 8(2):186–197. https://doi.org/10.1080/21505594.2016.1201250

Karababa M, Valentino E, Pardini G, Coste AT, Bille J, Sanglard D (2006) CRZ1, a target of the calcineurin pathway in *Candida albicans*. Mol Microbiol 59(5):1429–1451. https://doi.org/10.1111/j.1365-2958.2005.05037.x

Kaur R, Ma B, Cormack BP (2007) A family of glycosylphosphatidylinositol-linked aspartyl proteases is required for virulence of *Candida glabrata*. Proc Natl Acad Sci USA 104 (18):7628–7633. https://doi.org/10.1073/pnas.0611195104

Kim JH, Chan KL, Faria NCG, Martins M de L, Campbell BC (2012) Targeting the oxidative stress response system of fungi with redox-potent chemosensitizing agents. Front Microbiol 3 (88):1–11. https://doi.org/10.3389/fmicb.2012.00088

Kitamura A, Higuchi S, Hata M, Kawakami K, Yoshida K, Namba K, Nakajima R (2009a) Effect of B-1,6-glucan inhibitors on the invasion process of *Candida albicans*: potential mechanism of their in vivo efficacy. Antimicrob Agents Chemother 53(9):3963–3971. https://doi.org/10.1128/AAC.00435-09

Kitamura A, Someya K, Hata M, Nakajima R, Takemura M (2009b) Discovery of a small-molecule inhibitor of B-1,6-glucan synthesis. Antimicrob Agents Chemother 53(2):670–677. https://doi.org/10.1128/AAC.00844-08

Kontoyiannis DP, Lewis RE, Osherov N, Albert ND, May GS (2003) Combination of caspofungin with inhibitors of the calcineurin pathway attenuates growth in vitro in Aspergillus species. J Antimicrob Chemother 51(2):313–316. https://doi.org/10.1093/jac/dkg090

Kraus PR, Fox DS, Cox GM, Heitman J (2003) The *Cryptococcus neoformans* MAP kinase Mpk1 regulates cell integrity in response to antifungal drugs and loss of calcineurin function. Mol Microbiol 48(5):1377–1387

Krysan DJ (2015) Toward improved anti-cryptococcal drugs: novel molecules and repurposed drugs. Fungal Genet Biol 78:93–98. https://doi.org/10.1016/j.fgb.2014.12.001

Kuhn DM, Mukherjee PK, Clark TA, Pujol C, Chandra J, Hajjeh RA, Warnock DW, Soll DR, Ghannoum MA (2004) *Candida parapsilosis* characterization in an outbreak setting. Emerg Infect Dis 10(6):1074–1081. https://doi.org/10.3201/eid1006.030873

Kumamoto C a (2005) A contact-activated kinase signals *Candida albicans* invasive growth and biofilm development. Proc Natl Acad Sci USA 102(15):5576–5581. https://doi.org/10.1073/pnas.0407097102

Kumar R, Shukla PK (2010) Amphotericin B resistance leads to enhanced proteinase and phospholipase activity and reduced germ tube formation in *Candida albicans*. Fungal Biol 114 (2–3):189–197. https://doi.org/10.1016/j.funbio.2009.12.003

Lamoth F, Juvvadi PR, Fortwendel JR, Steinbach WJ (2012) Heat shock protein 90 is required for conidiation and cell wall integrity in *Aspergillus fumigatus*. Eukaryot Cell 11(11):1324–1332. https://doi.org/10.1128/EC.00032-12

Latgé JP (1999) *Aspergillus fumigatus* and Aspergillosis. Clin Microbiol Rev 12(2):310–350. https://doi.org/10.1128/JCM.01704-14

Lelièvre L, Groh M, Angebault C, Maherault AC, Didier E, Bougnoux ME (2013) Azole resistant *Aspergillus fumigatus*: an emerging problem. Medecine et Maladies Infectieuses 43 (4):139–145. https://doi.org/10.1016/j.medmal.2013.02.010

Lettner T, Zeidler U, Gimona M, Hauser M, Breitenbach M, Bito A (2010) *Candida albicans* AGE3, the ortholog of the *S. cerevisiae* ARF-GAP-encoding gene GCS1, is required for hyphal growth and drug resistance. PLoS One 5(8):1–15. https://doi.org/10.1371/journal.pone. 0011993

Li W, Handschumachers RE (1993) Specific interaction of the cyclophilin-cyclosporin complex with the B subunit of calcineurin. J Biol Chem 268(19):14040–14044

Liu M, Du P, Heinrich G, Cox GM, Gelli A (2006) Cch1 mediates calcium entry in *Cryptococcus neoformans* and is essential in low-calcium environments. Eukaryot Cell 5(10):1788–1796. https://doi.org/10.1128/EC.00158-06

Liu S, Hou Y, Liu W, Lu C, Wang W, Sun S (2015) Components of the calcium-calcineurin signaling pathway in fungal cells and their potential as antifungal targets. Eukaryot Cell 14 (4):324–334. https://doi.org/10.1128/EC.00271-14

Lugarini C, Goebel CS, Condas LAZ, Muro MD, De Farias MR, Ferreira FM, Vainstein MH (2008) *Cryptococcus neoformans* isolated from Passerine and Psittacine bird excreta in the state of Paraná, Brazil. Mycopathologia 166(2):61–69. https://doi.org/10.1007/s11046-008-9122-3

Ma H, Hagen F, Stekel DJ, Johnston S a, Sionov E, Falk R, Polacheck I, Boekhout T, May RC (2009) The fatal fungal outbreak on Vancouver Island is characterized by enhanced intracellular parasitism driven by mitochondrial regulation. PNAS 106(31):12980–12985. https://doi.org/10. 1073/pnas.0902963106

Magnani T, Soriani FM, Martins VDP, Policarpo ACDF, Sorgi CA, Faccioli LH, Curti C, Uyemura SA (2008) Silencing of mitochondrial alternative oxidase gene of *Aspergillus fumigatus* enhances reactive oxygen species production and killing of the fungus by macrophages. J Bioenerg Biomembr 40(6):631–636. https://doi.org/10.1007/s10863-008- 9191-5

Marchetti O, Moreillon P, Glauser MP, Bille J, Sanglard D (2000) Potent synergism of the combination of fluconazole and cyclosporine in *Candida albicans*. Antimicrob Agents Chemother 44(9):2373–2381

Mednick AJ, Nosanchuk JD, Casadevall A (2005) Melanization of *Cryptococcus neoformans* affects lung inflammatory responses during cryptococcal infection. Infect Immun 73 (4):2012–2019. https://doi.org/10.1128/IAI.73.4.2012-2019.2005

Mitchell TG, Perfect JR (1995) Cryptococcosis in the era of AIDS-100 years after the discovery of *Cryptococcus neoformans*. Clin Microbiol Rev 8(4):515–548. https://doi.org/10.1016/S0190- 9622(97)80336-2

Mitsuyama J, Nomura N, Hashimoto K, Yamada E, Nishikawa H, Kaeriyama M, Kimura A, Todo Y, Narita H (2008) In vitro and in vivo antifungal activities of T-2307, a novel arylamidine. Antimicrob Agents Chemother 52(4):1318–1324. https://doi.org/10.1128/AAC. 01159-07

Miyazaki T, Kohno S (2014) ER stress response mechanisms in the pathogenic yeast *Candida glabrata* and their roles in virulence. Virulence 5(2):365–370. https://doi.org/10.4161/viru. 27373

Miyazaki T, Miyazaki Y, Izumikawa K, Kakeya H, Miyakoshi S, Bennett JE, Kohno S (2006) Fluconazole treatment is effective against a *Candida albicans* erg3/erg3 mutant in vivo despite in vitro resistance. Antimicrob Agents Chemother 50(2):580–586. https://doi.org/10.1128/ AAC.50.2.580

Miyazaki T, Yamauchi S, Inamine T, Nagayoshi Y, Saijo T, Izumikawa K, Seki M, Kakeya H, Yamamoto Y, Yanagihara K, Miyazaki Y, Kohno S (2010) Roles of calcineurin and Crz1 in antifungal susceptibility and virulence of *Candida glabrata*. Antimicrob Agents Chemother 54 (4):1639–1643. https://doi.org/10.1128/AAC.01364-09

Miyazaki T, Izumikawa K, Nagayoshi Y, Saijo T, Yamauchi S, Morinaga Y, Seki M, Kakeya H, Yamamoto Y, Yanagihara K, Miyazaki Y, Kohno S (2011) Functional characterization of the regulators of calcineurin in *Candida glabrata*. FEMS Yeast Res 11(8):621–630. https://doi.org/ 10.1111/j.1567-1364.2011.00751.x

Miyazaki T, Nakayama H, Nagayoshi Y, Kakeya H, Kohno S (2013) Dissection of Ire1 functions reveals stress response mechanisms uniquely evolved in *Candida glabrata*. PLoS Pathog 9 (1):1–20. https://doi.org/10.1371/journal.ppat.1003160

Moralez ATP, França EJG, Furlaneto-Maia L, Quesada RMB, Furlaneto MC (2013) Phenotypic switching in *Candida tropicalis*: association with modification of putative virulence attributes and antifungal drug sensitivity. Med Mycol 52:106–114. https://doi.org/10.3109/13693786. 2013.825822

Mores AU, Souza RD, Cavalca L, de Paula e Carvalho A, Gursky LC, Rosa RT, Samaranayake LP, Rosa EAR (2011) Enhancement of secretory aspartyl protease production in biofilms of *Candida albicans* exposed to sub-inhibitory concentrations of fluconazole. Mycoses 54 (3):195–201. https://doi.org/10.1111/j.1439-0507.2009.01793.x

Morio F, Pagniez F, Lacroix C, Miegeville M, Le pape P (2012) Amino acid substitutions in the *Candida albicans* sterol δ5,6-desaturase (Erg3p) confer azole resistance: characterization of two novel mutants with impaired virulence. J Antimicrob Chemother 67(9):2131–2138. https://doi. org/10.1093/jac/dks186

Muend S, Rao R (2008) Fungicidal activity of amiodarone is tightly coupled to calcium influx. FEMS Yeast Res 8(3):425–431. https://doi.org/10.1007/s11103-011-9767-z.Plastid

Narasipura SD, Chaturvedi V, Chaturvedi S (2005) Characterization of *Cryptococcus neoformans* variety gattii SOD2 reveals distinct roles of the two superoxide dismutases in fungal biology and virulence. Mol Microbiol 55(6):1782–1800. https://doi.org/10.1111/j.1365-2958.2005.04503.x

Netea MG, Joosten LAB, van der Meer JWM, Kullberg B-J, van de Veerdonk FL (2015) Immune defence against Candida fungal infections. Nat Rev Immunol 15(10):630–642. https://doi.org/ 10.1038/nri3897

Nishikawa JL, Boeszoermenyi A, Vale-Silva LA, Torelli R, Posteraro B, Sohn Y-J, Ji F, Gelev V, Sanglard D, Sanguinetti M, Sadreyev RI, Mukherjee G, Bhyravabhotla J, Buhrlage SJ, Gray NS, Wagner G, Näär AM, Arthanari H (2016) Inhibiting fungal multidrug resistance by disrupting an activator–mediator interaction. Nature 530(7591):485–489. https://doi.org/10.1038/ nature16963

Nosanchuk JD, Casadevall A (2006) Impact of melanin on microbial virulence and clinical resistance to antimicrobial compounds. Antimicrob Agents Chemother 50(11):3519–3528. https://doi.org/10.1128/AAC.00545-06

Odom A, Muir S, Lim E, Toffaletti DL, Perfect J, Heitman J (1997a) Calcineurin is required for virulence of *Cryptococcus neoformans*. EMBO J 16(10):2576–2589. https://doi.org/10.1093/ emboj/16.10.2576

Odom A, Del Poeta M, Perfect J, Heitman J (1997b) The immunosuppressant FK506 and its nonimmunosuppressive analog L-685,818 are toxic to *Cryptococcus neoformans* by inhibition of a common target protein. Antimicrob Agents Chemother 41(1):156–61. Available at: http:// www.ncbi.nlm.nih.gov/pubmed/8980772%5Cn; http://www.ncbi.nlm.nih.gov/pubmed/8980772

Onyewu C, Afshari NA, Heitman J (2006) Calcineurin promotes infection of the cornea by *Candida albicans* and can be targeted to enhance fluconazole therapy. Antimicrob Agents Chemother 50 (11):3963–3965. https://doi.org/10.1128/AAC.00393-06

Orsi CF, Colombari B, Ardizzoni A, Peppoloni S, Neglia R, Posteraro B, Morace G, Fadda G, Blasi E (2009) The ABC transporter-encoding gene AFR1 affects the resistance of *Cryptococcus neoformans* to microglia-mediated antifungal activity by delaying phagosomal maturation. FEMS Yeast Res 9(2):301–310. https://doi.org/10.1111/j.1567-1364.2008.00470.x

Pais P, Costa C et al (2016a) Membrane proteome-wide response to the antifungal drug clotrimazole in *Candida glabrata*: role of the transcription factor CgPdr1 and the Drug:H+ antiporters CgTpo1_1 and CgTpo1_2. Mol Cell Proteomics 15(1):57–72

Pais P, Pires C et al (2016b) Membrane proteomics analysis of the *Candida glabrata* response to 5-flucytosine: unveiling the role and regulation of the drug efflux transporters CgFlr1 and CgFlr2. Front Microbiol 7(2045):1–14

Park BJ, Wannemuehler KA, Marston BJ, Govender N, Pappas PG, Chiller TM (2009) Estimation of the current global burden of cryptococcal meningitis among persons living with HIV/AIDS. AIDS 23(4):525–530. https://doi.org/10.1097/QAD.0b013e328322ffac

Paul S, Diekema D, Moye-Rowley WS (2013) Contributions of *Aspergillus fumigatus* ATP-binding cassette transporter proteins to drug resistance and virulence. Eukaryot Cell 12(12):1619–1628. https://doi.org/10.1128/EC.00171-13

Perlin DS, Rautemaa-Richardson R, Alastruey-Izquierdo A (2017) The global problem of antifungal resistance: prevalence, mechanisms, and management. Lancet Infect Dis 3099(17):1–10. https://doi.org/10.1016/S1473-3099(17)30316-X

Perlroth J, Choi B, Spellberg B (2007) Nosocomial fungal infections: epidemiology, diagnosis, and treatment. Med Mycol 45(4):321–346. https://doi.org/10.1080/13693780701218689

Poláková S, Blume C, Zárate JA, Mentel M, Jørck-Ramberg D, Stenderup J, Piskur J (2009) Formation of new chromosomes as a virulence mechanism in yeast *Candida glabrata*. Proc Natl Acad Sci USA 106(8):2688–2693. https://doi.org/10.1073/pnas.0809793106

Polke M, Hube B, Jacobsen ID (2015) Candida survival strategies. In: Advances in applied microbiology. Elsevier, New York, pp 139–235. https://doi.org/10.1016/bs.aambs.2014.12.002

Polvi EJ, Li X, O'Meara TR, Leach MD, Cowen LE (2015) Opportunistic yeast pathogens: reservoirs, virulence mechanisms, and therapeutic strategies. Cell Mol Life Sci 72 (12):2261–2287. https://doi.org/10.1007/s00018-015-1860-z

Prieto D, Román E, Correia I, Pla J (2014) The HOG pathway is critical for the colonization of the mouse gastrointestinal tract by *Candida albicans*. PLoS One 9(1):1–12. https://doi.org/10.1371/journal.pone.0087128

Ray TL, Payne CD (1990) Comparative production and rapid purification of Candida acid proteinase from protein-supplemented cultures. Infect Immun 58(2):508–514

Rho J, Shin JH, Song JW, Park M-R, Kee SJ, Jang SJ, Park YK, Suh SP, Ryang DW (2004) Molecular investigation of two consecutive nosocomial clusters of *Candida tropicalis* candiduria using pulsed-field gel electrophoresis. J Microbiol (Seoul, Korea), 42(2):80–6. Available at: http://www.ncbi.nlm.nih.gov/pubmed/15357299

Richie DL, Hartl L, Aimanianda V, Winters MS, Fuller KK, Miley MD, White S, McCarthy JW, Latgé JP, Feldmesser M, Rhodes JC, Askew DS (2009) A role for the unfolded protein response (UPR) in virulence and antifungal susceptibility in *Aspergillus fumigatus*. PLoS Pathog 5 (1):1–17. https://doi.org/10.1371/journal.ppat.1000258

Roemer T, Krysan DJ (2014) Antifungal drug development: challenges, unmet clinical needs, and new approaches. Cold Spring Harb Perspect Med 4:1–14. https://doi.org/10.1101/cshperspect.a019703

Román E, Correia I, Salazin A, Fradin C, Jouault T, Poulain D, Liu FT, Pla J (2016) The Cek1-mediated MAP kinase pathway regulates exposure of α-1,2 and β-1,2-mannosides in the cell wall of *Candida albicans* modulating immune recognition. Virulence 7(5):558–577. https://doi.org/10.1080/21505594.2016.1163458

Romano J, Nimrod G, Ben-Tal N, Shadkchan Y, Baruch K, Sharon H, Osherov N (2006) Disruption of the *Aspergillus fumigatus* ECM33 homologue results in rapid conidial germination, antifungal resistance and hypervirulence. Microbiology 152(7):1919–1928. https://doi.org/10.1099/mic.0.28936-0

Romão D et al (2017) A new determinant of *Candida glabrata* virulence: the acetate exporter CgDtr1. Front Cell Infect Microbiol 7(473):1–10

Sandovsky-Losica H, Shwartzman R, Lahat Y, Segal E (2008) Antifungal activity against *Candida albicans* of nikkomycin Z in combination with caspofungin, voriconazole or amphotericin B. J Antimicrob Chemother 62(3):635–637. https://doi.org/10.1093/jac/dkn216

Sanglard D, Kuchler K, Ischer F, Pagani JL, Monod M, Bille J (1995) Mechanisms of resistance to azole antifungal agents in *Candida albicans* isolates from AIDS patients involve specific multidrug transporters. Antimicrob Agents Chemother 39(11):2378–2386. https://doi.org/10.1128/AAC.39.11.2378

Sanglard D, Ischer F, Monod M, Bille J (1997) Cloning of *Candida albicans* genes conferring resistance to azole antifungal agents: characterization of CDR2, a new multidrug ABC transporter gene. Microbiology 143:405–416. https://doi.org/10.1099/00221287-143-2-405

Sanglard D, Ischer O, Calabrese D, Majcherczyk PA, De Microbiologie I, Diseases I, Hospitalier C, Vaudois U (1999) The ATP binding cassette transporter gene CgCDR1 from *Candida glabrata* is involved in the resistance of clinical isolates to azole antifungal agents. Antimicrob Agents Chemother 43(11):2753–2765

Sanglard D, Ischer F, Bille J (2001) Role of ATP-binding-cassette transporter genes in high-frequency acquisition of resistance to azole antifungals in *Candida glabrata*. Antimicrob Agents Chemother 45(4):1174–1183. https://doi.org/10.1128/AAC.45.4.1174

Sanglard D, Ischer F, Marchetti O, Entenza J, Bille J (2003) Calcineurin A of *Candida albicans*: involvement in antifungal tolerance, cell morphogenesis and virulence. Mol Microbiol 48 (4):959–976. https://doi.org/10.1046/j.1365-2958.2003.03495.x

Sanguinetti M, Posteraro B, La Sorda M, Torelli R, Fiori B, Santangelo R, Delogu G, Fadda G (2006) Role of AFR1, an ABC transporter-encoding gene, in the in vivo response to fluconazole and virulence of *Cryptococcus neoformans*. Infect Immun 74(2):1352–1359. https://doi.org/10.1128/IAI.74.2.1352-1359.2006

Santos R et al (2017) The multidrug resistance transporters CgTpo1_1 and CgTpo1_2 play a role in virulence and biofilm formation in the human pathogen *Candida glabrata*. Cell Microbiol 19 (15):1–13

Schmalzle SA, Buchwald UK, Gilliam BL, Riedel DJ (2016) *Cryptococcus neoformans* infection in malignancy. Mycoses 59:542–552. https://doi.org/10.1111/myc.12496

Selmecki A, Forche A, Berman J (2006) Aneuploidy and isochromosome formation in drug-resistant *Candida albicans*. Science 313(5785):367–370. https://doi.org/10.1111/j.1747-0285.2012.01428.x.Identification

Shah AH, Singh A, Dhamgaye S, Chauhan N, Vandeputte P, Suneetha KJ, Kaur R, Mukherjee PK, Chandra J, Ghannoum M a, Sanglard D, Goswami SK, Prasad R (2014) Novel role of a family of major facilitator transporters in biofilm development and virulence of *Candida albicans*. Biochem J 460(2):223–235. https://doi.org/10.1042/BJ20140010

Shin JH, Chae MJ, Song JW, Jung S, Cho D, Kee SJ, Kim SH, Shin MG, Suh SP, Ryang DW (2007) Changes in karyotype and azole susceptibility of sequential bloodstream isolates from patients with *Candida glabrata* candidemia. J Clin Microbiol 45(8):2385–2391. https://doi.org/10.1128/JCM.00381-07

Shingu-Vazquez M, Traven A (2011) Mitochondria and fungal pathogenesis: drug tolerance, virulence, and potential for antifungal therapy. Eukaryot Cell 10(11):1376–1383. https://doi.org/10.1128/EC.05184-11

Song MH, Lee JW, Kim MS, Yoon JK, White TC, Floyd A, Heitman J, Strain AK, Nielsen JN, Nielsen K, Bahn YS (2012) A flucytosine-responsive Mbp1/Swi4-like protein, Mbs1, plays pleiotropic roles in antifungal drug resistance, stress response, and virulence of *Cryptococcus neoformans*. Eukaryot Cell 11(1):53–67. https://doi.org/10.1128/EC.05236-11

Srikanta D, Santiago-Tirado FH, Doering TL (2014) *Cryptococcus neoformans*: historical curiosity to modern pathogen. Yeast 31(2):47–60. https://doi.org/10.1002/yea.2997.Cryptococcus

Stathopoulos-gerontides A, Guo JJ, Cyert MS (1999) Yeast calcineurin regulates nuclear localization of the Crz1p transcription factor through dephosphorylation. Genes Dev 13:798–803

Steinbach WJ, Schell W a, Blankenship JR, Onyewu C, Heitman J, Perfect JR (2004) In vitro interactions between antifungals and immunosuppressants against *Aspergillus fumigatus*. Antimicrob Agents Chemother 48(5):1664–1669. https://doi.org/10.1128/AAC.48.5.1664

Steinbach WJ, Cramer RA, Perfect BZ, Asfaw YG, Sauer TC, Najvar LK, Kirkpatrick WR, Patterson TF, Benjamin DK, Heitman J, Perfect JR (2006) Calcineurin controls growth, morphology, and pathogenicity in *Aspergillus fumigatus*. Eukaryot Cell 5(7):1091–1103. https://doi.org/10.1128/EC.00139-06

Suliman HS, Appling DR, Robertus JD (2007) The gene for cobalamin-independent methionine synthase is essential in *Candida albicans*: a potential antifungal target. Arch Biochem Biophys 467(2):218–226. https://doi.org/10.1002/ana.22528.Toll-like

Sun S, Li Y, Guo Q, Shi C, Yu J, Ma L (2008) In vitro interactions between tacrolimus and azoles against *Candida albicans* determined by different methods. Antimicrob Agents Chemother 52 (2):409–417. https://doi.org/10.1128/AAC.01070-07

Tonthat NK, Juvvadi PR, Zhang H, Lee SC, Venters R, Spicer L, Steinbach WJ, Heitman J, Schumacher MA (2016) Structures of pathogenic fungal FKBP12s reveal possible self-catalysis function. mBio 7(2):1–11. https://doi.org/10.1128/mBio.00492-16

Tscherner M, Schwarzmüller T, Kuchler K (2011) Pathogenesis and antifungal drug resistance of the human fungal pathogen *Candida glabrata*. Pharmaceuticals 4:169–186. https://doi.org/10.3390/ph4010169

Tsukahara K, Hata K, Nakamoto K, Sagane K, Watanabe NA, Kuromitsu J, Kai J, Tsuchiya M, Ohba F, Jigami Y, Yoshimatsu K, Nagasu T (2003) Medicinal genetics approach towards identifying the molecular target of a novel inhibitor of fungal cell wall assembly. Mol Microbiol 48(4):1029–1042. https://doi.org/10.1046/j.1365-2958.2003.03481.x

Uppuluri P, Nett J, Heitman J, Andes D (2008) Synergistic effect of calcineurin inhibitors and fluconazole against *Candida albicans* biofilms. Antimicrob Agents Chemother 52 (3):1127–1132. https://doi.org/10.1128/AAC.01397-07

Vale-Silva L, Ischer F, Leibundgut-Landmann S, Sanglard D (2013) Gain-of-function mutations in PDR1, a regulator of antifungal drug resistance in candida glabrata, control adherence to host cells. Infect Immun 81(5):1709–1720. https://doi.org/10.1128/IAI.00074-13

Vale-Silva LA, Moeckli B, Torelli R, Posteraro B, Sanglard D (2016) Upregulation of the Adhesin gene mediated by PDR1 in *Candida glabrata* leads to enhanced host colonization. mSphere 1 (2):1–16. https://doi.org/10.1128/mSphere.00065-15.Editor

Van Duin D, Casadevall A, Nosanchuk JD (2002) Melanization of *Cryptococcus neoformans* and *Histoplasma capsulatum* reduces their susceptibilities to amphotericin B and caspofungin. Antimicrob Agents Chemother 46(11):3394–3400. https://doi.org/10.1128/AAC.46.11.3394

Van Duin D, Cleare W, Zaragoza O, Nosanchuk JD, Casadevall A (2004) Effects of voriconazole on *Cryptococcus neoformans*. Antimicrob Agents Chemother 48(6):2014–2020. https://doi.org/10.1128/AAC.48.6.2014

Vermitsky J-P, Edlind TD (2004) Azole resistance in *Candida glabrata*: coordinate upregulation of multidrug transporters and evidence for a Pdr1-like transcription factor. Antimicrob Agents Chemother 48(10):3773–3781. https://doi.org/10.1128/AAC.48.10.3773-3781.2004

Vermitsky J-P, Earhart KD, Smith WL, Homayouni R, Edlind TD, Rogers PD (2006) Pdr1 regulates multidrug resistance in *Candida glabrata*: gene disruption and genome-wide expression studies. Mol Microbiol 61(3):704–722. https://doi.org/10.1111/j.1365-2958.2006.05235.x

Verwer PEB, Van Duijn ML, Tavakol M, Bakker-Woudenberg IAJM, Van De Sande WWJ (2012) Reshuffling of *Aspergillus fumigatus* cell wall components chitin and β-glucan under the influence of caspofungin or nikkomycin Z alone or in combination. Antimicrob Agents Chemother 56(3):1595–1598. https://doi.org/10.1128/AAC.05323-11

Walker LA, Munro CA, De Bruijn I, Lenardon MD, McKinnon A, Gow NAR (2008) Stimulation of chitin synthesis rescues *Candida albicans* from echinocandins. PLoS Pathog 4(4):1–12. https://doi.org/10.1371/journal.ppat.1000040

Walker LA, MacCallum DM, Bertram G, Gow NAR, Odds FC, Brown AJP (2009) Genome-wide analysis of *Candida albicans* gene expression patterns during infection of the mammalian kidney. Fungal Genet Biol 46(2):210–219. https://doi.org/10.1016/j.fgb.2008.10.012

Walker SS, Xu Y, Triantafyllou I, Waldman MF, Mendrick C, Brown N, Mann P, Chau A, Patel R, Bauman N, Norris C, Antonacci B, Gurnani M, Cacciapuoti A, McNicholas PM, Wainhaus S, Herr RJ, Kuang R, Aslanian RG, Ting PC, Black TA (2011) Discovery of a novel class of orally active antifungal β-1,3-D-glucan synthase inhibitors. Antimicrob Agents Chemother 55 (11):5099–5106. https://doi.org/10.1128/AAC.00432-11

Wenzel RP, Gennings C (2005) Bloodstream infections due to Candida species in the intensive care unit: identifying especially high-risk patients to determine prevention strategies. Clin Infect Dis 41(Suppl 6):389–393

Wheeler RT, Kombe D, Agarwala SD, Fink GR (2008) Dynamic, morphotype-specific *Candida albicans* b-glucan exposure during infection and drug treatment. PLoS Pathog 4(12):1–12. https://doi.org/10.1371/journal.ppat.1000227

Wiederhold NP, Kontoyiannis DP, Prince RA, Lewis RE (2005) Attenuation of the activity of caspofungin at high concentrations against *Candida albicans*: possible role of cell wall integrity and calcineurin pathways. Antimicrob Agents Chemother 49(12):5146–5148. https://doi.org/10.1128/AAC.49.12.5146-5148.2005

Willger SD, Puttikamonkul S, Kim K-H, Burritt JB, Grahl N, Metzler LJ, Barbuch R, Bard M, Lawrence CB, Cramer RA (2008) A sterol-regulatory element binding protein is required for cell polarity, hypoxia adaptation, azole drug resistance, and virulence in *Aspergillus fumigatus*. PLoS Pathog 4(11):e1000200. https://doi.org/10.1371/journal.ppat.1000200

Williamson PR, Jarvis JN, Panackal AA, Fisher MC, Molloy SF, Loyse A, Harrison TS (2016) Cryptococcal meningitis: epidemiology, immunology, diagnosis and therapy. Nat Rev Neurol 13:1–12. https://doi.org/10.1038/nrneurol.2016.167

Wimalasena TT, Enjalbert B, Guillemette T, Plumridge A, Budge S, Yin Z, Brown AJP, Archer DB (2008) Impact of the unfolded protein response upon genome-wide expression patterns, and the role of Hac1 in the polarized growth, of *Candida albicans*. Fungal Genet Biol 45(9):1235–1247. https://doi.org/10.1016/j.fgb.2008.06.001

Winkelströter LK, Bom VLP, de Castro PA, Ramalho LNZ, Goldman MHS, Brown NA, Rajendran R, Ramage G, Bovier E, dos Reis TF, Savoldi M, Hagiwara D, Goldman GH (2015) High osmolarity glycerol response PtcB phosphatase is important for *Aspergillus fumigatus* virulence. Mol Microbiol 96(1):42–54. https://doi.org/10.1111/mmi.12919

Wisplinghoff H, Bischoff T, Tallent SM, Seifert H, Wenzel RP, Edmond MB (2004) Nosocomial bloodstream infections in US hospitals: analysis of 24,179 cases from a prospective nationwide surveillance study. Clin Infect Dis 39(3):309–317. https://doi.org/10.1086/421946

Wu TAO, Wright K, Hurst SF, Morrison CJ (2000) Enhanced extracellular production of aspartyl proteinase, a virulence factor, by *Candida albicans* isolates following growth in subinhibitory concentrations of fluconazole. Antimicrob Agents Chemother 44(5):1200–1208

Xie JL, Polvi EJ, Shekhar-Guturja T, Cowen LE (2014) Elucidating drug resistance in human fungal pathogens. Future Microbiol 9(4):523–542. https://doi.org/10.2217/fmb.14.18

Xu T, Feng Q, Jacob MR, Avula B, Mask MM, Baerson SR, Tripathi SK, Mohammed R, Hamann MT, Khan IA, Walker LA, Clark AM, Agarwal AK (2011) The marine sponge-derived polyketide endoperoxide plakortide F acid mediates its antifungal activity by interfering with calcium homeostasis. Antimicrob Agents Chemother 55(4):1611–1621. https://doi.org/10.1128/AAC.01022-10

Yamada-Okabe T, Yamada-Okabe H (2002) Characterization of the CaNAG3, CaNAG4, and CaNAG6 genes of the pathogenic fungus *Candida albicans*: possible involvement of these genes in the susceptibilities of cytotoxic agents. FEMS Microbiol Lett 212(1):15–21. https://doi.org/10.1016/S0378-1097(02)00726-7

Yoshimoto H, Saltsman K, Gasch AP, Li HX, Ogawa N, Botstein D, Brown PO, Cyert MS (2002) Genome-wide analysis of gene expression regulated by the calcineurin/Crz1p signaling pathway in *Saccharomyces cerevisiae*. J Biol Chem 277(34):31079–31088. https://doi.org/10.1074/jbc.M202718200

Yu Q, Wang H, Cheng X, Xu N, Ding X, Xing L, Li M (2012) Roles of Cch1 and Mid1 in morphogenesis, oxidative stress response and virulence in *Candida albicans*. Mycopathologia 174(5–6):359–369. https://doi.org/10.1007/s11046-012-9569-0

Yu Q, Ding X, Xu N, Cheng X, Qian K, Zhang B, Xing L, Li M (2013) In vitro activity of verapamil alone and in combination with fluconazole or tunicamycin against *Candida albicans* biofilms. Int J Antimicrob Agents 41(2):179–182. https://doi.org/10.1016/j.ijantimicag.2012.10.009

Yu SJ, Chang YL, Chen YL (2015) Calcineurin signaling: lessons from Candida species. FEMS Yeast Res 15(4):1–7. https://doi.org/10.1093/femsyr/fov016

Printed in the United States
By Bookmasters